ret: red

13

)

ELECTRONIC WAVE FORMING AND PROCESSING CIRCUITS

ELECTRONIC WAVE FORMING AND PROCESSING CIRCUITS

HAI HUNG CHIANG

Biomedical Engineering Department
Chung Yuan Christian University
Taiwan, Republic of China

A Wiley-Interscience Publication
JOHN WILEY & SONS
New York / Chichester / Brisbane / Toronto / Singapore

Library of Congress Cataloging in Publication Data:

Chiang, Hai Hung.
  Electronic wave forming and processing circuits.

  "A Wiley-Interscience publication."
  Includes index.
  1. Pulse circuits.   2.   Television circuits.
  3. Pulse height analyzers.   4.   Spectrometers. I. Title.
  TK7868.P8C45   1985   621.3815'34   85-17825
  ISBN 0-471-82826-2

Printed in the United States of America

10 9 8 7 6 5 4 3 2 1

PREFACE

The information in all electronic systems is carried by waves. Therefore wave forming and processing circuits have been utilized extensively in the designs of contemporary electronic devices or equipment, and the practical applications of these circuits are still being expanded.

In light of the importance of the wave forming and processing circuits in advanced electronic system designs, this book begins by illustrating various fundamental techniques of wave forming and processing and concludes by presenting useful analyses for typical wave forming and/or processing systems.

The first part of the book comprises the following fundamental topics: (1) introduction to electronic waveforms; (2) linear waveshaping circuits; (3) semiconductor–diode switching; (4) transistor switching; (5) transistor and IC comparator circuits; (6) astable and monostable multivibrators and blocking oscillators; (7) bistable multivibrators; (8) logic gates and logic families; (9) digital counters, registers, and data converters; (10) linear and approximate ramp generators; and (11) modulation, demodulation, and time-division multiplexing. This part can be used as a text or major reference for a one-semester course on pulse circuits at the senior or first-year graduate level.

The second part of the book employs the fundamental techniques given in the first part to analyze three widely used wave processing systems: television (in communication electronics), pulse-height analyzers (in applied physics), and time-of-flight mass spectrometers (in applied chemistry). This part provides typical system design examples, and thus it can be utilized as a guide for advanced electronic system designs. Overall, the book contains the fundamental techniques and the practical applications of wave forming and processing circuits; thus, it should serve as a useful handbook for engineers in the electronics industry and for experimental physicists and chemists in universities and national laboratories. Design examples, references, questions, and problems are given to assist the reader to fully understand the contents. A solution manual will be available upon request from the publisher.

Finally, the author would like to express appreciation to his colleagues, Professor H. S. Chang and Dr. B. N. Hung of Chung Yuan University, for their encouragement, and to his son, Dr. R. T. Chiang of General Electric Company, for his comments and suggestions.

HAI HUNG CHIANG

Chung-Li, Taiwan
Republic of China
January 1986

CONTENTS

6. ASTABLE AND MONOSTABLE MULTIVIBRATORS AND BLOCKING OSCILLATORS   189

7. BISTABLE MULTIVIBRATORS OR FLIP-FLOPS   234

8. BASIC LOGIC GATES AND LOGIC FAMILIES   260

ELECTRONIC WAVE FORMING AND PROCESSING CIRCUITS

1

INTRODUCTION TO ELECTRONIC WAVEFORMS

1.1 ANALOG AND DIGITAL SIGNALS PRODUCED BY DIFFERENT CIRCUITS

Waveforms are defined in terms of amplitude and time-interval measurements. Various signals have their own different waveforms. Before discussing waveforms, let us review the difference between the analog and digital or pulse signals.

The word *analog* comes from the word *analogous* (meaning "similar to"). Since all electronic circuits use voltages and currents to perform their work, analog circuits use currents and voltages that are analogous to whatever quantity is being described, measured, generated, and/or controlled. Thus these currents and voltages are the analog signals that pertain to data in the form of continuous variable physical quantities. The broadcasting radio AM or FM signal and TV composite signal are all the analog signals to which the AM or FM receiver and TV receiver respond, respectively.

Digital signals are quantities that are restricted to two values. They may appear as levels or pulses. Pulse signals change from one level to the other, remain at the new level for a fixed period of time, and then automatically return to the first level.

In order to illustrate the analog and digital concepts, let us assume an LED (light emitting diode) used as a light source. An LED will emit light if enough current is passed through it. This will occur if it is forward biased by the more positive voltage to the *P* material, as shown in Figure 1.1-1*a*. If the LED is reverse biased, as in Figure 1.1-1*b*, then no current flows and therefore no light is emitted. Suppose we want the brightness of the light source to be analogous to the amount of current flowing in the circuit. Then the brighter the light, the

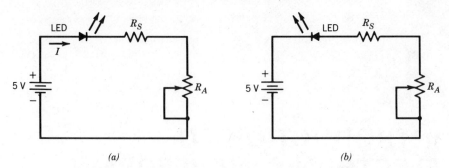

(a)   *(b)*

FIGURE 1.1-1 Analog LED circuit: (*a*) forward biased; (*b*) reverse biased.

more current that flows; the dimmer the light, the less current that flows. Now look at the forward-biased LED circuit of Figure 1.1-1*a*. The resistor R_A is used to alter the current through the LED and thereby vary the emitted light. The resistor R_S is employed to limit the current to a safe value that will not damage the LED whenever R_A is set to $0\,\Omega$. When R_A is turned to maximum resistance, there is very little current allowed to flow and the LED emits very little light. Theoretically, if we looked at the LED and saw maximum (100%) brightness (or 75% of maximum brightness, or 32.2622% of maximum brightness), we would conclude that this is analogous to maximum (100%) current (or 75% of maximum current, or 32.2622% of maximum current). Like most analog circuits, however, the differences between theoretical operation and actual operation can be large and disturbing. For instance, it is highly probable that 75% (or 32.2622%) of maximum current does not cause exactly 75% (or 32.2622%) of the maximum amount of light to be emitted. We would certainly have difficulty distinguishing the difference between 30% brightness and 33% brightness using only our eyes. The accuracy of an analog output is often less than desirable.

The simple digital circuit of Figure 1.1-2 is an alternative to the analog

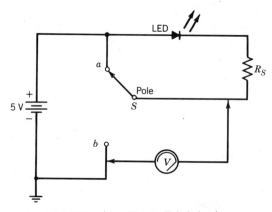

FIGURE 1.1-2 Simple digital circuit.

circuit shown in Figure 1.1-1*a*. When the pole of the switch *s* is connected to contact *a*, the voltmeter *V* will read + 5 V. However, the LED is shorted, and so no light is emitted from it. When the pole is connected to contact *b*, the voltmeter V will read 0 V. There is a path for current flow and the LED emits light. If we think of the voltmeter as a measuring device for the input voltage to this circuit and of our eyes as a measuring device for the output light of this circuit, we can summarize our observations in Table 1.1-1. The circuit of Figure 1.1-2 is behaving like a NOT circuit; that is, a low input (no voltage applied) causes a high output (maximum light emitted) and a high input (maximum voltage applied) causes a low output (no light emitted). Far more common are digital circuits that use voltages for both inputs and outputs. Thus to make the definition of a NOT operation as general as possible, rewrite Table 1.1-1 in the more popular form of Table 1.1-2. Compare Table 1.1-2*a* to *b*. Note that a 0 corresponds to a low (L) and a 1 corresponds to a high (H). There is no need for concern about the accuracy of values between these two extremes.

Notice that a perfect analog circuit could provide an infinite number of meaningful outputs, such as 100% of maximum, 75%, 33.2664%, 33.2663%, and so on. Realistically we cannot obtain perfect accuracy over the range from 0 to 100% of maximum output. The digital operation described in Table 1.1-2*b* has only two outputs. The output of 0 represents 0% of maximum output. The output of 1 represents 100% of maximum output. Since any digital input or output is restricted to the values 0 and 1, these digital devices must be working in radix 2. This is because these two values are the complete character set of this radix. Another name for this radix is *binary*.

TABLE 1.1-1
Conclusions from Figure 1.1-2

Input (Voltage at Pole Contact)	Output (Light from LED)
At *b*, no voltage measured	Maximum light emitted
At *a*, maximum voltage measured	No light emitted

TABLE 1.1-2
Conclusions from Figure 1.1-2 Rewritten

Input	Output	Input	Output
L	H	0	1
H	L	1	0

(*a*) NOT operation described with highs and lows	(*b*) NOT operation described with 0's and 1's

1.2   SINUSOIDAL WAVEFORM

The sine wave shown in Figure 1.2-1 is a repeating cycle of voltage (or current) with a sinusoidal relationship to time. It is represented by the equation $v = V_{max} \sin \omega t$ (or $i = I_{max} \sin \omega t$). The sine wave is the easiest waveform to generate, distribute, and use. All other periodic waveforms can be synthesized from sine-wave components.

The distribution and use of sine-wave ac involve inductance and capacitance. Unless sine waves are used, inductance and capacitance will result in waveform changes that prevent effective operation of a system. These waveform changes occur since the relationships between inductive or capacitive voltages and currents are dependent on rate of change ($v = L \, di/dt$ and $i = C \, dv/dt$). A sine wave has a rate-of-change curve that is also a sine wave; once a sine wave, always a sine wave.

In the electronics area, the sinusoidal waveform is produced by oscillators. There are many types of commonly used oscillators. LC oscillators usually operate at high frequencies, which can be calculated by the formula $f \simeq 1/(2\pi\sqrt{LC})$. The quartz-crystal-controlled oscillators usually operate at high frequencies with high-frequency stability. The Wien-bridge op-amp oscillators operate at frequencies up to 1 MHz, which are determined by the RC time constant of the series RC arm and the parallel RC arm.

The *average value* (V_{av}) of a sine wave is the average magnitude of a half-cycle that begins with zero magnitude; thus $V_{av} = 2V_{max}/\pi$ or $0.637 \, V_{max}$. The sine wave of voltage or current in a network will result in a certain power level. The steady dc voltage or current that will result in the same power is the *rms* (root-mean-square) or effective value of the sine wave; it is given by $V_{rms} = V_{max}/\sqrt{2}$ or $0.707 \, V_{max}$.

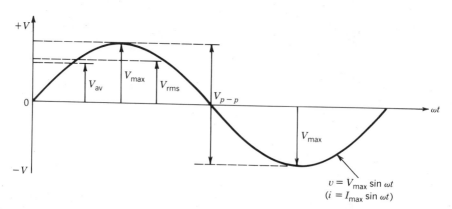

FIGURE 1.2-1   Sine-wave magnitude values.

1.3 PULSE-WAVEFORM CHARACTERISTICS

Square (Rectangular) Waveform

A *step change* occurs when a dc voltage suddenly changes from one level to another. The change might be positive or negative. A *square* or *rectangular waveform* consists of successive cycles of positive step changes followed by negative step changes. An asymmetrical square wave with two unequal half-periods is usually called the *pulse waveform.*

Ideal Pulse Waveform

An ideal pulse waveform is shown in Figure 1.3-1. The pulses are positive with respect to ground. The pulse amplitude is simply the voltage level of the top of the pulse measured from ground. The first edge of the pulse at $t = 0$ is called the *leading edge.* The second edge is called the *trailing edge.* The period T is the time measured from the leading edge of one pulse to the leading edge of the next pulse. The pulse repetition frequency (PRF) or pulse repetition rate (PRR) is equal to $1/T$ pulses per second (pps) (or Hz). The time measured from the leading edge to the trailing edge of one pulse is referred to as the *pulse width* (t_p) *or pulse duration.* The time between pulses is known as the *space width.* The proportion of the period occupied by the pulse is defined as the *duty cycle:*

$$\text{duty cycle} = \left(\frac{t_p}{T}\right) \times 100\%. \tag{1.3-1}$$

Pulse Waveform with Rise Time, Fall Time, and Tilt

Actually the top of the pulse is not perfectly flat, as shown in Figure 1.3-2. The amplitude of the trailing edge is less than that of the leading edge. There is a definite rise time t_r and fall time t_f at the leading and trailing edges of the pulse. In these cases, the pulse width (t_p) is defined as the average pulse width, and it is

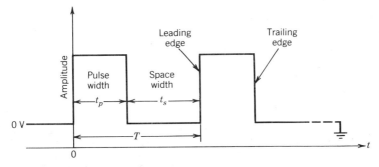

FIGURE 1.3-1 Characteristics of an ideal pulse waveform.

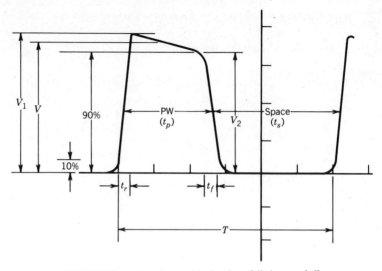

FIGURE 1.3-2   Waveform with rise time, fall time, and tilt.

normally measured at half the average amplitude. The space width (t_s) is measured at the same amplitude as the pulse width. Thus the sum of pulse width and
space width is equal to the period T:

$$t_p + t_s = T. \tag{1.3-2}$$

In Figure 1.3-2, V_1 is the maximum pulse amplitude, V_2 is the minimum
amplitude, and V is the average pulse amplitude; hence

$$V = \frac{V_1 + V_2}{2}. \tag{1.3-3}$$

The *rise time* is the time required for voltage to go from 10 to 90% of the
average amplitude. The *fall time* is the time required for the pulse to fall from
90 to 10% of the average amplitude. The *tilt* or *slope* of a waveform is defined in
terms of the average amplitude:

$$\text{tilt} = \frac{V_3}{V} \times 100\%$$

$$= \frac{V_1 - V_2}{V} \times 100\%. \tag{1.3-4}$$

The pulse waveform shown in Figure 1.3-3 is symmetrical above and below
ground level. In this case V_1 and V_2 are each measured with respect to ground,
and the tilt is given by Equation (1.3-4).

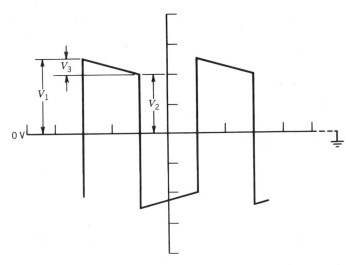

FIGURE 1.3-3 Actual waveform symmetrical above and below ground.

1.4 SAWTOOTH WAVEFORM PRODUCED BY A BASIC RAMP GENERATOR

A voltage that increases or decreases at a constant rate with respect to time ($dv/dt = $ constant α), has a graph that is a positive or negative ramp. When one ramp is much steeper than the other, the waveform is usually called the *sawtooth waveform*, as illustrated in Figure 1.4-1c.

In the basic ramp generator of Figure 1.4-1, the transistor is cut off between the input pulses and saturated during the positive pulse time. The capacitor C slowly charges between the pulses and rapidly discharges during the pulse time. As output v_o rises, capacitor C charges until its voltage at the end of 10 ms is 5 V ($= 25$ V/5). The ramp is nearly linear.

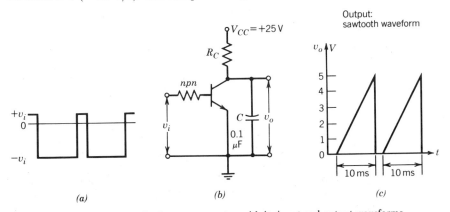

FIGURE 1.4-1 A basic ramp generator with its input and output waveforms.

Ramp generators are widely used to provide constant-velocity electron-beam deflection in cathode-ray tubes. The beam motion causes a spot of light to move steadily across the screen of the tube, usually so fast that it appears to be a line of light. This line of light is called a *trace*.

A linear ramp can be generated by a Miller integrator or a bootstrap circuit (see Chapter 10).

1.5 FREQUENCY-MODULATED (FM) WAVEFORM

Comparison of Amplitude-Modulated Waveform and Frequency-Modulated Waveform

Modulation is the process of superimposing the information contained within a frequency band onto an information carrier. The commonly used modulation methods are *amplitude modulation* (AM) and *frequency modulation* (FM). In amplitude modulation the magnitude of the carrier is varied in accordance with the modulating signal. Figure 1.5-1 depicts a carrier that is amplitude modulated by a square wave.

In frequency modulation the amplitude is kept constant while incorporating the signal into variations of the carrier frequencies. FM reduces most natural and man-made electrical noises in the form of amplitude-modulated signals. Figure 1.5-2 depicts a carrier that is frequency modulated by a square wave.

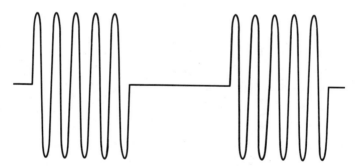

FIGURE 1.5-1 Carrier amplitude modulated by square wave.

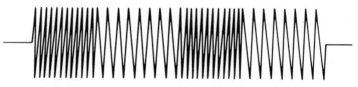

FIGURE 1.5-2 Carrier frequency modulated by square wave.

FM Waveform Produced by a Frequency Modulator

A basic frequency modulator uses a variable capacitance in parallel with the oscillator tank. This variable capacitance may be provided by a varactor diode or by a reactance circuit, as shown in Figure 1.5-3. The actual and equivalent circuits of the reactance transistor are indicated in Figures 1.5-3a and b, respectively. The applied voltage v produces a current i in the reactance network. The component values used in the reactance network are such that $X_C \gg R$. The input resistance to the base is h_{ie} or $(1 + \beta)r'_e$; it is much larger than R and can be neglected in determining i.

Thus the current i is determined solely by X_C:

$$i = \frac{v}{-jX_C} = j\frac{v}{X_C} = j2\pi fCv. \qquad (1.5\text{-}1)$$

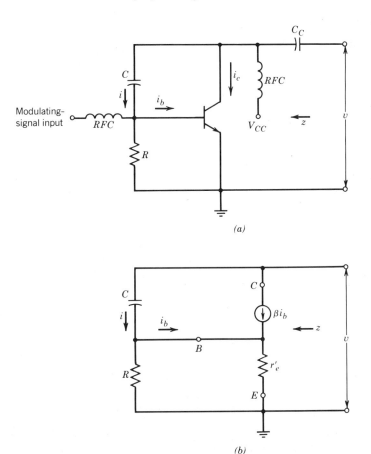

(a)

(b)

FIGURE 1.5-3   (a) Actual and (b) equivalent circuits of a basic reactance transistor. For some practical purposes, the *npn* transistor may be replaced by a *pnp* device.

The voltage from base to ground is

$$iR = j2\pi f\, RCv, \qquad (1.5\text{-}2)$$

and the base current is

$$i_b = \frac{j2\pi f\, RCv}{(1 + \beta)r'_e}. \qquad (1.5\text{-}3)$$

The collector current is

$$i_c = \beta i_b = j\,\frac{\beta}{1 + \beta}\, 2\pi f\, \frac{RC}{r'_e}\, v \simeq j2\pi f\, \frac{RC}{r'_e}\, v. \qquad (1.5\text{-}4)$$

The impedance presented by the transistor to the external voltage is

$$Z = \frac{v}{i_c} = \frac{v}{j2\pi f(RC/r'_e)v} = -j\,\frac{1}{2\pi f\, C_{eq}}, \qquad (1.5\text{-}5)$$

where

$$C_{eq} = \frac{RC}{r'_e}. \qquad (1.5\text{-}5a)$$

This equivalent capacitance C_{eq} is often described as an injected capacitance, since the circuit places a parallel capacitance across the source of v. C_{eq} is inversely proportional to r'_e, and r'_e is determined from

$$\frac{25\text{ mV}}{I_E} \leqslant r'_e \leqslant \frac{50\text{ mV}}{I_E}. \qquad (1.5\text{-}6)$$

As far as C_{eq} is concerned, the use for C_{eq} is at radio frequencies in the oscillator. Although Equation (1.5-6) is based on I_E, a dc value, r'_e can be varied at a rate determined by the modulating signal.

A system of a frequency-modulated square or triangular waveform can be made up using a voltage controlled oscillator (VCO).

1.6   PULSE-AMPLITUDE-MODULATED (PAM) WAVEFORM

Pulse-amplitude modulation (PAM) is the simplest type of pulse modulation. As the name implies, the amplitude of each pulse is made proportional to the instantaneous amplitude of the modulating signal. A one-shot multivibrator can be used for PAM, as shown in Figure 1.6-1. Normally Q_3 is on, and Q_2 is

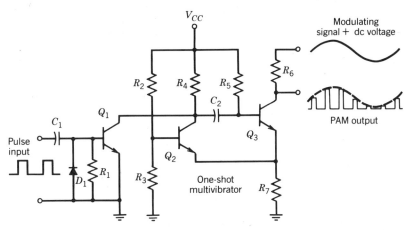

FIGURE 1.6-1   The PAM waveform produced by a one-shot multivibrator with a trigger stage.

off. When Q_3 is on, the output voltage is the saturation level of the Q_3 collector. The base of the trigger stage Q_1 is connected to a clamper consisting of C_1, R_1, and diode D_1. The modulating signal source superimposed on a proper dc voltage is supplied through the Q_3 load resistor R_6. The positive trigger pulse inverted by the Q_1 stage is applied to the Q_3 base. When Q_3 is switched off for the pulse time, the output voltage is equal to the modulating signal level. When Q_3 is saturated between the pulses, the output voltage approximates the emitter voltage across resistor R_7.

The PAM can be demodulated by a simple low-pass RC filter, as shown in Figure 1.6-2. The PAM waveform consists of the fundamental modulating frequency and a number of HF components that determine the shape of the pulses. Since the impedance of capacitor C is inversely proportional to the frequency, the low-pass filter output is the signal frequency with perhaps a very small pulse-frequency component. Surely more than one filter stage can be used to remove the pulse frequency completely.

If a radio frequency is pulse-amplitude modulated instead of simply amplitude modulated, much less power is required for the transmission of information since the transmitter is actually switched off between pulses.

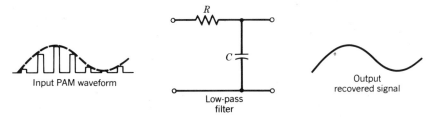

FIGURE 1.6-2   Demodulation of PAM accomplished by passing the PAM waveform through a low-pass RC filter.

1.7 HARMONIC CONTENT OF SAWTOOTH AND SQUARE WAVEFORMS

Spectrum of a Sinusoidal or Cosine Waveform

A signal waveform can be represented by a series of sine and/or cosine waves. Such a representation is termed the *spectrum* of the signal. The spectrum of a sine wave [$v = V_{max} \sin 2\pi ft = V_{max}\cos(2\pi ft - \pi/2)$] or cosine wave ($v = V_{max} \cos 2\pi ft$) is simply a straight line of height V_{max} positioned at f on the frequency axis, as shown in Figure 1.7-1*b*.

Complex Repetitive Waveforms Represented by Fourier Series

Any waveform, other than the sine or cosine wave, which repeats itself at regular intervals, is called a *complex repetitive wave*. By the mathematical operation known as Fourier analysis, complex repetitive waveforms (such as sawtooth or square waves) can be analyzed to determine their harmonic content.

The sawtooth waveform can be represented by the Fourier series

$$v = \frac{V}{2} - \frac{V}{\pi}(\sin \omega t + \tfrac{1}{2} \sin 2\omega t + \tfrac{1}{3} \sin 3\omega t + \cdots), \qquad (1.7\text{-}1)$$

where the first term is the zero-frequency or dc component. Any waveform that is not symmetrical in area about the time axis will have a zero-frequency component. The series in Equation (1.7-1) has an infinite number of terms, but it can be seen that the amplitude of each term decreases as $1/n$. It will also be seen that the Fourier series contains odd and even harmonics (i.e., components at frequencies $2f$, $3f$, $4f$, $5f$, etc.) in addition to the fundamental (at f) and dc component. The sawtooth waveform and its frequency spectrum are shown in Figure 1.7-2.

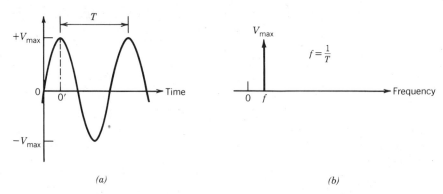

(a)   *(b)*

FIGURE 1.7-1 *(a)* A sinusoidal waveform; if the zero-time origin is started at 0' instead of 0, the wave can be described by $v = V_{max}\cos 2\pi ft$; *(b)* spectrum of a sinusoidal (or cosine) waveform.

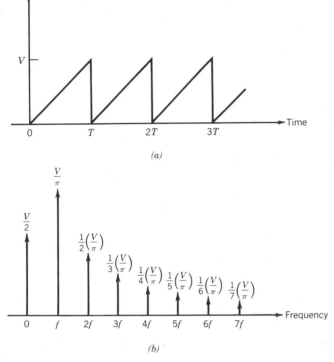

FIGURE 1.7-2 (*a*) A sawtooth waveform; (*b*) its frequency spectrum.

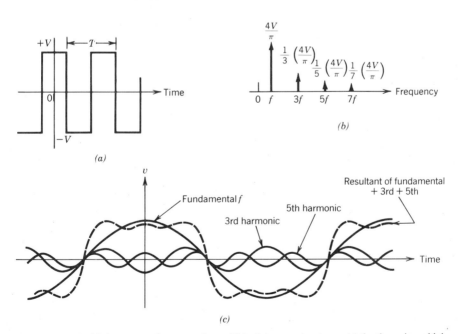

FIGURE 1.7-3 (*a*) A square-voltage waveform; (*b*) its frequency spectrum; (*c*) the three sinusoidal components add to approximate a square wave.

13

The square waveform shown in Figure 1.7-3a can be represented by the Fourier series

$$v = \frac{4V}{\pi}(\cos \omega t - \tfrac{1}{3}\cos 3\omega t + \tfrac{1}{5}\cos 5\omega t - \tfrac{1}{7}\cos 7\omega t + \cdots) \quad (1.7\text{-}2)$$

(see also Appendix 2). Note that the symmetry of the square wave about the time axis is similar to a cosine wave and as a result the series (1.7-2) contains only cosine terms. It can be seen that the series contains only odd harmonics (i.e., components at frequencies 3f, 5f, 7f, etc.) in addition to the fundamental (at f). The spectrum for the square wave is shown in Figure 1.7-3b. The first three components of the square-wave spectrum are shown in Figure 1.7-3c. Adding these produces the resultant waveform shown dotted, and it will be seen that this approaches the square-wave shape.

Distortion on Square Waves

The tilt on the top and bottom of a square waveform would result if the low-frequency components were not passed by the circuit. Both long rise times and tilt result when the involved circuitry has neither a low-enough nor a high-enough frequency response for the applied square wave. When circuits over-emphasize some of the HF harmonics, overshoots are produced as shown in Figure 1.7-4b.

The highest harmonic frequency that can be reproduced is the upper cutoff frequency (= upper 3-dB frequency) f_H. The rise time in terms of f_H is given by

$$t_r = \frac{0.35}{f_H} = \frac{0.35}{f_2}. \quad (1.7\text{-}3)$$

This equation is derived from Equation (2.2-4).

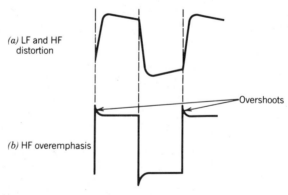

(a) LF and HF distortion

Overshoots

(b) HF overemphasis

FIGURE 1.7-4 Square-wave distortion: (a) low- and high-frequency distortion; (b) overemphasis of high frequencies.

If a square wave is applied as input to an amplifier with a lower cutoff frequency (= lower 3-dB frequency) of f_L greater than zero, then the tilt is present on the output waveform. It is found that

$$\text{Fractional tilt} = \pi \frac{f_L}{f} = \pi \frac{f_1}{f}. \tag{1.7-4}$$

This equation is derived from Equation (2.4-10).

In many cases, the rise time nearly equal to one-tenth of pulse width (or $t_r = t_p/10$) might be used as a guide for acceptable high-frequency distortion.

REFERENCES

1. Ridsdale, R. E., *Electric Circuits*, 2nd ed., McGraw-Hill, New York, 1983, Chap. 11.
2. Chirlian, P. M., *Basic Network Theory*, McGraw-Hill, New York, 1969, Chap. 11.
3. Roddy, D. and J. Coolen, *Electronic Communications*, 2nd ed., Reston Publishing Company, Reston, Virginia, 1981, Chap. 2.
4. Nilsson, J. W., *Electric Circuits*, Addison-Wesley, Reading, Massachusetts, 1983, Chaps. 18–19.
5. Caviglia, D. D., Design and Construction of an Arbitrary Waveform Generator, *IEEE Transactions on Instrumentation and Measurement*, Vol. IM-32, No. 3, September 1983, pp. 398–403.

QUESTIONS

1-1. Define analog, digital, and pulse signals.

1-2. Draw a sketch to show the waveforms of the capacitor voltage v_C and resistor voltage v_R when a dc voltage is applied to the *RC* series circuit.

1-3. Sketch the shapes of a pulse waveform, a sawtooth waveform, and a transient or brief nonrepetitive waveform.

1-4. For a pulse waveform, define leading edge, trailing edge, period, pulse repetition rate, pulse width, and duty cycle.

1-5. Define rise time, fall time, and tilt for a pulse waveform.

1-6. Why should the output voltage be a small fraction of the V_{CC} supply in the basic ramp generator of Figure 1.4-1?

1-7. Draw a waveform to indicate that the carrier is amplitude modulated by a sinusoidal waveform.

1-8. Draw a waveform to indicate that the carrier is frequency modulated by a sinusoidal waveform.

1-9. Briefly explain how the waveform of pulse-amplitude modulation is produced.

PROBLEMS

1-1. (a) Write voltage equations to express that the cosine waveform leads the sinusoidal waveform by 90°. (b) Draw a sketch to show this phase relationship.

1-2. Sketch the following pulse waveforms: (a) positive-going pulse with peak-to-peak amplitude of 10 V riding on a -2-V level and (b) negative-going pulse with peak-to-peak amplitude of 12 V riding on a -3-V level.

1-3. Derive the expressions for the R_2/R_4 ratio and the oscillating frequency from the Wien-bridge oscillator shown in Figure P1-3.

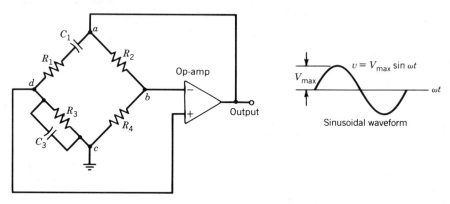

FIGURE P1-3

1-4. In the square-wave generator of Figure P1-4, assume $R_{B1} = R_{B2} = R = 100 \text{ k}\Omega$ and $C_1 = C_2 = C = 0.01 \ \mu\text{F}$. Find the frequency f.

   Hint. $f \simeq 0.725/RC$; symmetrical square-wave generated.

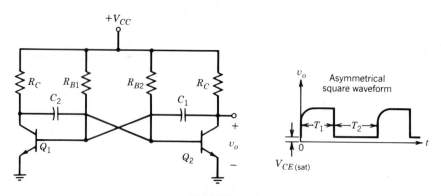

FIGURE P1-4

1-5. If the output pulse waveform from an amplifier has a rise of 0.5 μs, calculate the upper 3-dB frequency of the amplifier.

   Answer. 700 kHz

1-6. A pulse waveform has a pulse repetition frequency (PRF) of 2 kHz and a duty cycle of 5%. (a) Find the frequency of the highest harmonic required for accurate reconstruction of the waveform. (b) If the 2-kHz pulse is to be amplified by a circuit having a high frequency limit of 1 MHz, determine the minimum pulse width and duty cycle that can be reproduced accurately.

Answer. (a) 140 kHz; (b) 0.7%

1-7. Calculate the bandwidth required to amplify a 1.2 kHz square wave if the rise time of the output does not exceed 0.2 μs and 3% tilt is acceptable.

Answer. $f_H = 1.75$ MHz; $f_L = 11.46$ Hz.

1-8. A 1-kHz square-wave output from an amplifier has $t_r = 0.25$ μs and tilt = 3.5%. Calculate the upper and lower 3-dB frequencies of the amplifier.

Answer. $f_H = 1.4$ MHz; $f_L = 11.1$ Hz.

1-9. A 5-kHz square wave is an input applied to an amplifier. Determine the rise time and tilt that may be expected on the square-wave output with a bandwidth extending from 10 Hz to 500 kHz.

Answer. $t_r = 0.7$ μs; tilt = 6.3%

2

THEORY AND APPLICATION OF LINEAR WAVESHAPING CIRCUITS

2.1 *RC* NETWORK WITH SINE-WAVE INPUT

High-Pass Filter

A *high-pass network* is a high-pass filter that attenuates low frequencies and causes a flat-topped pulse to decay. Its simplest form is the *RC* network shown in Figure 2.1-1*a*. If v_i is a sine-wave generator of frequency f, the output of the network as a function of frequency (see Figure 2.1-1*b*) is

$$v_o = \frac{v_i}{\sqrt{1 + (f_1/f)^2}} \quad \text{and} \quad \theta = \arctan \frac{f_1}{f}, \tag{2.1-1}$$

where

$$\frac{1}{\sqrt{1 + (f_1/f)^2}} = \text{magnitude of gain } |A|[1], \tag{2.1-2}$$

$$f_1 = f_L = \text{lower 3-dB frequency} = \text{cutoff frequency} = 1/2\pi RC,$$

$$\theta = \text{angle}[2] \text{ by which the output leads the input.}$$

---

[1] From the phasor diagram,

$$|A| = \frac{R}{\sqrt{R^2 + X_c^2}} = \frac{1}{\sqrt{1 + (X_c/R)^2}} = \frac{1}{\sqrt{1 + (f_1/f)^2}}$$

(assuming $R = 1/2\pi f_1 C$).
[2] $\tan \theta = X_c/R = (1/2\pi f)/(1/2\pi f_1) = f_1/f$.

At the frequency f_1, the magnitude of the capacitive reactance is equal to the resistance and the gain is 0.707. This drop in signal level corresponds to a signal reduction of 3 dB.[3] The maximum possible gain is equal to 1.

From Equation (2.1-1), $|A| \simeq f/f_1$ for $f/f_1 < 1$ (below cutofff) or $f_1/f > 1$. Thus

$$A_{dB} = 20 \log \frac{f}{f_1} = -20 \log \frac{f_1}{f}. \qquad (2.1\text{-}3)$$

For $f/f_1 \geqslant 1$ (above or at cutoff), or $f_1/f \leqslant 1$,

$$|A| = \frac{1}{\sqrt{1 + (f_1/f)^2}} \simeq 1 \text{ and } A_{dB} = 20 \log 1 = 0 \text{ dB}. \qquad (2.1\text{-}4)$$

The response of the network shown in Figure 2.1-1*b* is listed in Table 2.1-1.

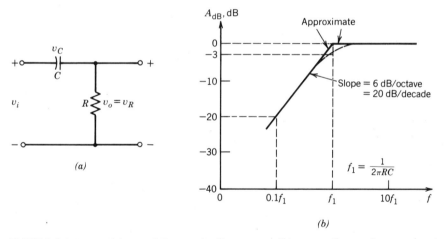

FIGURE 2.1-1 (*a*) High-pass *RC* network; (*b*) response of the network to a sine-wave input.

TABLE 2.1-1

Formula	$f/f_1 < 1, A_{dB} = -20 \log(f_1/f)$				$f/f_1 \geqslant 1, A_{dB} = 0$				
f/f_1	0.1	0.2	0.4	0.8	1	2	4	10	100
f_1/f	10	5		2.5	1.25				
A_{dB}	-20	$-20 \times 0.7 = -14$	-8	-2	0	0	0	0	0

Remark: For example, if the frequency ratio = 0.2/0.1 = 2:1, the A_{dB} difference = $-14 - (-20) = 6$.

[3]dB = $10 \log P_2/P_1 = 20 \log V_2/V_1$.

From the data listed, we see the slope of the curve is 6 dB/octave or 20 dB/decade as shown. An *octave* is defined as a 2:1 frequency ratio; a decade is defined as a 10:1 frequency ratio.

Low-Pass Filter

A *low-pass network* is a low-pass filter that attenuates high frequencies, causing a rounding of the leading edge of a square pulse. Its simplest form is the *RC* network shown in Figure 2.1-2*a*. If v_i is a sine-wave generator of frequency f, the output of the network as a function of frequency (see Figure 2.1-2*b*) is

$$v_o = \frac{v_i}{\sqrt{1 + (f/f_2)^2}} \quad \text{and} \quad \phi = -\arctan\frac{f}{f_2} \tag{2.1-5}$$

where

$$\frac{1}{\sqrt{1 + (f/f_2)^2}} = \text{magnitude of gain } |A|. \tag{2.1-6}$$

$f_2 = f_H$ = upper 3-dB frequency = cutoff frequency = $1/2\pi RC$;

ϕ = angle by which the output leads the input.

The gain falls to 0.707 of its low-frequency value at frequency f_2. For $f/f_2 \leqslant 1$ (below or at cutoff), $|A| \simeq 1$ and $A_{dB} = 0$. For $f/f_2 > 1$ (above cutoff), $|A| \simeq f_2/f$ and $A_{dB} = 20 \log (f_2/f) = -20 \log(f/f_2)$.

Simple Active High-Pass and Low-Pass Filters

A simple active high-pass filter shown in Figure 2.1-3 is a high-pass *RC* network connected to the noninverting input terminal of the *voltage* (unity) *follower* (an

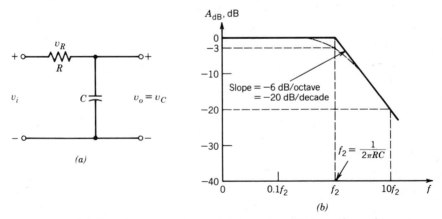

FIGURE 2.1-2   (*a*) Low-pass *RC* network; (*b*) response of the network to a sine wave.

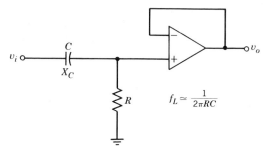

FIGURE 2.1-3 A simple active high-pass filter.

op-amp with its output connected to the inverting input). A voltage follower provides a voltage gain (amplification) of unity with no phase reversal; it acts like an emitter follower and is primarily used for matching impedance purposes. When the frequency of the input voltage v_i is less than the cutoff frequency $f_L(\simeq 1/2\pi RC)$, the capacitive reactance X_c is large, so that most of the applied voltage is dropped across capacitor C, and the voltage across resistor R is very small. Since the circuit is a voltage follower, its output voltage v_o is also very small. When the input frequency is higher than f_L, reactance X_c is low; thus most of the applied voltage is dropped across resistor R, and output v_o is nearly equal to input v_i. The resulting slope of the frequency-response curve is 20 dB per decade.

A simple active low-pass filter is shown in Figure 2.1-4. Its operation can be explained referring to the preceding discussion of the low-pass RC network and the voltage follower. With $R = 10\,\text{k}\Omega$ and $C = 0.1\,\mu\text{F}$, the cutoff frequency is

$$f_H \simeq \frac{1}{2\pi RC} = \frac{1}{6.28(10 \times 10^3)(0.1 \times 10^{-6})} = 159 \text{ Hz.}$$

The slope of the frequency-response curve is $-20\,\text{dB}$ per decade.

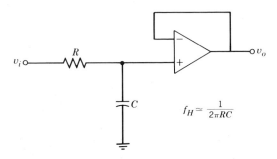

FIGURE 2.1-4 A simple active low-pass filter.

2.2 *RC* NETWORK WITH STEP-FUNCTION (STEP-VOLTAGE) INPUT

High-Pass Filter (Figure 2.1-1*a*)

The *unit step function* U(t) is defined as having the value zero for all negative times and the value unity for all positive times. A *step* (or *step-function*) *voltage* is one that maintains the value zero for all times $t < 0$ and maintains the value V for all times $t > 0$. The response of the network to the step-voltage input is exponential, with a time constant $\tau \equiv RC$, and the output voltage v_o is of the form

$$v_o = V_f + (V_i - V_f)e^{-t/\tau}, \tag{2.2-1}$$

where V_f and V_i are the final and initial output voltages, respectively. If we assume that the capacitor is initially uncharged, then the output voltage must jump to V immediately after $t = 0$ (at $t = 0+$), since the capacitor voltage cannot change instantaneously. For $t > 0$ the input is a constant, and since the capacitor blocks the dc component of the input, the final output voltage is zero, or $V_f = 0$. Then Equation (2.2-1) becomes

$$v_o(= v_R) = Ve^{-t/RC}. \tag{2.2-2}$$

The output is 0.607 of its initial value at 0.5 *RC*, 0.368 at 1*RC*, 0.135 at 2 *RC*,

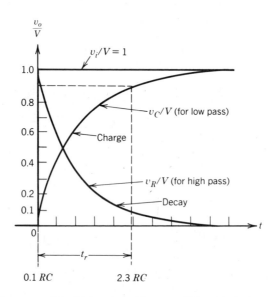

FIGURE 2.2-1 Response of the high-pass and low-pass *RC* networks to a step-voltage input.

0.05 at 3 *RC*, 0.018 at 4 *RC*, and 0.007 at 5 *RC*. Equation (2.2-2) can be written

$$\frac{v_R}{V} = e^{-t/RC},$$

which is of the form $y = e^{-x}$.

Low-Pass Filter (Figure 2.1-2*a*)

The response of the low-pass network to a step-function input is exponential with a time constant *RC*. Because the capacitor voltage v_C cannot change instantaneously, the output starts from zero and rises toward the steady-state value *V*, as indicated in Figure 2.2-1. Thus, according to Kirchhoff's voltage law,

$$v_C = v_i - v_R;$$

and from Equation (2.2-2), we have

$$v_o(= v_C) = V(1 - e^{-t/RC}), \qquad (2.2\text{-}3)$$

or $v_C/V = 1 - e^{-t/RC}$, which is of the form $y = 1 - e^{-x}$. If the capacitor initial charge voltage $V_{CI} \neq 0$, then

$$v_C = V - (V - V_{CI})e^{-t/RC}. \qquad (2.2\text{-}3a)$$

As mentioned previously, the pulse rise time t_r is the time interval between the 10 and 90% amplitude points on the waveform (see Figure 2.2-1). The time required for v_C to reach 10% of its final value is 0.1 *RC*, and the time required for it to reach 90% of its final value is 2.3 *RC*. The difference between these two values is the rise time t_r of the network and is given by

$$t_r = 2.2\, RC = \frac{2.2}{2\pi f_2} = \frac{0.35}{f_2} = \frac{0.35}{f_H}, \qquad (2.2\text{-}4)$$

where $f_2 = f_H = 1/2\pi RC$, by which the high-frequency 3-dB point is given.

2.3   *RC* NETWORK WITH PULSE INPUT

High-Pass Filter

If the pulse in Figure 2.3-1*a* is applied to the circuit of Figure 2.1-1*a*, the response for times that are less than the pulse duration t_p is the same as that for the step-voltage input. At the end of the pulse, the input falls abruptly by the amount *V*

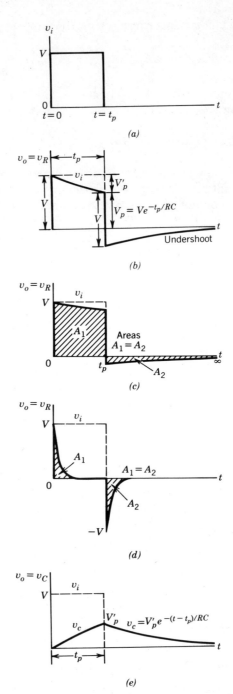

FIGURE 2.3-1 (*a*) A pulse applied to the circuits of Figures 2.1-1*a* and 2.1-2*a*; (*b*) exponential decay of the output pulse v_R; (*c*) output pulse v_R if $RC \gg t_p$; (*d*) output v_R if $RC \ll t_p$; (*e*) response of the low-pass RC network to a pulse input.

and, since the capacitor voltage cannot change instantaneously, the output must also drop by V. Thus, immediately after $t = t_p$ (or at $t = t_p +$), $v_0 = V_p - V$ ($= v_R$ at $t_p +$); v_0 becomes negative and then decays exponentially to zero (at $t = \infty$, ideally), as shown in Figure 2.3-1b. For $t > t_p$, v_o is given by

$$v_o(= v_R) = V(e^{-t_p/RC} - 1)e^{-(t-t_p)/RC}, \qquad (2.3\text{-}1)$$

which is of the form $y = -ae^{-x}$. For all values of the ratio RC/t_p there must always be an undershoot, and the area below the axis will always equal the area above (Problem 2-2). Because the input and output are separated by the blocking capacitor C, the dc or average level of the output signal is zero for the high-pass RC network. If $RC \gg t_p$, the capacitor charges slowly, and there is only a slight tilt to the output pulse and the undershoot is very small (see Figure 2.3-1c). If $RC \ll t_p$, the capacitor charges rapidly, and the output consists of a positive spike or pip of amplitude V at the beginning of the pulse and a negative spike of the same size at the end of the pulse, as shown in Figure 2.3-1d. The output negative spike is actually the capacitor voltage that discharges via the input and output terminals at $t = t_p$.

Low-Pass Filter

The response of the low-pass network to a pulse input for times t less than the pulse width t_p is the same as that for a step-function input and is given by Equation (2.2-3). At $t = t_p$, the voltage is V_p' and the output must decrease to zero from this value with a time constant RC, as shown in Figure 2.3-1e. Note that the output will always extend beyond the pulse width t_p, since whatever charge has accumulated on C during the pulse duration cannot leak off instantaneously. If it is desired to minimize this distortion, then the rise time must be small relative to the pulse width. A pulse shape will be preserved if the 3-dB frequency f_2 is approximately equal to the reciprocal of the pulse width t_p or if the rise time is $t_r \simeq 0.35\, t_p \,(= 2.2/2\pi f_2)$. Thus, to pass a 0.4-$\mu$s pulse reasonably well requires a circuit with an upper 3-dB frequency of the order of 2.5 MHz ($= 1/0.4\,\mu$s) or a rise time of 0.14 μs ($= 0.35 \times 0.4\,\mu$s).

2.4   RC NETWORK WITH SQUARE-WAVE INPUT

Periodic Waveform

For any periodic waveform, the average value is

$$V_{\text{av}} = \frac{A}{T}, \qquad (2.4\text{-}1)$$

where A is the area under the curve and T is the period. The effective value of the waveform is

$$V_{rms} = \sqrt{\frac{A'}{T}}, \qquad (2.4\text{-}2)$$

where A' is the area under the square of the waveform.

High-Pass Filter

Figure 2.4-1a shows a square wave that maintains itself at one constant level V' for a time T_1 and at another constant level V'' for a time T_2 and that is repetitive with a period $T = T_1 + T_2$. For this square-wave input or any other periodic-input waveform, the average level of the steady-state output signal from the circuit of Figure 2.1-1a is always zero and is independent of the dc level of the input. The output must consequently extend in both the positive and negative directions with respect to the zero-voltage axis, and the area of the part of the waveform above the zero axis must equal the area below the zero axis. When the input changes discontinuously by an amount V, the output changes discontinuously by an equal amount and in the same direction. During any finite time interval when the input maintains a constant level, the output decays exponentially toward zero voltage. In the limiting case, where RC/T_1 and RC/T_2 are both arbitrarily large in comparison with unity, the output waveform will be identical to the input, except that the dc component will be lacking (see Figures 2.4-1a and b). At the other extreme, if RC/T_1 and RC/T_2 are both very small in comparison with unity, the output will consist of alternate positive and negative peaks (see Figure 2.4-1d), and the peak-to-peak amplitude of the output will be twice the peak-to-peak amplitude of the input (see Figure 2.4-1a).

More generally, the response to a square wave must appear as in Figure 2.4-1c where

$$V'_1 - V_2 = V, \qquad V'_1 = V_1 e^{-T_1/RC}, \qquad (2.4\text{-}3)$$

$$V_1 - V'_2 = V, \qquad V'_2 = V_2 e^{-T_2/RC}. \qquad (2.4\text{-}4)$$

For a symmetrical square wave, $T_1 = T_2 = T/2$, $V_1 = -V_2$, and $V'_1 = -V'_2$. Under this condition the equations in (2.4-3) are identical with those in (2.4-4). Then from Equation (2.4-4),

$$V_1 - V_2 e^{-T/2RC} = V_1 + V_1 e^{-T/2RC} = V,$$

$$V_1 = \frac{V}{(1 + e^{-T/2RC})}; \qquad (2.4\text{-}5)$$

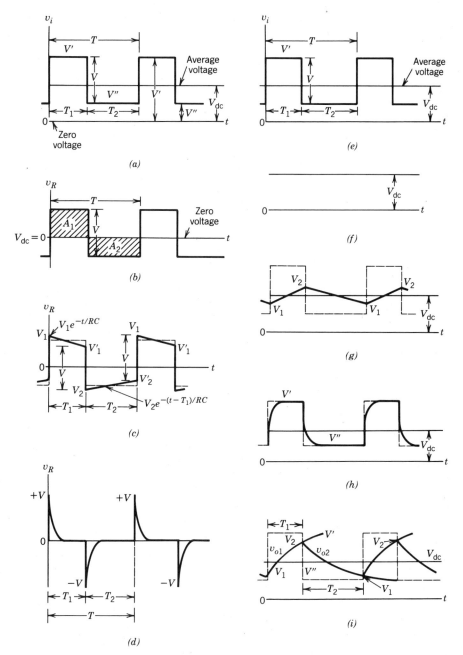

FIGURE 2.4-1 (*a*) Square wave applied to the high-pass *RC* network; (*b*) output waveform identical to the input if $RC \gg T$; (*c*) general output-waveform (v_R) if $RC \gg T$; (*d*) peaking of square wave if $RC \ll T$; (*e*) square wave applied to the low-pass *RC* network; (*f*) v_C is V_{dc} while v_R is a square-wave if $RC \gg T$; (*g*) v_C waveform if $RC \gg T$; (*h*) v_C waveform if $RC \ll T$; (*i*) v_C waveform if *RC* is comparable to *T.*

and from Equation (2.4-3),

$$V'_1 = V_1 \, e^{-T/2RC};$$

substituting V_1 from Equation (2.4-5), we have

$$V'_1 = \frac{V}{1 + e^{-T/2RC}} \, e^{-T/2RC}$$

or

$$V'_1 = \frac{V}{1 + e^{T/2RC}}. \tag{2.4-6}$$

For $T/2RC \ll 1$,

$$e^{-T/2RC} = 1 - T/2RC + \frac{1}{2}\left(\frac{-T}{2RC}\right)^2 - \cdots \simeq 1 - T/2RC; \tag{2.4-7}$$

substituting into Equation (2.4-5), we have

$$V_1 = \frac{V}{2 - T/2RC} = \frac{V}{2}\left(\frac{1}{1 - T/4RC}\right) = \frac{V}{2}\left[1 \div \left(1 - \frac{T}{4RC}\right)\right] \simeq \frac{V}{2}\left(1 + \frac{T}{4RC}\right)$$

or

$$V_1 \simeq \frac{V}{2}\left(1 + \frac{T}{4RC}\right). \tag{2.4-8}$$

In the same manner, Equation (2.4-6) becomes

$$V'_1 \simeq \frac{V}{2}\left(1 - \frac{T}{4RC}\right). \tag{2.4-9}$$

Then the exponential portions of the output are approximately linear, as indicated in Figure 2.4-2. The percentage tilt on the waveform is

$$P_{\text{tilt}} = \frac{V_1 - V'_1}{V/2} \times 100\% \simeq \frac{T}{2RC} \times 100\% \tag{2.4-10}$$

or

$$P_{\text{tilt}} \simeq \pi \frac{f_1}{f} \times 100\% = \pi \frac{f_L}{f} \times 100\%, \tag{2.4-11}$$

where $f_1 = f_L = 1/2\pi RC$, by which the low-frequency 3-dB point is given,
f = frequency of the applied square wave = $1/T$.

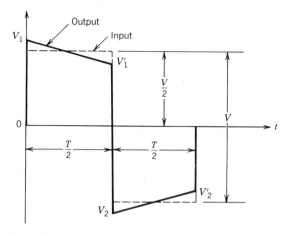

FIGURE 2.4-2   Linear tilt of a symmetrical square wave when $RC \gg T$. Time constant RC is much greater than T, so that areas of the parts of v_R and v_i waveforms above (or below) zero-voltage axis are nearly equal.

Low-Pass Filter

A square wave applied to a low-pass RC network (Figure 2.1-2a) is shown in Figure 2.4-1e. If the time constant RC is small relative to the period T of the input square wave, the output will appear as in Figure 2.4-1h, which is a reasonable reproduction of the input. If the time constant RC is comparable to the period T, the output will be as in Figure 2.4-1i. If $RC \gg T$, the output will consist of exponential sections that are essentially linear (see Figure 2.4-1g).

By referring to Equation (2.2-1), one can give the equations for the rising portion and the falling portion in Figure 2.4-1i. They are

$$v_{o1} = V' + (V_1 - V')e^{-t/RC} \tag{2.4-12}$$

and

$$v_{o2} = V'' + (V_2 - V'')e^{-(t-T_1)/RC}. \tag{2.4-13}$$

2.5   *RC* NETWORK WITH EXPONENTIAL INPUT

Exponential Waveform

The exponential waveform with a time constant RC is given by Equation (2.2-3). Similarly the exponential waveform with a time constant τ is given by

$$v_i = V(1 - e^{-t/\tau}), \tag{2.5-1}$$

which is plotted in Figure 2.5-1.

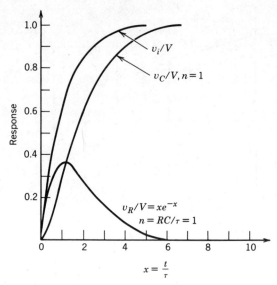

FIGURE 2.5-1 Response of the high-pass and low-pass RC networks to an exponential input. The curves for v_R/V and v_C/V when $n > 1$ can be drawn according to the data obtained from Equations (2.5-6) and (2.5-11).

High-Pass Filter

In any RC network, $v_i = it/C + v_R$. Differentiating this equation gives

$$\frac{dv_i}{dt} = \frac{i}{C} + \frac{dv_R}{dt} \quad \text{or} \quad \frac{dv_i}{dt} = \frac{v_R}{RC} + \frac{dv_R}{dt}. \tag{2.5-2}$$

Suppose the input of a high-pass RC network is an exponential waveform given by Equation (2.5-1); then Equation (2.5-2) becomes

$$\frac{V}{\tau} e^{-t/\tau} = \frac{v_R}{RC} + \frac{dv_R}{dt}. \tag{2.5-3}$$

We define n and x by

$$n \equiv \frac{RC}{\tau} \quad \text{and} \quad x \equiv \frac{t}{\tau}.$$

Multiplying Equation (2.5-3) by $e^{t/RC}$ yields

$$\frac{V}{\tau} e^{-t/\tau + t/RC} = \frac{v_R}{RC} e^{t/RC} + e^{t/RC} \frac{dv_R}{dt},$$

or

$$\frac{V}{\tau} e^{[(\tau - RC)/\tau RC]t} = \frac{d}{dt}(v_R e^{t/RC}).$$ (2.5-4)

If $n = RC/\tau = 1$, then

$$\frac{V}{\tau} = \frac{d}{dt}(v_R e^{t/RC}).$$

Integrating, we obtain

$$\int_0^t \frac{V}{\tau} dt = \int d(v_R e^{t/RC}) \quad \text{or} \quad \frac{Vt}{\tau} = v_R e^{t/RC} = v_R e^{t/\tau}.$$

Substituting t/τ by x, we obtain

$$v_R = Vxe^{-x}, \quad \text{if } \tau = RC.$$ (2.5-5)

A set of data from Equation (2.5-5) is listed in Table 2.5-1, and the response of a high-pass *RC* network to an exponential input is indicated in Figure 2.5-1.

If $n \neq 1$, the solution[1] of Equation (2.5-4) is

$$v_R = \frac{nV}{n-1}(e^{-x/n} - e^{-x}).$$ (2.5-6)

---

[1] From Equation (2.5-4),

$$\frac{V}{\tau} e^{[(\tau - RC)/\tau RC]t} dt = d(v_R e^{t/RC}).$$

Integrating,

$$\frac{V}{\tau} \int e^{[(\tau - RC)/\tau RC]t} dt = \frac{V}{\tau} \frac{\tau RC}{\tau - RC} \int d e^{[(\tau - RC)/\tau RC]t}$$

$$= \int d(V_R e^{t/RC})$$

$$\frac{VRC}{\tau - RC} e^{[(\tau - RC)/\tau RC]t} = v_R e^{t/RC} + K.$$ (2.5-7)

Substituting the initial condition $v_R = 0$ at $t = 0$, we obtain

$$K = \frac{VRC}{\tau - RC}.$$

Now returning to Equation (2.5-7), we have

$$\frac{VRC}{\tau - RC} e^{[(\tau - RC)/\tau RC]t} = v_R e^{t/RC} + \frac{VRC}{\tau - RC}.$$

Substituting *RC* by $n\tau$ and t by $x\tau$ and then solving for v_R, we obtain Equation (2.5-6).

TABLE 2.5-1
A Set of Data from Equations (2.5-5) and (2.5-10)

x	e^{-x}	$1 - e^{-x} = v_i/V$	$xe^{-x} = v_R/V$	$1 - e^{-x} - xe^{-x} = v_C/V$
0.2	0.818	0.182	0.163	0.019
0.5	0.606	0.394	0.30	0.094
1	0.370	0.630	0.37	0.260
2	0.135	0.865	0.27	0.595
3	0.050	0.950	0.15	0.800
4	0.018	0.982	0.07	0.912
5	0.0067	0.993	0.03	0.963

If n is a large number, the capacitor C will charge slowly. The curve for v_R/V when $n > 1$ can be drawn according to the data obtained from Equation (2.5-6).

Low-Pass Filter

With the exponential input, the output of the low-pass RC network (Figure 2.1-2a) is given by

$$v_O = v_C = V(1 - e^{-x}) - Vxe^{-x} = V(1 - e^{-x} - xe^{-x}) \qquad (2.5\text{-}8)$$

if $n = 1$ and

$$v_o = v_C = V\left[1 - e^{-x} - \frac{n}{n-1}(e^{-x/n} - e^{-x})\right] \qquad (2.5\text{-}9)$$

if $n \neq 1$. Thus,
for $n = 1$,

$$\frac{v_C}{V} = 1 - (1 + x)e^{-x}, \qquad (2.5\text{-}10)$$

and for $n \neq 1$,

$$\frac{v_C}{V} = 1 + \frac{1}{n-1}e^{-x} - \frac{n}{n-1}e^{-x/n}. \qquad (2.5\text{-}11)$$

The response of the low-pass circuit to the exponential input is plotted in Figure 2.5-1.

2.6 THE RC NETWORK USED AS A BASIC EXPONENTIAL PULSE GENERATOR

Generating an Exponential Waveform

As described previously, an exponential waveform may be obtained from the RC network when a step voltage is applied. If an automatic two-position switch is used in this RC network, capacitor C will charge on one position when a dc voltage is applied and will discharge through resistor R on the other position. When the capacitor voltage (V) is discharged, the output voltage v_o (across R) will be an exponential waveform, as given by the expression

$$v_o = Ve^{-t/RC}. \tag{2.6-1}$$

The inverted waveform of v_o/V is similar to that of v_i/V in Figure 2.5-1.

A Practical Exponential-Waveform Generator

The exponential waveform is used an an input to instruments for testing linearity and waveshaping. Figure 2.6-1 is the circuit of the exponential-waveform generator. The output voltage from the power supply is either positive or negative, which is controlled by the Polarity switch. The Attenuation switch provides two factors, 1 and 10, with the 10-step control. For the factor 1 (or 10), 1 (or 0.1) V/step is given. The vibrating mercury relay operates at 60 Hz and is used to control the charge and discharge rate of the decay-time capacitors, so that the exponential output of the RC network is at the rate of 60 Hz. The decay time is mainly determined by the decay-time capacitor, 1-kΩ resistor, and load.

2.7 RC NETWORK WITH RAMP INPUT

Slope of the Ramp

A ramp or sweep voltage (V) is a waveform that is zero for $t < 0$ and that increases linearly with time for $t > 0$. Thus $v = \alpha t$, where α is the linearly rising rate or the slope of the ramp, and $\alpha = v/t = dv/dt = $ constant, as shown in Figure 2.7-1.

High-Pass Filter

For the time constant RC much longer than the period T, the capacitor (Figure 2.1-1a) charges slowly, so that at $t = T$, the response only has a small deviation from linearity, as shown in Figure 2.7-1a.

For $RC \ll T$, the capacitor charges rapidly. Except near $t = 0$, the charging current $i = C\, dv_i/dt = C\alpha = $ constant. Thus $v_R = iR = \alpha RC = $ constant.

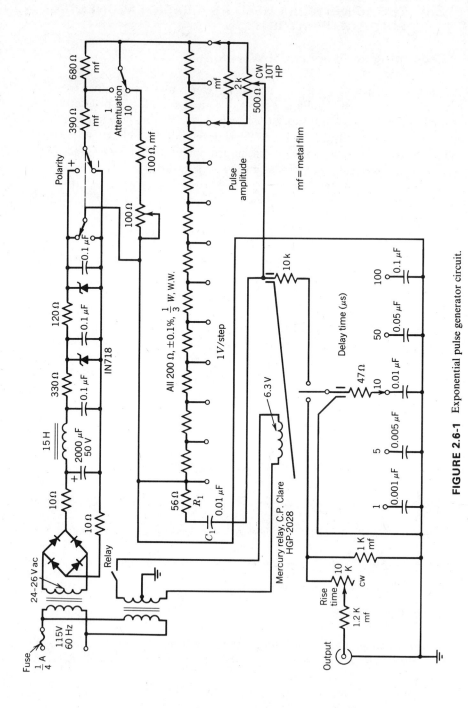

FIGURE 2.6-1 Exponential pulse generator circuit.

34

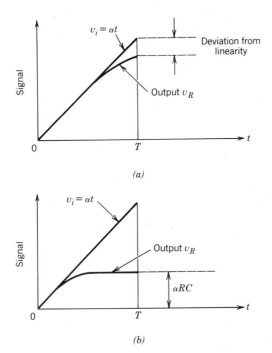

(a)

(b)

FIGURE 2.7-1 (*a*) Response of a high-pass *RC* network to a ramp voltage for $RC \gg T$; (*b*) response to a ramp voltage for $RC \ll T$.

As we know, when $RC \ll t$, $v_R \simeq \alpha RC$, and when $RC \gg t$ or $t \rightarrow 0$, $v_R \simeq 0$; then, in the general case, the output voltage across the resistor is given by

$$v_R = \alpha RC(1 - e^{-t/RC}). \tag{2.7-1}$$

Clearly this output is governed by Equation (2.5-2), which becomes

$$\alpha = \frac{v_R}{RC} + \frac{dv_R}{dt}. \tag{2.7-2}$$

Equation (2.7-2) has the solution[1] given by Equation (2.7-1). For $t \ll RC$, the

---

[1] Multiplying Equation (2.7-2) by $e^{t/RC}$ yields

$$\alpha e^{t/RC} = \left(\frac{v_R}{RC}\right) e^{t/RC} + e^{t/RC}\left(\frac{dv_R}{dt}\right).$$

$$\int \alpha e^{t/RC} \, dt = \int d(v_R e^{t/RC}); \quad \alpha RC \int d(e^{t/RC}) = \int d(v_R e^{t/RC}); \quad \alpha RC e^{t/RC} = v_R e^{t/RC} + K;$$

since $v_R = 0$ at $t = 0$, $K = \alpha RC$; substituting, we have $\alpha RC e^{t/RC} = V_R e^{t/RC} + \alpha RC$; solving for v_R, we obtain Equation (2.7-1).

exponential in Equation (2.7-1) may be replaced by the series

$$e^{-t/RC} = 1 - \frac{t}{RC} + \frac{1}{2!}\left(\frac{t}{RC}\right)^2 - \frac{1}{3!}\left(\frac{t}{RC}\right)^3 + \cdots$$

and the result is given by

$$v_R = \alpha t \left(1 - \frac{t}{2RC} + \cdots\right). \tag{2.7-3}$$

The ramp input and output for $RC \gg T$ is shown in Figure 2.7-1a. As a measure of the departure from linearity, the transmission error at $t = T$ is defined by

$$e_t = \frac{v_i - v_R}{v_i} \simeq \frac{T}{2RC} = \pi f_1 T, \tag{2.7-4}$$

where $f_1 = 1/2\pi RC$; for example, if we want to pass a 1-msec sweep with less than 0.1% deviation from linearity, Equation (2.7-4) yields $RC > 0.5$ sec or $f_1 < 0.3$ Hz.

Low-Pass Filter

For $v_i = \alpha t$, the output across the capacitor (Figure 2.1-2a) is given by

$$v_C = v_i - v_R = \alpha t - \alpha RC(1 - e^{-t/RC})$$

or

$$v_C = \alpha(t - RC) + \alpha RC e^{-t/RC}. \tag{2.7-5}$$

If $RC \ll T$, the output becomes $v_C \simeq \alpha(t - RC)$, and its waveform appears as in Figure 2.7-2a, where it is seen that the output follows the input but is delayed

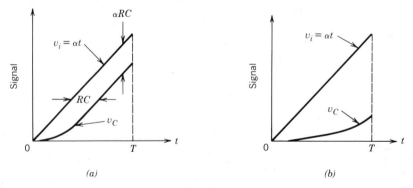

(a)   *(b)*

FIGURE 2.7-2 Response of low-pass RC network to a ramp input: (a) $RC/T \ll 1$; (b) $RC/T \gg 1$.

by one time constant RC from the input (except near the origin where there is distortion). The transmission error e_t is defined as the difference between input and output divided by the input at $t = T$. For $RC/T \ll 1$, we find

$$e_t \simeq \frac{RC}{T} = \frac{1}{2\pi f_2 T},$$

(2.7-6)

where $f_2 = 1/2\pi RC$.

If $RC/T \gg 1$, the capacitor charges slowly, and thus the output is very distorted, as in Figure 2.7-2b. Replacing the exponential in Equation (2.7-5) by a series, we obtain

$$v_C \simeq \frac{\alpha t^2}{2RC}.$$

(2.7-7)

This quadratic response indicates that the circuit acts as an integrator, since

$$v_C = \frac{1}{RC} \int_0^t \alpha t \, dt = \frac{\alpha t^2}{2RC}.$$

(2.7-8)

2.8 *RC* DIFFERENTIATORS AND INTEGRATORS

The High-Pass Filter as a Differentiator

A *differentiator* is a network in which the output is proportional to the derivative of the input signal. One of its simplest forms is a high-pass RC network. If, as in Figure 2.1-1a, the time constant is very small compared with the time required for the input signal to make an appreciable change, then the network is referred to as a differentiator. In this case, the capacitor charges rapidly, and thus the voltage drop across R will be small relative to the drop across C. We may consider that the total input v_i appears across C, so that the current is determined entirely by C. Therefore the current is $C dv_i/dt$, and the output signal across R is

$$v_o = RC \frac{dv_i}{dt}.$$

(2.8-1)

Differentiating a square wave gives a waveform that is uniformly zero except at the points of discontinuity. At these points, precise differentiation would yield impulses of infinite amplitude, zero width, and alternating polarity. However, in the actual waveform provided by the RC differentiator, the amplitude of the peaks never exceeds V, as shown in Figure 2.4-1d. Such an error exists because at the time of the discontinuity, the voltage across R is not negligible compared with that across C.

For the sweep input $v_i = \alpha t$, the output signal is

$$v_o = RC\frac{dv_i}{dt} = \alpha RC.$$

This result is true except near $t = 0$ (see Figure 2.7-1b). The error near $t = 0$ is again due to the fact that in this region the voltage across R is not negligible compared with that across C.

Consider how to obtain a criterion for good differentiation in terms of steady-state analysis. If a sine wave $v_i = V_m \sin \omega t$ is applied to the circuit of Figure 2.1-1a, the output will be a sine wave shifted by a leading angle θ such that

$$\tan \theta = \frac{X_c}{R} = \frac{1}{\omega RC}, \tag{2.8-2}$$

and the output will be proportional to $\sin(\omega t + \theta)$. In order to obtain

$$RC\frac{dv_i}{dt} = V_m \omega RC \cos \omega t,$$

the angle θ must equal 90°. If $\omega RC = 0.01$, then $1/\omega RC = 100$ and $\theta = 89.4$, which is sufficiently close to 90° for most purposes. Therefore, if $\omega RC \ll 1$, the output is approximately the expected value

$$v_0 = V_m \omega RC \cos \omega t = V_m \omega RC \sin (\omega t + 90°).$$

Double Differentiation

Assume that the inverting amplifier of Figure 2.8-1 operates linearly and that its output impedance is small compared with the impedance of R_2 and C_2, so that this combination does not load the amplifier. Let R_1 be the parallel combination of R and the input impedance of the amplifier. If the time constants

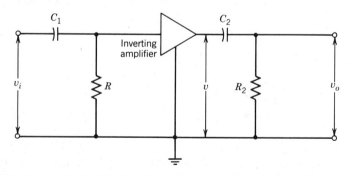

FIGURE 2.8-1 Double-RC-clipped amplifier performing a double differentiation.

R_1C_1 and R_2C_2 are small compared with the period of the input signal, then this double-*RC*-clipped amplifier performs approximately a second-order differentiation.

Consider the exponential waveform $v_i = V(1 - e^{-t/\tau})$ applied to the circuit of Figure 2.8-1. If $R_1C_1 = R_2C_2 = \tau$, the output[1] is given by

$$v_o = -AVx(1 - x/2)e^{-x}, \qquad (2.8\text{-}3)$$

where A is the magnitude of the amplifier gain and $x \equiv t/\tau$. This result is plotted in Figure 2.8-2. The initial slope of v_o is $-AV/\tau$, since V/τ is the initial slope of v_i. Note that the output waveform has as much area above the time axis as it does below, which is a fact of importance in pulse spectrometry. However this area balance will fail if the amplifier is driven out of its linear range.

The Low-Pass Filter as an Integrator

An *integrator* is a network in which the output signal is proportional to the mathematical integral of the input signal. One of the most commonly used forms is a low-pass *RC* network. If, as in Figure 2.1-2a, the time constant is very large compared with the time required for the input signal to make an appreciable change, the network is referred to as an integrator. In this case, the capacitor C

---

[1] The output v of the inverting amplifier (Figure 2.8-1) is the negative of the waveform Vxe^{-x} in Figure 2.5-1 and is given by $v = (-AVt/\tau)e^{-t/\tau}$, where A is the magnitude of the amplifier gain. Now, from Equation (2.5-2), $dv/dt = v_o/\tau + dv_o/dt$ or

$$\left(-\frac{AV}{\tau}\right)\left(1 - \frac{t}{\tau}\right)e^{-t/\tau} = \frac{v_o}{\tau} + \frac{dv_o}{dt}. \qquad (2.8\text{-}3a)$$

Multiplying Equation (2.8-3a) by $e^{t/\tau}$ yields

$$\left(-\frac{AV}{\tau}\right)\left(1 - \frac{t}{\tau}\right) = \left(\frac{v_o}{\tau}\right)e^{t/\tau} + \left(\frac{dv_o}{dt}\right)e^{t/\tau}.$$

Therefore

$$\left(-\frac{AV}{\tau}\right)\left(1 - \frac{t}{\tau}\right)dt = d(v_o e^{t/\tau});$$

integrating,

$$\left(-\frac{AVt}{\tau}\right)\left(1 - \frac{t}{2\tau}\right) = v_o e^{t/\tau} + K.$$

$K = 0$ since $v_o = 0$ at $t = 0$. Thus $v_o = (-AVt/\tau)(1 - t/2\tau)e^{-t/\tau}$, or

$$v_o = (-AVx)\left(1 - \frac{x}{2}\right)e^{-x}. \qquad (2.8\text{-}3)$$

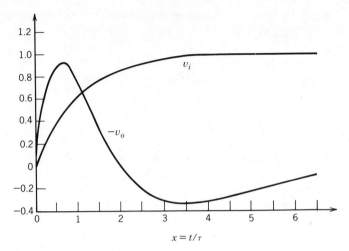

FIGURE 2.8-2 Response of a double differentiator to an exponential input. The numerical values correspond to $A = 4$ and $V = 1$.

charges slowly and the voltage drop across C is very small relative to the drop across R. We may consider that the total input v_i appears across R. Therefore the current is v_i/R, and the output signal across C is

$$v_o = \frac{1}{C} \int i\,dt = \frac{1}{RC} \int v_i\,dt. \qquad (2.8\text{-}4)$$

If $v_i = \alpha t$, the output is $\alpha t^2/2RC$, as given by Equation (2.7-8). When time increases, the output will change from a quadratic to a linear function of time, as shown in Figure 2.7-2a. The integral of a constant is a linear function, and this agrees with the curves of Figure 2.4-1g, which correspond to $RC/T \gg 1$.

The integration of a sinusoidal waveform is illustrated in Figure 2.8-3. From time zero at the peak of the sinusoidal input, the wave is divided into sections of equal widths. The sine wave is represented by a series of pulses of varying amplitudes. The first and second pulses cause linear increases in capacitor voltage Δv_1 and Δv_2 from time 0 to t_1 and t_1 to t_2, respectively. Δv_2 is less than Δv_1 since the pulse amplitude during t_1 to t_2 is smaller. Similarly the third and fourth pulses produce linear voltage increases at decreasing rates. Negative pulses 5 to 8 linearly decrease the capacitor voltage. If we extend the waveforms, we can see that integration of the sinusoidal input produces a negative cosine waveform output from capacitor C of the RC network with $RC \gg T$ (the period of the sine wave).

Loading Effects on Differentiator and Integrator

One loading problem that often occurs is the case of a transistor that is to be switched on by the filter (differentiator or integrator) output. In order to avoid

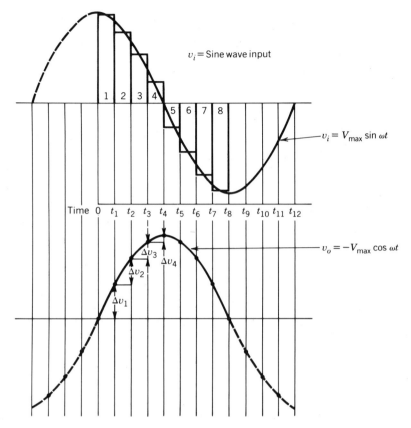

FIGURE 2.8-3 Integration of a sinusoidal input resulting in a negative cosine waveform output from capacitor C of the RC network when $RC \gg T$ (the period of the sine wave).

the loading effects on differentiating and integrating circuits, a resistor R_B is connected between the filter output and the transistor base input, as shown in Figures 2.8-4 and 2.8-5.

In Figure 2.8-4, the total input resistance offered by the transistor and R_B is

$$R_i = R_B + \frac{V_{BE}}{I_B}.$$

The differentiator resistance (during positive peaks when an *npn* transistor is used) now becomes

$$R' = R \| R_i.$$

During negative-going spikes, Q_1 is biased off, I_B ceases to flow, and the dif-

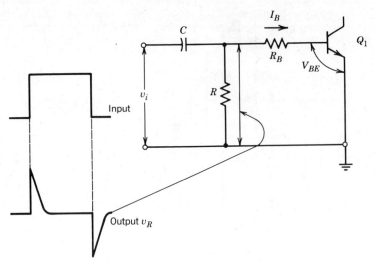

FIGURE 2.8-4 Using resistor R_B to avoid the transistor loading effect on the differentiator. Since $R \ll R_B$, waveform is not clipped. If R_B is shorted, the positive spike of v_R will be clipped.

ferentiator resistance becomes R. Under this condition, Q_1 has a negligible effect on the differentiator.

The similar loading effect may present a problem with the integrator. In Figure 2.8-5, the capacitor current I must be made very much larger than I_B if the transistor is to have a negligible effect on the integrator. Alternatively, an emitter follower or a voltage follower may be used to minimize the loading effect on the integrator or differentiator.

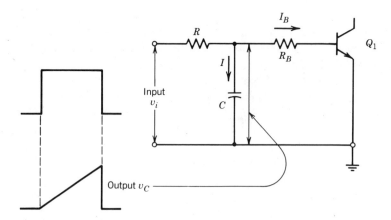

FIGURE 2.8-5 Using resistor R_B to avoid the transistor loading effect on the integrator. Since $I \gg I_B$, waveform is not clipped. If R_B is shorted, the v_C peak will be clipped.

2.9 *RC* DIFFERENTIATOR AND INTEGRATOR USED FOR SEPARATION OF HORIZONTAL AND VERTICAL SYNC PULSES IN A TV RECEIVER

The high-pass and low-pass *RC* filters are used to separate horizontal and vertical sync pulses from the stripped sync waveform in a TV receiver. As shown in Figure 2.9-1*a*, horizontal sync pulses are passed by a differentiator. Incoming pulses have a width of approximately 5 μs, and the differentiator has an *RC* time constant of about 1 or 1.5 μs. The peak voltage of the output spikes can be utilized by succeeding circuits. Vertical sync pulses are effectively rejected by the differentiator.

The vertical sync pulse is serrated at half-line intervals, so that complete rejection does not take place in the differentiator. Each serration produces a differentiated output. The reason that serrations are employed is to provide a means of maintaining sync lock in the horizontal deflection system during the passage of the vertical sync pulse. If serrations were not provided, the horizontal

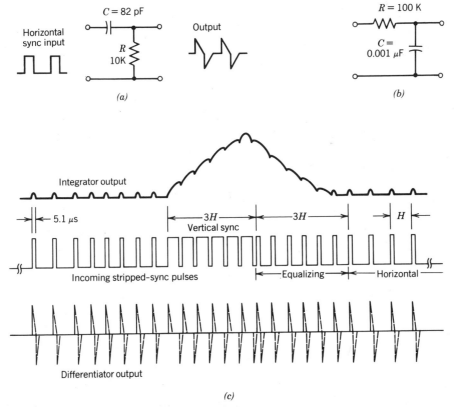

FIGURE 2.9-1 Separation of vertical and horizontal sync pulses: (*a*) simple differentiating circuit; (*b*) simple integrating circuit; (*c*) incoming stripped-sync pulses with differentiated and integrated stripped-sync waveforms.

oscillator would "wander" during the time that the vertical sync pulse occurs; under this condition, the top of the picture would appear bent. Figure 2.9-1b and c shows a simple integrating circuit and the incoming stripped-sync pulses with differentiated and integrated stripped-sync waveforms.

The separation of vertical sync pulses from horizontal sync pulses is accomplished by means of an RC integrator (e.g., $RC = 100 \, \mu s$). In the vertical sync pulse, the serration width is 2.5 μs, and the section width is 29 μs. Hence the width of vertical sync pulse is 6(29 + 2.5) or 190 μs. The horizontal sync pulse with a width of only 5 μs will produce a very small charge on the integrating capacitor.

Equalizing pulses are necessary because interlaced scanning is employed; that is, the start of the vertical sync pulse is shifted by half a line from the last horizontal sync pulse from odd to even fields.

Because the serrations in vertical sync pulse tend to produce irregularities in the output waveform from a simple integrator, two-section (or three-section) integrators are generally used.

The TV synchronization can be explained further as follows. A sync pulse is transmitted at the start of each forward scan. The pulse triggers the scanning generator in the TV receiver so that the beam in the cathode-ray tube (CRT) starts a new scanning line at exactly the same time that a new scanning line is started in the camera. To minimize flicker, the basic frame of 525 lines is broken down into two fields, each of which contains $262\frac{1}{2}$ lines. This is the interlaced scanning method. The horizontal line frequency is 262.5 × 60 or 15,750 Hz. The field frequency is 60 Hz, and the frame frequency is 30 Hz. The line period H is 1/15,750 or 63.5 μs. The field period is 1/60 or 16,667 μs.

2.10   COMPENSATED ATTENUATORS

An attenuator is used to reduce the amplitude of a signal waveform. The compensated attenuator shown in Figure 2.10-1a is an RC bridge, where C_1 is adjustable. The final adjustment for compensation is made experimentally by the method of square-wave testing. When the bridge is at balance,

$$\frac{1/\omega C_1}{1/\omega C_2} = \frac{R_1}{R_2},$$

$$C_1 = \frac{R_2 C_2}{R_1} \equiv C_p, \qquad (2.10\text{-}1)$$

and no current flows in the branch connecting the point x to the point y. When the output is calculated, the branch xy may be omitted. Attentuation is the ratio $a = R_2/(R_1 + R_2)$, and so the output is equal to av_i independent of the

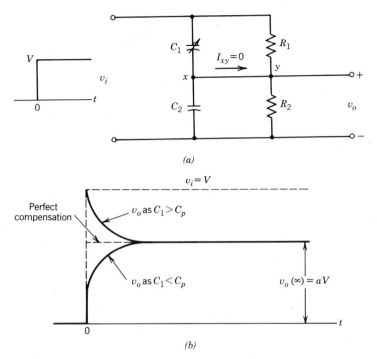

FIGURE 2.10-1 (*a*) Compensated attenuator operated as a bridge; (*b*) response of the attenuator to a step-voltage input.

frequency:

$$v_o = av_i = \frac{R_2}{R_1 + R_2} v_i. \tag{2.10-2}$$

In the bridge of Figure 2.10-1*a*, C_1 and C_2 in series are shunted by R_1 and R_2, respectively. Assume that the attenuator compensation is incorrect for a step-voltage input of magnitude V. As the input changes abruptly by V at $t = 0$, the voltages across C_1 and C_2 must also change discontinuously. Since the voltage across a capacitor cannot change instantaneously if the current remains finite, we conclude that an infinite current must flow in the $C_1 - C_2$ path at $t = 0$ for an infinitesimal time. Hence a finite charge

$$q = \int_{0-}^{0+} i \, dt$$

is delivered to each capacitor. At $t = 0+$ (immediately after $t = 0$),

$$V = \frac{q}{C_1} + \frac{q}{C_2} = \frac{(C_1 + C_2)q}{C_1 C_2}. \tag{2.10-3}$$

Therefore the output voltage at $t = 0+$ is

$$v_o(0+) = \frac{q}{C_2} = \frac{C_1}{C_1 + C_2} V.$$

(2.10-4)

Since a capacitor acts as an open circuit under steady-state conditions for an applied dc voltage, the final output voltage at $t = \infty$ is determined by the resistors R_1 and R_2. Hence

$$V_o(\infty) = \frac{R_2}{R_1 + R_2} V.$$

(2.10-5)

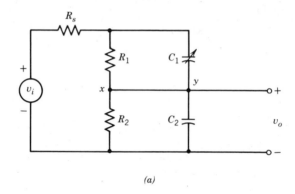

(a)

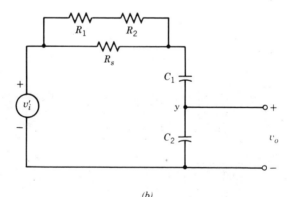

(b)

FIGURE 2.10-2 (a) Compensated attenuator including source impedance R_s; (b) Thevenin equivalent circuit where $V_i' = (R_1 + R_2)V/(R_s + R_1 + R_2)$. This equivalent circuit is obtained from the circuit in (a) when the branch xy is omitted. The equivalent voltage V_i' and the equivalent resistance $(R_1 + R_2)\|R_s$ are derived from Thevenin's Theroem which is stated as follows: Any linear network may, with respect to a pair of terminals, be replaced by a series combination of a voltage generator (source) and an impedance. The voltage generator equals the open-circuit voltage between the two terminals. The value of the equivalent impedance is equal to that impedance seen looking back into the network with all energy sources replaced by their respective internal impedance.

Because of the Thevenin resistance $R = R_1 R_2/(R_1 + R_2)$ in parallel with $C = C_1 + C_2$, the decay of the output from initial to final value takes place exponentially with a time constant $\tau = RC$. The responses of the attenuator for $C_1 = C_p \equiv R_2 C_2/R_1$, $C_1 > C_p$, and $C_1 < C_p$ are indicated in Figure 2.10-1b. Perfect compensation is obtained if $v_o(0+) = v_o(\infty)$, equivalent to $C_1 = R_2 C_2/R_1 = C_p$.

In the preceding analysis we have implicitly assumed a generator with zero source impedance (R_s). In practice, the infinite current cannot exist due to $R_s \neq 0$. Thus the ideal step response can no longer be obtained from the practical compensated attenuator shown in Figure 2.10-2. However an improvement in rise time does result if the compensated attenuator is used. If $R_s \ll R_1 + R_2$, as is usually the case, the input to the attenuator will be an exponential of time constant $\tau = R_s C_1 C_2/(C_1 + C_2)$. A cathode-ray oscilloscope probe is an example of the bridge attenuator.

2.11 *RL* AND *RLC* CIRCUITS

RL Circuits

Assume that the resistor R and capacitor C of the networks shown in Figures 2.1-1a and 2.1-2a are replaced by a resistor R' and an inductor L, respectively. Then, if the time constant L/R' equals the time constant RC, all the results obtained from the RC networks remain unchanged.

The small air-core coil is used in small-time-constant applications. The peaking circuit shown in Figure 2.11-1 illustrates how a square wave is converted into pulses by means of a small time constant L/R. The output impedance R is the drain resistance r_d of the FET and the open-circuit voltage gain $A = g_m r_d$,

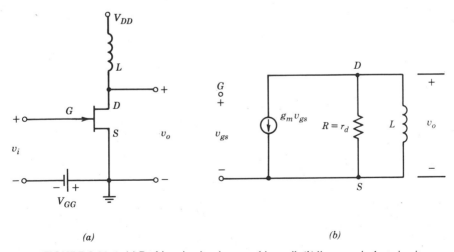

(a)   *(b)*

FIGURE 2.11-1 *(a)* Peaking circuit using a peaking coil; *(b)* linear equivalent circuit.

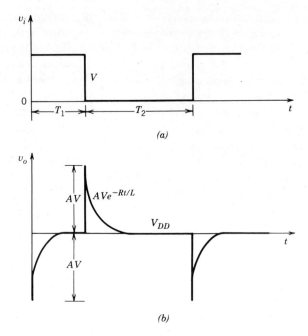

(a)

(b)

FIGURE 2.11-2 Input v_i and output v_o for the circuit of Figure 2.11-1, with $R = r_d$. $L/R \ll T_1 < T_2$.

where g_m is the transconductance. The peaking coil L acts as an open circuit at the time of an abrupt change in voltage. For an FET with the output open circuited, the change in drain voltage equals A times the gate-voltage change. Therefore, as indicated in Figure 2.11-2, the peak of the output pulse equals AV, where V is the jump in voltage of the input signal. The output voltage falls or rises exponentially with a time constant L/R toward V_{DD}, the quiescent voltage.

RLC Circuits

In the RLC circuit of Figure 2.11-3, the LC parallel combination and resistor R are connected in series. Current i is the sum of i_1 and i_2. The voltages across L and C are equal, or

$$\frac{(i - i_1)t}{C} = L \frac{di_1}{dt}.$$

(2.11-1)

Differentiating Equation (2.11-1) with respect to time, we have

$$\frac{i - i_1}{C} = L \frac{d^2 i_1}{dt^2}.$$

(2.11-2)

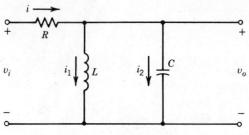

FIGURE 2.11-3   An *RLC* circuit.

Hence

$$i - i_1 = LC \frac{d^2 i_1}{dt^2}.$$

(2.11-3)

Kirchhoff voltage law (KVL) yields

$$v_i = iR + L \frac{di_1}{dt},$$

from which

$$i = \frac{v_i}{R} - \frac{L}{R} \frac{di_1}{dt}.$$

(2.11-4)

Substituting Equation (2.11-4) into Equation (2.11-3), we obtain

$$\frac{v_i}{R} - \frac{L}{R} \frac{di_1}{dt} - i_1 = LC \frac{d^2 i_1}{dt^2},$$

$$\frac{d^2 i_1}{dt^2} + \frac{1}{RC} \frac{di_1}{dt} + \frac{1}{LC} i_1 = \frac{v_i}{RLC}.$$

(2.11-5)

The homogeneous equation of Equation (2.11-5) is

$$\frac{d^2 i_1}{dt^2} + \frac{1}{RC} \frac{di_1}{dt} + \frac{1}{LC} i_1 = 0.$$

(2.11-6)

Assume that one of the particular solutions of Equation (2.11-6) is $i_1 = e^{st}$. Substituting this into Equation (2.11-6), we find that the characteristic equation of Equation (2.11-6) is

$$s^2 + \frac{1}{RC} s + \frac{1}{LC} = 0.$$

(2.11-7)

Hence

$$s = -\frac{1}{2RC} \pm \left[\frac{1}{4R^2C^2} - \frac{1}{LC}\right]^{1/2}. \qquad (2.11\text{-}8)$$

Defining k and T_0 by

$$k \equiv (1/2R)(L/C)^{1/2} \equiv \text{damping constant},$$

$$T_0 \equiv 2\pi(LC)^{1/2} \equiv \text{resonant or undamped period},$$

we obtain

$$\frac{1}{2RC} = \frac{1}{2R}\left(\frac{L}{LC^2}\right)^{1/2} = 2\pi \frac{1}{2\pi(LC)^{1/2}} \times \frac{1}{2R}\left(\frac{L}{C}\right)^{1/2} = \frac{2\pi k}{T_o} \qquad (2.11\text{-}9)$$

and

$$\left[\frac{1}{4R^2C^2} - \frac{1}{LC}\right]^{1/2} = \frac{j}{(LC)^{1/2}}\left[1 - \frac{1}{4R^2}\frac{L}{C}\right]^{1/2} = j\frac{2\pi}{T_0}[1 - k^2]^{1/2}. \qquad (2.11\text{-}10)$$

Substituting Equations (2.11-9) and (2.11-10) into Equation (2.11-8), we obtain

$$s = -\frac{2\pi k}{T_0} \pm j\frac{2\pi k}{T_0}(1 - k^2)^{1/2}. \qquad (2.11\text{-}11)$$

Equation (2.11-11) indicates the roots of characteristic Equation (2.11-7). The solution of v_i, v_o, i, and i_1 are in the form e^{st} since the exponential function still conserves its form after it is differentiated or integrated.

Ringing Circuit

A *ringing circuit* is an *RLC* circuit that has nearly undamped oscillations. If the damping factor k is small, the circuit will ring for many cycles. If the damping is small enough, the response approaches an undamped sine wave. The initial magnetic energy stored in the inductor is converted into electric energy in the capacitor at the end of one quarter cycle. Thus

$$\frac{LI^2}{2} = \frac{CV_{max}^2}{2} \qquad \text{or } V_{max} = I\left(\frac{L}{C}\right)^{1/2}. \qquad (2.11\text{-}12)$$

The following example illustrates the generation of a burst of oscillations by a switch circuit. Such a burst may be useful for timing purposes if its frequency is suitably chosen. The basic technique is shown in Figure 2.11-4. While the

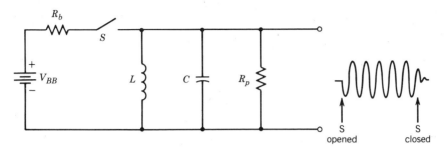

FIGURE 2.11-4 Switch-excited *RLC* circuit (high *Q*) generates a burst of oscillations.

switch is closed, a current $I = V_{BB}/R_b$, corresponding to a stored energy $LI^2/2$, is maintained in the inductor. When the switch is opened, this current begins to charge the capacitor, initiating the alternating exchange of magnetic and electrostatic energy in the circuit. The peak capacitor voltage V_{max} is such that the energy $CV_{max}^2/2$ is the same (neglecting dissipation) as that initially in the inductor: $V_{max} = I(L/C)^{1/2}$. While the switch is open, the oscillations are damped by R_p, with $Q(=\omega C R_p) = R_p(C/L)^{1/2}$. When the switch is later closed, the damping due to $R_b(\ll \sqrt{L/C})$ quickly brings the circuit back to its quiescent condition.

A practical ringing circuit is shown in Figure 2.11-5. The transistor used as a switch is originally in a saturated state, so that the initial current in the inductor is determined by V_{CC} and R_C. At $t = 0$ the transistor is turned off by the negative pulse input, and the current that was flowing in the emitter now flows into the capacitor, tending to charge it negatively; thus the *RLC* circuit rings, giving a sinusoidal output waveform that starts with a negative half cycle, as shown in the figure. The high-*Q* circuit excited by the transistor switch operates as a ringing oscillator, which can be used as a timing standard for oscilloscopic measure-

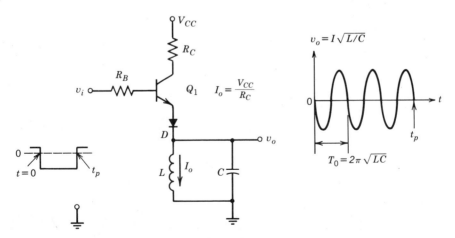

FIGURE 2.11-5 A practical ringing circuit used to generate a pulsed train of sinusoidal oscillations.

ments. The diode D is used to keep the emitter-base junction of the transistor from breaking down, because it must otherwise withstand a voltage at least as great as the peak-to-peak voltage of the output. At the end of input pulse the transistor is turned back on, and the circuit is damped by the resistors R_C and R_B effectively acting in parallel.

2.12   PULSE TRANSFORMERS

Iron-Core Transformers

Iron-core transformers are used in steady-state ac circuits, in audio and rf circuits, and in the transmission and shaping of pulses with nanosecond and microsecond pulse duration. Some of the more common applications of pulse transformers include changing the amplitude of a pulse, inverting the polarity of a pulse, providing dc isolation between pulse source and the load, and coupling pulses from the pulse-generating circuit to various circuit points.

Figure 2.12-1 shows the schematic diagram of a transformer, including the source v_i and the load R_L. The primary, secondary, and mutual inductances are L_p, L_s, and M, respectively. The coefficient of coupling K between primary and secondary inductances is defined by $K = M/(L_p L_s)^{1/2}$. In a perfect or ideal transformer, $K = 1$, and

$$\frac{v_o}{v_i} = \frac{i_p}{i_s} = \sqrt{\frac{L_s}{L_p}} = \frac{N_s}{N_p} = \frac{R_L}{R_p} = n, \qquad (2.12\text{-}1)$$

where v_o = output voltage,
$\quad\ \ v_i$ = input voltage,
$\quad\ \ i_p$ = primary current,
$\quad\ \ i_s$ = secondary current,

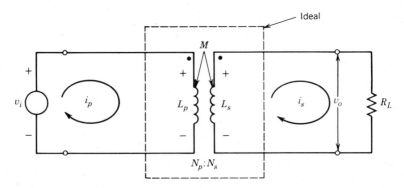

FIGURE 2.12-1 Schematic diagram of a transformer. The coil ends with dots have the same polarity at any instant.

N_p = primary number of turns,
N_s = secondary number of turns,
n = transformation ratio,
$R_p = R_L/n^2$ = effective load resistance reflected to primary side.

In designing a small 60-Hz power transformer, the turns per volt is usually determined by $T/V = 5.78/A$, where A is the cross-sectional area of 4% silicon-steel core, expressed in in^2. In this case, $B_{max} = 65{,}000$ lines/in^2 has been taken as the maximum permissible flux density. The cross-sectional area limitation is A (in^2) $\geqslant 0.16\sqrt{V_p I_p}$, where V_p and I_p are the primary voltage and current, respectively. A copper wire of cross-sectional area equal to 700 circular mils is permitted to carry current up to 1 A. The transformer efficiency ($= P_o/P_i$) is usually 85%. The actual secondary turns = (ideal secondary turns N_s)/(Eff)$^{1/2}$ = $1.11 N_s$. The number of primary turns is determined as in an ideal case.

A pulse transformer behaves as a reasonable approximation to an ideal transformer when used in connection with the fast waveforms it is intended to handle. The ideal transformer will be important since it will be part of the equivalent circuit that we set up for the practical pulse transformer.

Practical Pulse Transformers

Figure 2.12-2a shows the practical-pulse-transformer equivalent circuit, which accounts for the main effects of flux leakage, total shunt capacitance, and winding resistances. The series inductance L_L is the leakage inductance, which is presented at the terminals of the primary when the secondary is short circuited. The shunt inductance L_M is the magnetizing inductance, which is presented at the terminals of the primary when the secondary is open circuited. L_M represents that portion of the primary inductance that actually contributes to the transformer action. Usually $L_L \ll L_M$. Pulse transformers are said to be closely coupled when they only have a small amount of leakage. L_L is typically up to 100 μH, while L_M is typically 10 mH to 100 mH. The resistances R_p and R_s are the ohmic values of the primary and secondary windings, respectively. The capacitance C is the total effective shunt capacitance (including the effective interwinding capacitance and the capacitance reflected from the secondary capacitance). The value of C depends on many factors, including the geometry of the core and the number of turns per winding. The pulse transformers normally employ turns ratios of less than 10 in order to minimize the undesired effects of C. Usually C is up to 100 pF.

The circuit of Figure 2.12-2a can be simplified by reflecting the secondary resistances back to the primary, as shown in Figure 2.12-2b. Before discussing the response of practical pulse transformers, let us review several things about the circuit of Figure 2.12-2b: For a step or pulse input, the rising or falling edges contain the high-frequency components, while the flat, constant portions contain the low-frequency components. Hence the pulse transformer must respond to a wide frequency range. At very high frequencies, the impedances of L_L and C

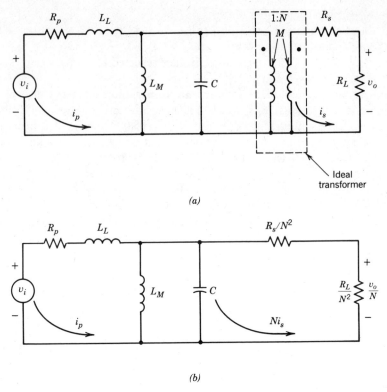

(a)

(b)

FIGURE 2.12-2 (a) Equivalent circuit for a practical pulse transformer; (b) same circuit simplified by reflecting the secondary resistances back to the primary. Core loss is negligible in good pulse transformers.

in Figure 2.12-2b will be very high and very low, respectively. At very low frequencies, the impedances of L_M will be very low. Therefore, as far as the output is concerned, the equivalent circuit acts as a band-pass circuit. A simple procedure to follow in analyzing this circuit is to divide it into a rise-time (or high-frequency) analysis and a flat-top (or low-frequency) analysis. A typical equi-

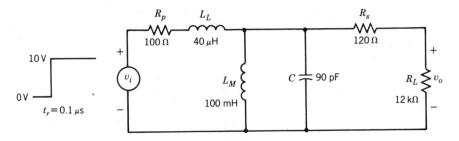

FIGURE 2.12-3 Pulse transformer (N_s/N_p) equivalent circuit used to illustrate the high-frequency analysis and low-frequency analysis.

valent circuit used to illustrate these analyses is shown in Figure 2.12-3. The input is a 10-V step with $t_r = 0.1\ \mu s$. The turns ratio is $N_s : N_p = 1:1$. $R_s = 120\ \Omega \ll 12\ k\Omega = R_L$.

The Rise-Time Response

Consider the situation in Figure 2.12-3. Since $t_r = 0.1\ \mu s$, $f_H = 0.35/t_r = 3.5$ MHz must be considered. At 3.5 MHz,

$$X_{L_M} = 2\pi f L_M = (6.28)(3.5 \times 10^6)(100 \times 10^{-3}) \simeq 2.2\ M\Omega;$$

$$X_c = \frac{1}{2\pi f C} = \frac{1}{6.28 \times 3.5 \times 10^6 \times 90 \times 10^{-12}} = 506\ \Omega.$$

Hence L_M and $R_L + R_s$ can be neglected. The simplified circuit for high-frequency analysis is shown in Figure 2.12-4a.

This simplified circuit is a series RLC circuit. At $t = 0$, KVL yields

$$V = Ri + L\frac{di}{dt} + \frac{1}{C}\int i\,dt. \qquad (2.12\text{-}2)$$

Differentiating Equation (2.12-2) with respect to time,

$$L\frac{d^2 i}{dt^2} + R\frac{di}{dt} + \frac{i}{C} = 0. \qquad (2.12\text{-}3)$$

This is the homogeneous equation. Substituting the trial solution $i = ke^{st}$ into Equation (2.12-3), we obtain

$$s^2 + \frac{R}{L}s + \frac{1}{LC} = 0. \qquad (2.12\text{-}4)$$

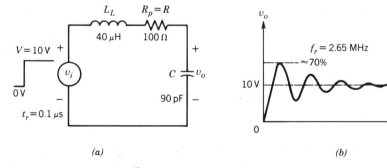

(a)   (b)

FIGURE 2.12-4 (a) High-frequency equivalent circuit used to calculate the rise-time response; (b) response to rising edge of input.

This is the characteristic equation. Its roots are

$$s = -\frac{R}{2L} \pm \sqrt{\left(\frac{R}{2L}\right)^2 - \frac{1}{LC}} = -\frac{R}{2L} \pm \frac{1}{L}\sqrt{\frac{R^2}{4} - \frac{L}{C}}. \qquad (2.12\text{-}5)$$

The borderline case occurs when

$$\frac{R^2}{4} = \frac{L}{C}, \qquad (2.12\text{-}6)$$

or

$$R_{\text{crit}} = R = 2\sqrt{\frac{L}{C}}. \qquad (2.12\text{-}6a)$$

Let

$$\omega_r = 2\pi f_r = \frac{1}{\sqrt{LC}}. \qquad (2.12\text{-}7)$$

When $R^2/4 > L/C$, the circuit is overdamped. When $R^2/4 < L/C$, the circuit is underdamped. The circuit Q is

$$Q = \frac{\omega r L}{R} = \frac{L/\sqrt{LC}}{R} = \frac{1}{R}\sqrt{\frac{L}{C}}. \qquad (2.12\text{-}8)$$

In the circuit of Figure 2.12-4a,

$$Q = \frac{1}{R}\sqrt{\frac{L_L}{C}} = 6.67$$

Since

$$\frac{R_p^2}{4} = 2500 < \frac{L_L}{C} = 444{,}444,$$

the circuit is underdamped, and the output will contain ringing oscillations in response to the rising edge of the step. The ringing frequency is

$$f_r = \frac{1}{2\pi\sqrt{LC}} \simeq 2.65 \text{ MHz}. \qquad (2.12\text{-}9)$$

If the circuit were valid for all frequencies, output v_o would be expected to settle at 10 V. A sketch of the initial portion of v_o is shown in Figure 2.12-4b. The v_o

waveform exhibits an overshoot of around 7 V (70%) during the first ringing cycle. This overshoot can be reduced by decreasing the circuit Q, which can be accomplished by increasing the series resistance R_p. However the increase in R_p will cause a slower rise time in v_o. Hence a compromise has to be made; a Q value of about 0.8 gives an overshoot of 9% and a rise time equal to approximately one-fourth of the ringing period.

The Flat-Top Response and Composite Total v_o Waveform

The response during the flat top of the pulse or step voltage is given by the low-frequency equivalent circuit of Figure 2.12-5a, which is obtained from Figure

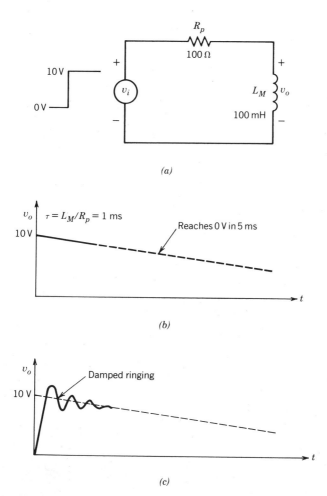

FIGURE 2.12-5 (a) Low-frequency equivalent circuit; (b) output response to flat-top of the step-voltage input; (c) complete output waveform.

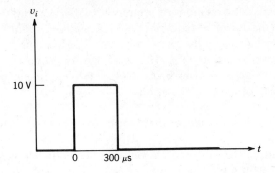

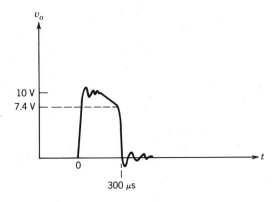

FIGURE 2.12-6   Response to a 10-V pulse of width $t_p = 300\ \mu$s.

2.12-3 by neglecting the effect of L_L, C, and $R_L + R_s$. Basically the circuit of Figure 2.12-5a is a high-pass RL network. Its output v_o will be exponentially decaying from 10 V to zero, as shown in Figure 2.12-5b.

The complete v_o waveform shown in Figure 2.12-5c is a combination of the results of Figures 2.12-4b and 2.12-5b.

Figure 2.12-6 shows the response to an input pulse of width $t_p = 300\ \mu$s. The output waveform has ringing oscillations on both the rising and falling edges due to the high-frequency response of the pulse transformer; the tilt in the output pulse occurs because of the low-frequency response of the transformer. In this case, $\tau = L_M/R_p = 100\ \text{mH}/100\ \Omega = 1\ \text{ms}$, and $t_p = 0.3\ \text{ms} = 0.3\tau$; hence the exponential portion goes through 0.3τ during the pulse-width interval. This results in a tilt of

$$1 - e^{-t_p/\tau} = 1 - e^{-0.3} = 1 - 0.74 = 26\%$$

from 10 V down to 7.4 V.

2.13 DELAY LINES

Introduction

Delay lines are passive four-terminal networks that have the property that a signal applied to the input terminals appears at the output terminals at the end of a delay time t_d. In many cases transmission lines are used to obtain the required t_d and so they are usually termed delay lines. Delay times ranging from a few nanoseconds to about $1000\ \mu s$ are obtainable with electromagnetic delay lines, which are either of lumped-parameter form or distributed-parameter form. The applications of delay lines are numerous. For example, an oscilloscope used for observing fast waveforms has a built-in delay line in the vertical channel so that the input signal, which also triggers the sweep, is delayed slightly before being applied to the vertical-output amplifier. If the sweep was not allowed to start before the signal was applied, the first portion of the waveform might not be visible on the scope face. Other applications of delay lines occur in precise time measurement, pulse shaping, radar systems, digital computers, and so on.

Lumped-Parameter Delay Lines

Lumped-parameter delay-lines are made up of actual individual circuit elements (usually inductors and capacitors), as shown in Figure 2.13-1. In Figure 2.13-1a, the portion enclosed in the dotted lines is the basic T-section; these T-sections in cascade are included in the delay line. Figures 2.13-1b and c are the simplification of part a and the symbolic representation of a delay line, respectively.

The characteristic impedance of the delay line shown in Figure 2.13-1 is given by

$$Z_o = \sqrt{\frac{L}{C}} \qquad (2.13\text{-}1)$$

and is purely resistive. Z_o is independent of the number of T-sections.

The delay line terminated in its characteristic impedance is shown in Figure 2.13-2. When a resistor equal to Z_o is placed across the output terminals of the delay line, the input impedance of the delay line equals z_o. This characteristic is ideally independent of the number of T-sections employed; however, when the number of sections increases, the series resistances associated with each inductor become important; thus the input impedance is more complex. This characteristic is only valid at frequencies below one-fourth of the cutoff frequency of the line f_c, which is given by

$$f_c = \frac{1}{\pi \sqrt{LC}}. \qquad (2.13\text{-}2)$$

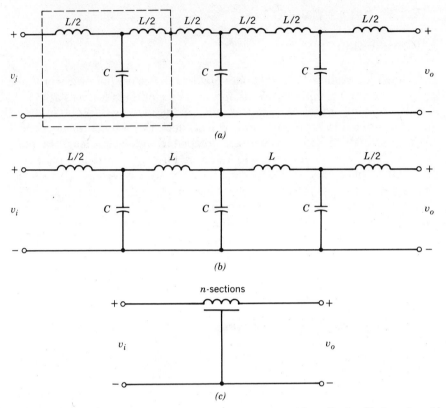

FIGURE 2.13-1 (*a*) Three-T-section lumped-parameter delay line; (*b*) simplified version; (*c*) common symbol used to represent delay lines of any type.

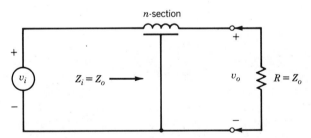

FIGURE 2.13-2 Delay line terminated in its characteristic impedance Z_o, showing $Z_i = Z_o$ at $f < 1/\pi\sqrt{LC}$.

Hence, for a delay line terminated with its characteristic impedance (Z_o), the input impedance (Z_i) equals the purely resistive impedance Z_o at frequencies below $f_c/4$. Under these conditions, the input source supplied real power $v_i^2/Z_o = v_o^2/R$. At frequencies above $f_c/4$, the input impedance is partly resistance and partly reactance; in this case the output will have some degree of distortion. When the delay line is terminated with $R \neq Z_o$, the output will be distorted. Even when $R = Z_o$, there will be some slight attenuation and distortion due to the resistance of the inductors.

The delay time t_d produced by the delay line depends on the number of T-sections included. The value of t_d for a single section is given by

$$t_d = \sqrt{LC}. \tag{2.13-3}$$

For n sections, the delay time is

$$t_d = n\sqrt{LC}. \tag{2.13-4}$$

Since the amount of distortion increases as more sections are added, it is difficult to obtain very long delay times. In practice, t_d is typically up to 1 ms with lumped-parameter delay lines.

Distributed-Parameter Delay Lines

The distributed-parameter delay line is merely a transmission line, which is simply a means for connecting one circuit to another. For a transmission line to function as a delay line, it must be terminated in its characteristic impedance Z_o. If C and L are the capacitance and inductance, respectively, of a unit length of the line, then the characteristic impedance (resistive) is given by $Z_o = (L/C)^{1/2}$. For an ideal transmission line, an arbitrary waveform impressed on the input terminals will appear without distortion at the output terminals after a delay time t_d. If the line is terminated in the characteristic impedance Z_o, no reflection will take place when the signal reaches the end of the line.

For a uniform lossless line, the delay per meter T is given by

$$T = \sqrt{LC} = CZ_o = \frac{L}{Z_o}. \tag{2.13-5}$$

For air dielectric,

$$T = (3 \times 10^8)^{-1} \text{ s/m} = 3.3 \text{ ns/m} \simeq 1 \text{ ns/ft};$$

taking a nominal value $Z_o = 100 \, \Omega$, we have, for air,

$$C \simeq 0.33 \text{ pF/cm} \simeq 10 \text{ pF/ft};$$

$$L = 3.3 \text{ nH/cm} \simeq 100 \text{ nH/ft}.$$

For a medium of relative dielectric constant ε_r, the delay is $3.3\varepsilon_r^{1/2}$ ns/m.

Delay-Line Pulse Shaping

The termination of the delay line is matched if the load equals the characteristic impedance $Z_o = R_o$. Also the input end of the line is matched if the impedance of the signal source equals Z_o. Mismatching a delay line at one end causes delayed reflections to appear at the other. These can be used for pulse shaping. Their outstanding feature is that the profile of the reflection is similar to that of the leading edge of the input within the limitations of the rise time of the line itself. Use of a delay line can transform the initial step into a neat rectangular pulse. This is termed *delay-line clipping* or *delay-line differentiation*. The pulse width produced in a delay line is dependent only on the line delay T, a relatively constant and reliable parameter.

Two chief types of mismatch for the line are a short circuit and an open circuit. The first returns a reflection with inverted voltage, wiping out the voltage existing on the line (for a step-function input) and carrying back the news that the line is really a short circuit. The second inverts the current in the reflection: this cancels the existing line current and carries the news of the open-circuit termination back to the input. In either case the input end of the line must be properly matched if further reflections are to be avoided.

If source and load can be idealized (impedance much smaller or much larger than $R_o = Z_o$), the clipping circuits shown in Figure 2.13-3 can be assembled. The shorted lines are used in shunt positions and the open ones are used in series. In each case the line looks like an impedance R_o until the reflection has had time to return, after time $2T$. The output pulse amplitudes are therefore half those of the input, as indicated, and the clipped pulse widths are $2T$. When source or load has finite impedance, this fact must be taken into account for matching the line. Resistive impedances may often be incorporated as part of the matching resistor shown in Figure 2.13-3.

Practical Delay-Line Clipping (Differentiation)

A delay line can be used in several different ways to convert a step function into a square wave of controlled duration. One way is to impress the signal on a short-circuited delay line from a source whose impedance matches the characteristic impedance of the delay line (Figure 2.13-4). After a time interval equal to twice the propagation time of the line, the signal returns from the short-circuited end to cancel the signal at the input end. The accuracy of the cancellation depends upon the accuracy of the impedance match and the phase distortion and attenuation in the delay line. If care is taken to choose the terminating network that will compensate for the inevitable lumped capacitance at the driving end and that will compensate, at least partially, for the phase distortion, the residual ringing signals observed with commercially available helically wound delay cables can be successfully reduced to a few percent of the signal size. The "rear porch" that results from attenuation in the line (Figure 2.13-5a) can be effectively eliminated by passing the signal through an appropriate RC

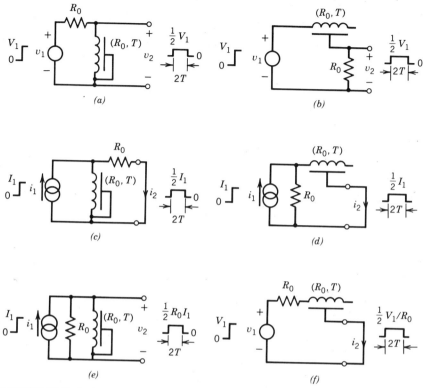

FIGURE 2.13-3 Delay-line clipping circuits with idealized source and load impedances. Delay-line characteristic impedance $Z_0 = R_o$.

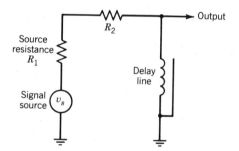

FIGURE 2.13-4 Short-circuited delay-line clipping configuration. R_2 is chosen so that $R_1 + R_2$ is equal to the delay-line impedance.

differentiating circuit. However the top of the resultant signal (Figure 2.13-5b) is no longer perfectly flat.

There is one way of eliminating the rear porch so that the output signal will retain its flat top, without resorting to additional differentiation. The technique involves the use of a delay line terminated in its characteristic impedance at the receiving end and, preferably but not necessarily, at the sending end. The delayed signal is subtracted from the input signal in a difference amplifier. The input signal may be slightly attenuated at the difference-amplifier input to

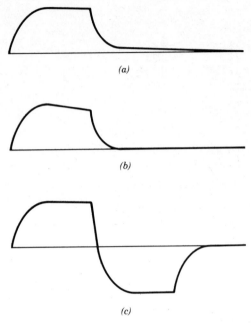

FIGURE 2.13-5 Single delay-line differentiated signal: (a) showing rear porch due to delay-line attenuation; (b) with rear porch removed by RC differentiation; (c) double-delay-line differentiated signal (bipolar pulse).

compensate for the attenuation of the delayed signal in the delay line. Such an arrangement is shown in Figure 2.13-6. The delay line is terminated at the receiving end by R_1 and at the sending end by the parallel combination of $(R_1 + R_2)$ and R_3. The tap on R_3 is adjusted to compensate for attenuation in the line. Since the delay line is terminated at both ends, much poorer matches can be tolerated without producing excessive ringing.

Applying delay-line clipping at two places in the amplifier results in a bipolar signal (see Figure 2.13-5c) in which the positive and negative areas are

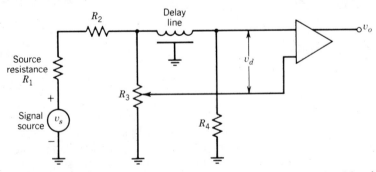

FIGURE 2.13-6 Alternative delay-line differentiating circuit. The delayed signal is subtracted from the input signal in the difference amplifier.

equal. With a bipolar signal in nuclear pulse spectrometry, the base-line[1] location is independent of the counting rate, and both pileup overloading and low-frequency noise can be minimized. However, with double-delay-line differentiation (clipping), the midband noise is increased by an additional factor of $\sqrt{3}$ over that obtained with single-delay-line clipping and by an overall factor of $\sqrt{6}$ over single or double RC differentiation.

2.14   LAPLACE TRANSFORMS APPLIED TO LINEAR NETWORK ANALYSIS

Introduction to Laplace Transforms

The *Laplace transform* provides a systematic method for finding the response of linear circuits to transient input signals. Suppose the input signal (voltage or current) is $f_1(t)$ and we want to find the output $f_2(t)$ of the circuit. The essence of the method is, first, to take the Laplace transform of $f_1(t)$ and, then, to form the product of this and the transfer function of the circuit. The inverse transform of this product then gives the output signal $f_2(t)$. The taking of the transform and of the inverse transform can often be done by recourse to standard tables. The transfer function of the circuit can then usually be written down, so the Laplace transform can often be used to give the output $f_2(t)$ very easily.

The Laplace transform is a very important and very effective tool for studying linear time-invariant networks. In these networks all elements except the independent sources are linear and *time invariant*—that is, characteristics of the time-invariant elements do not vary with time. The importance of the Laplace transform arises from the following facts: (1) The Laplace transform reduces the solution of linear differential equations to the solution of linear algebraic equations; (2) Laplace transform theory uses the concept of network function; and (3) the Laplace transform exhibits the close relation that exists between the time-domain behavior of a network (say, as waveforms on the scope) and its sinusoidal steady-state behavior.

The key idea of the Laplace transform is that to a time function f defined on $(0^-, \infty)$ it associates a function F of the complex frequency s. The transform is constructed as follows: $f(t)$ is multiplied by the factor e^{-st}, and the resulting function of t, $f(t)e^{-st}$, is integrated between 0^- and ∞, giving the defining integral

$$\int_{0^-}^{\infty} f(t)e^{-st}\, dt.$$

Since the integral is definite (the limits 0^- and ∞ are fixed), it does not depend on t but only on the parameter s; thus it is a function of the complex frequency s.

---

[1] The base line is the datum from which pulse heights are measured.

We call $F(s)$ the function defined by this integral; that is,

$$\mathscr{L}[f(t)] = F(s) \equiv \int_{0^-}^{\infty} f(t)e^{-st}\,dt = \lim_{\Delta t_i \to 0} \sum_{i=0}^{\infty} f(t_i)e^{-st_i}\,\Delta t_i. \quad (2.14\text{-}1)$$

We say that $F(s)$ is the Laplace transform of $f(t)$. The operator notation $\mathscr{L}[f(t)]$ stands for $F(s)$. The Laplace transform $F(s)$ is a function of the complex variable s. The integral (2.14-1) does not converge for all values of s. In most practical situations, we can find a real number σ_0, such that the integral converges for all values of s satisfying $\mathrm{Re}\,s > \sigma_0$; $s = \sigma + j\omega$.

A transformation of this type is useless unless the inverse transformation can be found—that is, unless $f(t)$ can be recovered from $F(s)$. The mathematical expression for finding the inverse Laplace transformation is

$$f(t) = \mathscr{L}^{-1}[F(s)] = \frac{1}{2\pi j} \int_{\sigma - j\infty}^{\sigma + j\infty} F(s)e^{st}\,ds. \quad (2.14\text{-}2)$$

Laplace Transforms of Unit Step, Impulse, and Pulse

Unit-Step Function. Let $f(t) = u(t)$, the unit-step function, which is defined as

$$u(t) = \begin{cases} 0 & \text{for } t < 0 \\ 1 & \text{for } t > 0 \end{cases} \quad (2.14\text{-}3)$$

and is illustrated in Figure 2.14-1a. The Laplace transform is given by

$$\mathscr{L}[u(t)] = F(s) = \int_{0^-}^{\infty} 1e^{-st}\,dt. \quad \text{For } \mathrm{Re}\,s = \sigma > 0. \quad (2.14\text{-}4)$$

Integration yields

$$\mathscr{L}[u(t)] = \left[-\frac{1}{s}e^{-st}\right]_{0^-}^{\infty} = -\frac{1}{s}(0-1) = \frac{1}{s}. \quad (2.14\text{-}5)$$

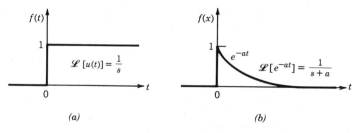

(a)   *(b)*

FIGURE 2.14-1 Laplace transforms (a) unit step $\mathscr{L}[u(t)] = 1/s$; (b) exponential pulse $\mathscr{L}[e^{-at}] = 1/(s+a)$.

Exponential Pulse. The exponential pulse shown in Figure 2.14-1*b* is expressed by the function

$$f(t) = \begin{cases} 0 & \text{for } t < 0 \\ e^{-at} & \text{for } t \geqslant 0 \end{cases}, \tag{2.14-6}$$

where a is any real or complex number. The Laplace transform of e^{-at} is given by

$$F(s) = \int_{0^-}^{\infty} e^{-at} e^{-st} \, dt = \int_{0^-}^{\infty} e^{-(s+a)t} \, dt = \frac{e^{-(s+a)t}}{-(s+a)} \bigg|_{0^-}^{\infty}.$$

Hence

$$\mathscr{L}[e^{-at}] = \frac{1}{s+a} \quad \text{for Re } (s+a) > 0. \tag{2.14-7}$$

Similarly

$$\mathscr{L}[e^{at}] = \frac{1}{s-a} \quad \text{for Re } (s-a) > 0. \tag{2.14-8}$$

Unit Impulse. The unit impulse function is introduced on a limit basis. Consider the function of time $f_\delta(t)$ drawn in Figure 2.14-2*a*. This flat-topped pulse is zero for $t < 0$ and is also zero for $t > t_0$. For $0 < t \leqslant t_0$ its value is constant at $1/t_0$. Hence the area under the curve of $f_\delta(t)$ is unity. This is independent of the value of t_0. Now consider that t_0 becomes smaller and smaller. The width of the pulse decreases, but its height increases so that the area remains unity. Thus the impulse function is defined as

$$\delta(t) = \lim_{t \to 0} f_\delta(t). \tag{2.14-9}$$

The letter delta denotes the impulse function, which is known as the *delta function*. The *impulse function* is a pulse of infinite height that is zero everywhere but at the origin. The area of the pulse is unity. $\mathscr{L}[\delta(t)] = 1$.

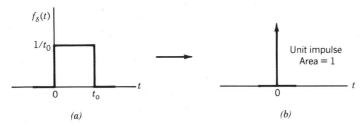

(a)   *(b)*

FIGURE 2.14-2 *(a)* A pulse of duration t_0 s and height $1/t_0$. As t_0 approaches zero, this becomes *(b)*, the unit-impulse function $\delta(t)$.

Flat-Topped Pulse. The pulse $f_p(t)$ shown in Figure 2.14-3a can be obtained by taking the sum of the unit-step and the negative unit-step functions of Figure 2.14-3b:

$$f_p(t) = u(t) - u(t - t_1). \qquad (2.14\text{-}10)$$

In order to obtain the transform $F_p(s)$, we need to know the Laplace transform of a function of time that is the same as another function, except that it is delayed in time. To relate the Laplace transforms of these two time functions let us consider that

$$\mathcal{L}[u(t)f(t)] = F(s) \qquad (2.14\text{-}11)$$

and we wish to find $\mathcal{L}[u(t - t_1)f(t - t_1)]$; that is, the original function $f(t)$ is delayed by t_1 s. From the definition of the Laplace transform we have

$$\mathcal{L}[u(t - t_1)f(t - t_1)] = \int_{0^-}^{\infty} u(t - t_1)f(t - t_1)e^{-st}\,dt = \int_{t_1}^{\infty} f(t - t_1)e^{-st}\,dt.$$

$$(2.14\text{-}12)$$

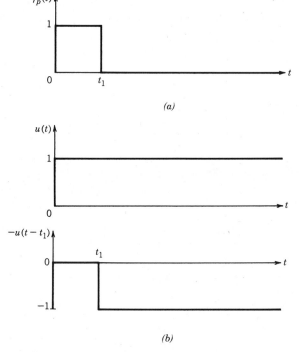

(a)

(b)

FIGURE 2.14-3 *(a)* A flat-topped pulse; *(b)* two unit-step functions whose sum provides the pulse of *(a)*.

Substituting $t - t_1 = \tau$ in Equation (2.14-12) we obtain

$$\mathscr{L}[u(t - t_1)f(t - t_1)] = \int_{0^-}^{\infty} f(\tau)e^{-s(\tau + t_1)}\, d\tau.$$

However e^{-st_1} is a constant as far as the integration is concerned, so

$$\mathscr{L}[u(t - t_1)f(t - t_1)] = e^{-st_1} \int_{0^-}^{\infty} f(\tau)e^{-s\tau}\, d\tau. \qquad (2.14\text{-}13)$$

The integral is just the definition of $\mathscr{L}[f(t)]$. Thus

$$\mathscr{L}[u(t - t_1)f(t - t_1)] = e^{-st_1}F(s). \qquad (2.14\text{-}14)$$

Therefore, to obtain the Laplace transform of a function that has been delayed by t_1, we multiply the transform of the undelayed function by e^{-st_1}. The Laplace transform of the unit-step function is $1/s$. Then, applying Equation (2.14-14), we obtain

$$F(s) = F_p(s) = \mathscr{L}[u(t)] - \mathscr{L}[u(t - t_1)] = \frac{1}{s}(1 - e^{-t_1 s}). \qquad (2.14\text{-}15)$$

Laplace Transform of the Derivative

Suppose we have a function $f(t)$ whose Laplace transform is $F(s)$. Let us obtain $\mathscr{L}[(d/dt)f(t)]$ in terms of $F(s)$. If we substitute in Equation (2.14-1), we have

$$\mathscr{L}\left[\frac{df(t)}{dt}\right] = \int_{0^-}^{\infty} \frac{df(t)}{dt} e^{-st}\, dt. \qquad (2.14\text{-}16)$$

Now we integrate by parts, using the formula

$$\int_0^{\infty} u\, dv = uv\Big|_0^{\infty} - \int_0^{\infty} v\, du. \qquad (2.14\text{-}17)$$

In Equation (2.14-16), we let

$$u = e^{-st} \quad \text{and} \quad dv = \frac{df(t)}{dt}\, dt = df(t). \qquad (2.14\text{-}18)$$

Hence

$$du = -se^{-st}\, dt \quad \text{and} \quad v = f(t). \qquad (2.14\text{-}19)$$

Then, substituting in Equation (2.14-17), we obtain

$$\mathscr{L}\left[\frac{df(t)}{dt}\right] = [f(t)e^{-st}]_0^\infty + \int_0^\infty sf(t)e^{-st}\,dt. \qquad (2.14\text{-}20)$$

Assume that Re s is large enough so that $\lim_{t\to\infty} f(t)e^{-st} = 0$. Then

$$[f(t)e^{-st}]_0^\infty = -f(0). \qquad (2.14\text{-}21)$$

In addition, s is a constant as far as the integration is concerned. Thus

$$\int_0^\infty sf(t)e^{-st}\,dt = s\int_0^\infty f(t)e^{-st}\,dt = sF(s); \qquad (2.14\text{-}22)$$

that is, the integral is just s times $\mathscr{L}[f(t)]$. Therefore

$$\mathscr{L}\left[\frac{df(t)}{dt}\right] = sF(s) - f(0). \qquad (2.14\text{-}23)$$

This result states that the Laplace transform of the derivative of a function is s times the Laplace transform of the function minus the initial value of the function. Since $f(0^-)$, not $f(o^+)$, will always be known, we use Equation (2.14-1) as the definition of the Laplace transform, and so the expression that we shall use for the transform of the derivative is

$$\mathscr{L}\left[\frac{df(t)}{dt}\right] = sF(s) - f(0^-). \qquad (2.14\text{-}24)$$

Usually we write the lower limit of integration as 0, with the understanding that this means 0^-.

Determination of Transfer Functions and Output Functions

In general, the transfer functions of networks relate an output voltage or current to the input voltage or current that causes it. The numerator of the transfer function is the response, and the denominator is the input (generator) quantity.

Low-Pass *RC* Network Driven by a Voltage Source.   The ratio of the output function to that of the input for the network of Figure 2.14-4 is the transfer function expressed by

$$G(s) = \frac{V_o(s)}{V_i(s)}, \qquad (2.14\text{-}25)$$

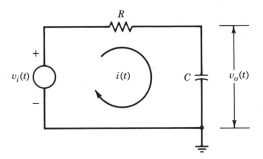

FIGURE 2.14-4 Low-pass RC network.

which is derived as follows. The voltage equation for the circuit is

$$v_i(t) = Ri(t) + v_o(t) = RC\frac{dv_o(t)}{d(t)} + v_o(t),$$

or

$$v_i(t) = \tau\frac{dv_o(t)}{d(t)} + v_o(t), \qquad (2.14\text{-}26)$$

where $\tau = RC$. The Laplace transform of Equation (2.14-26) is

$$\int_{0-}^{\infty} v_i(t)e^{-st}\,dt = \tau\int_{0-}^{\infty}\frac{dv_o(t)}{d(t)}e^{-st}\,dt + \int_{0-}^{\infty}v_o(t)e^{-st}\,dt,$$

or

$$V_i(s) = (\tau s + 1)V_o(s). \quad [\text{Note that } v_o(0) = 0.]$$

Hence the transfer function is

$$G(s) = \frac{V_o(s)}{V_i(S)} = \frac{1}{\tau s + 1}. \qquad (2.14\text{-}27)$$

Any network with a transfer function of the mathematical form of Equation (2.14-27) is a low-pass network.

If the input voltage is a unit-step function $u(t)\{\mathscr{L}[u(t)] = 1/s\}$, then the Laplace transform of the output is

$$V_o(s) = G(s)V_i(s) = \frac{1/\tau}{(s + 1/\tau)s}. \qquad (2.14\text{-}28)$$

To obtain the output v_o, take the inverse Laplace transform of $V_o(s)$. Now expand $V_o(s)$ into partial fractions:

$$\frac{1/\tau}{(s + 1/\tau)s} = \frac{A}{s + 1/\tau} + \frac{B}{s}. \tag{2.14-29}$$

Multiply both sides by $s + 1/\tau$;

$$\frac{1/\tau}{s} = A + B\frac{s + 1/\tau}{s}.$$

In the limit as $s \to -1/\tau$, we obtain $A = -1$.
Multiply both sides of Equation (2.14-29) by s:

$$\frac{1/\tau}{s + 1/\tau} = A\frac{s}{s + 1/\tau} + B.$$

In the limit as $s \to 0$, we obtain $B = 1$.
Thus

$$V_o(s) - \frac{1/\tau}{(s + 1/\tau)s} = -\frac{1}{s + 1/\tau} + \frac{1}{s}.$$

Since

$$\mathscr{L}^{-1}\left[\frac{1}{s + 1/\tau}\right] = e^{-t/\tau}, \quad \mathscr{L}^{-1}\left[\frac{1}{s}\right] = 1, t > 0,$$

we have

$$v_o(t) = \mathscr{L}^{-1}\left[\frac{1/\tau}{(s + 1/\tau)s}\right] = -\mathscr{L}^{-1}\left[\frac{1}{s + 1/\tau}\right] + \mathscr{L}^{-1}\left[\frac{1}{s}\right],$$

or

$$v_o = 1 - e^{-t/\tau}. \tag{2.14-30}$$

If the input step is V for $t > 0$ rather than $u(t)$, then the output becomes

$$v_o = V(1 - e^{-t/\tau}) = V(1 - e^{-t/RC}). \tag{2.2-3}$$

RC **Integrator.** The transfer function of the network shown in Figure 2.14-5 is expressed by

$$G(s) = \frac{I_o(s)}{I_i(s)},$$

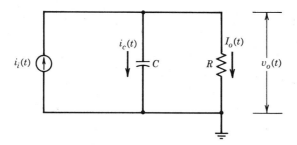

FIGURE 2.14-5   Integrating RC network.

which is derived as follows. The current equation for the circuit is

$$i_i(t) = i_c(t) + i_o(t),$$

or

$$i_i(t) = C \frac{dv_o(t)}{dt} + \frac{v_o(t)}{R}, \qquad (2.14\text{-}31)$$

where $v_c(0) = 0$. The Laplace transform of Equation (2.14-31) is

$$I_i(s) = sCV_o(s) + \frac{V_o(s)}{R}. \qquad (2.14\text{-}32)$$

Solving for $V_o(s)$,

$$V_o(s) = \frac{R}{\tau s + 1} I_i(s), \qquad (2.14\text{-}33)$$

where $\tau = RC$. Since

$$I_o(s) = \frac{V_o(s)}{R} \quad \text{or} \quad V_o(s) = RI_o(s),$$

then

$$I_o(s) = \frac{1}{\tau s + 1} I_i(s).$$

Therefore

$$G(s) = \frac{I_o(s)}{I_i(s)} = \frac{1}{\tau s + 1}. \qquad (2.14\text{-}34)$$

The condition for the network to be an integrator is determined from

$$(\tau s + 1)I_o(s) = I_i(s).$$

In the time domain,

$$i_i(t) = i_o(t) + \tau \frac{di_o(t)}{dt}.$$

The condition for integrating is $i_i(t) \gg i_o(t)$,

$$i_i(t) \simeq \tau \frac{di_o(t)}{d(t)},$$

or

$$i_o(t) \simeq \frac{1}{\tau} \int_0^t i_i(t)dt \tag{2.14-35}$$

The condition may be satisfied if R and C are large, so that $\tau \gg T$, the entire period of the input waveform.

High-Pass *RC* Network Driven by Voltage Source.   The transfer function of the network shown in Figure 2.14-6 is derived as follows. The voltage equation for this circuit is

$$v_i(t) = v_o(t) + v_c(t). \tag{2.14-36}$$

Since

$$v_c(t) = \frac{1}{C} \int_0^t i(t)dt = \frac{1}{RC} \int_0^t Ri(t)dt = \frac{1}{\tau} \int_0^t v_o(t)dt,$$

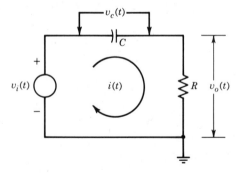

FIGURE 2.14-6 High-pass network.

where $\tau = RC$, then Equation (2.14-36) can be written as

$$v_i(t) = v_0(t) + \frac{1}{\tau} \int_0^t v_o(t)dt. \tag{2.14-37}$$

Differentiating Equation (2.14-37) with respect to time,

$$\frac{dv_i(t)}{dt} = \frac{dv_o(t)}{dt} + \frac{v_o(t)}{\tau},$$

or

$$v_o(t) = \tau \frac{d}{dt}[v_i(t) - v_o(t)]. \tag{2.14-38}$$

Taking the Laplace transform of Equation (2.14-38),

$$V_o(s) = \tau s V_i(s) - \tau s V_o(s). \tag{2.14-39}$$

Rearranging Equation (2.14-39),

$$(\tau s + 1)V_o(s) = \tau s V_i(s).$$

Thus the transfer function is

$$G(s) = \frac{V_o(s)}{V_i(s)} = \frac{\tau s}{\tau s + 1}. \tag{2.14-40}$$

Any network with a transfer function of the mathematical form of Equation (2.14-40) is a high-pass network.

If the input is a step voltage $V(t > 0)\{\mathscr{L}[V] = V/s\}$, then the Laplace transform of the output is

$$V_o(s) = G(s)V_i(s) = \frac{V}{s + 1/\tau}. \tag{2.14-41}$$

Since $\mathscr{L}^{-1}\left[\dfrac{1}{s + 1/\tau}\right] = e^{-t/\tau}$ we have

$$v_o(t) = V\mathscr{L}^{-1}\left[\frac{1}{s + 1/\tau}\right] = Ve^{-t/\tau}$$

or

$$v_o = Ve^{-t/\tau}. \tag{2.2-2a}$$

RC **Differentiator.**   The transfer function of the network shown in Figure 2.14-7 is expressed as

$$G(s) = \frac{I_o(s)}{I_i(s)},$$

which is derived as follows. The current equation for the circuit is

$$i_i(t) = \frac{v_o(t)}{R} + i_o(t) = \frac{v_o(t)}{R} + C\frac{dv_o(t)}{dt}. \tag{2.14-42}$$

Assuming $v_o(0) = 0$, then

$$I_i(s) = \frac{V_o(s)}{R} + sCV_o(s). \tag{2.14-43}$$

Hence

$$V_o(s) = \frac{R}{\tau s + 1} I_i(s), \tag{2.14-44}$$

where $\tau = RC$. Since $I_o(s) = sCV_o(s)$, then

$$I_o(s) = \frac{\tau s}{\tau s + 1} I_i(s),$$

and therefore

$$G(s) = \frac{I_o(s)}{I_i(s)} = \frac{\tau s}{\tau s + 1}. \tag{2.14-45}$$

The condition necessary for the network to be a differentiator is determined from

$$(\tau s + 1)I_o(s) = \tau s I_i(s).$$

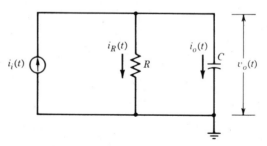

FIGURE 2.14-7 Differentiating *RC* network.

Rearranging yields

$$I_o(s) = \tau s[I_i(s) - I_o(s)]; \tag{2.14-46}$$

$$\mathcal{L}^{-1}[I_o(s)] = \tau \mathcal{L}^{-1}[sI_i(s) - sI_o(s)].$$

Thus

$$i_o(t) = \tau \frac{d}{dt}[i_i(t) - i_o(t)]. \tag{2.14-47}$$

If $i_i \gg i_o$, then

$$i_o(t) \simeq \tau \frac{di_i(t)}{dt}. \tag{2.14-48}$$

This condition can be satisfied if C and R are small.

High-Pass RL Network Driven by a Current Source. The current equation for the circuit shown in Figure 2.14-8 is

$$i_i(t) = \frac{V_o(t)}{R} + \frac{1}{L}\int_0^t v_o(t)dt. \tag{2.14-49}$$

Differentiating Equation (2.14-49),

$$\frac{di_i(t)}{dt} = \frac{1}{R}\left[\frac{dV_o(t)}{dt}\right] + \frac{V_o(t)}{L},$$

or

$$\frac{di_i(t)}{dt} = \frac{di_o(t)}{dt} + \frac{R}{L}i_o(t). \tag{2.14-50}$$

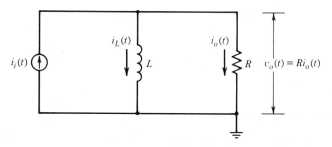

FIGURE 2.14-8 High-pass RL network.

Rearranging Equation (2.14-50),

$$i_o(t) = \frac{L}{R} \frac{d}{dt} [i_i(t) - i_0(t)] = \tau \frac{d}{dt} [i_i(t) - i_o(t)], \qquad (2.14\text{-}51)$$

where $\tau = L/R$ = time constant. The Laplace transform of Equation (2.14-51) is

$$i_o(s) = \tau s I_i(s) - \tau s I_o(s). \qquad (2.14\text{-}52)$$

Hence the transfer function is

$$G(s) = \frac{I_o(s)}{I_i(s)} = \frac{\tau s}{\tau s + 1}. \qquad (2.14\text{-}53)$$

From Equation (2.14-51), for the network to be a differentiator, $i_i(t) \gg i_o(t)$. Therefore

$$i_o(t) \simeq \tau \frac{di_i}{dt}(t). \qquad (2.14\text{-}54)$$

REFERENCES

1. Johnson, D. E. and J. R. Johnson, *Introductory Electric Circuit Analysis*, Prentice-Hall, Englewood Cliffs, New Jersey, 1981.

2. Tuttle, D. F., *Circuits*, McGraw-Hill, New York, 1977, Chap. 2.

3. Tocci, J. R., *Fundamentals of Pulse and Digital Circuits*, 2nd ed., Charles E. Merrill, Columbus, Ohio, 1977, Chaps. 1–5.

4. Pettit, J. M. and M. M. Mcwhorter, *Electronic Switching, Timing, and Pulse Circuits*, 2nd ed., McGraw-Hill, New York, 1970, Chaps. 2 and 7.

5. Millman, J. and H. Taub, *Pulse, Digital, and Switching Waveforms*, McGraw-Hill, New York, 1965, Chaps. 2 and 3.

6. Nilsson, J. W., *Electric Circuits*, Addison-Wesley, Reading, Massachusetts, 1983, Chaps. 7–8 and 15–16.

QUESTIONS

2-1. What is a high-pass network? What is a low-pass network?

2-2. Define the unit-step function.

2-3. Sketch the shape of the output waveform from a high-pass RC network with a pulse input when the RC time constant is much less than the pulse width.

2-4. Sketch the shape of the output waveform from an *RC* integrator when the input is a cosine wave.

2-5. Explain why the output of an *RC* differentiator represents the differential of the input waveform.

2-6. Explain why the output of an *RC* integrator represents the integration of the input waveform.

2-7. Sketch the shape of the output waveform from an *RC* differentiator when the input is a ramp voltage.

2-8. Explain how the loading problems that occur with a differentiator and an integrator may be overcome.

2-9. Explain how the horizontal and vertical sync pulses are separated in a TV receiver.

2-10. Describe the TV synchronization.

2-11. In practice, perfect compensation of the *RC* bridge attenuator is impossible. Why?

2-12. Explain how a switch-excited *RLC* circuit (high *Q*) generates a burst of oscillations.

2-13. Discuss the response of a pulse transformer to the input pulse.

2-14. Describe the delay-line differentiation.

PROBLEMS

2-1. What will the current i_c be in the circuit of Figure P2-1 at $t = 10$ ms after switch *S* is closed if the capacitor is initially uncharged?

   Hint.   Draw the Thevenin equivalent circuit and then calculate i_c.

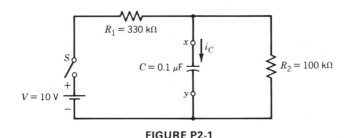

FIGURE P2-1

2-2. Prove by direct integration that the area under the curve of Figure 2.3-1*b* is zero.

2-3. Prove that for any periodic input waveform the average level of the steady-state output signal from the circuit of Figure 2.1-1*a* is always zero.

2-4.   Determine the amplitude of the capacitor voltage waveform for the low-pass RC ($=1$ kΩ $\times$ 1 μF) network with an input 6-V pulse of duration equal to 0.4 ms.

Answer. $v_c = 2$ V at $t = 0.4$ ms.

2-5.   A 0.1-μF capacitor is charged from a 10-V source through a 20-kΩ resistor. If the capacitor has an initial charge of -5 V, calculate its voltage after 10 ms.

Answer.   4.48 V.

2-6.   The signal applied to an RC circuit is a symmetrical square wave of amplitude equal to 20 V peak to peak and half-period equal to 4 ms. The resistor and capacitor are of 3.3 kΩ and 1 μF. Calculate the capacitor voltage at 15 ms from $t = 0$ and sketch the input and output waveforms.

Answer. $v_c = 6.16$ V at $t = 15$ ms.

2-7.   The signal applied to an RC integrator is a 10-V pulse with a duration of 1 ms. $R = 20$ kΩ; $C = 10\mu$F. Find the level of v_C at the end of the pulse. The initial voltage on C is assumed to be zero.

Answer.   50 mV.

2-8.   Draw the input and output waveforms of an RC differentiating circuit to illustrate the differential of a sine wave to be a cosine. At the peak of the sine wave ($t = 0$), the differentiator output voltage is zero. Why?

2-9.   Sketch the output waveform from a differentiator when the input is a cosine wave.

2-10.   The signal applied to an RC differentiator is a triangular waveform with a peak of 10 V. The duration for positive ramp or negative ramp is 100 ms. The resistor and capacitor are of 2 kΩ and 0.5 μF. Calculate the amplitude of the differentiated output and sketch the waveform of the output.

Answer. ± 0.1 V.

2-11.   A 10-Hz symmetrical square wave whose peak-to-peak amplitude is 4 V is impressed on a high-pass RC network whose lower 3-dB frequency is 5 Hz. Calculate and sketch the output waveform. In particular, what is the peak-to-peak output amplitude?

Answer.   6.62 V.

2-12.   A square wave whose peak-to-peak value is 2 V extends ± 1 V with respect to ground. The half-period is 0.1 sec. This voltage is applied to an RC differentiating circuit whose time constant is 0.2 sec. What are the steady-state maximum and minimum values of the output voltage?

Answer. $V_1 = -V_2 = 1.244$ V; $V_1' = -V_2' = 0.756$ V.

2-13.   (a) Design a simple circuit of an exponential pulse generator; its output waveform is similar to that of Figure 2.5-1. (b) Plot the curves in Figure 2.5-1, which represent v_R/V and v_C/V for $n = RC/\tau = 10$.

2-14. The bridge-type compensated attenuator shown in Figure 2.10-1 is used as an oscilloscope input test probe. $R_1 = R_2 = 1\ \text{M}\Omega$, and $C_2 = 50\ \text{pF}$. To obtain perfect compensation the bridge should be adjusted to balance so that $R_1 C_1 = R_2 C_2$ or $v_o(0) = v_o(\infty)$) when a step voltage is applied. Calculate and plot to scale the output waveform for (a) $C_1 = 75\ \text{pF}$, (b) $C_1 = 50\ \text{pF}$, and (c) $C_1 = 25\ \text{pF}$. The input is a 20-V step.

Hint. At low frequencies (corresponding to $t \to \infty$), $v_o = v_i/2$. At high frequencies (corresponding to $t \to 0$), $v_o = [C_1/(50\text{p} + C_1)]v_i$. The time constant $= 0.5\ \text{M}\ (C_1 + 50\ \text{p})$, which can be found if we look back from the output terminal (with the input short circuited).

Answer. (a) $v_o(0+) = 12\ \text{V}$; (b) $v_o(0+) = v_o(\infty) = 10\ \text{V}$; (c) $v_o(0+) = 6.6\ \text{V}$.

2-15. In the scope probe bridge of Figure 2.10-1, $R_1 = 4.7\ \text{M}\Omega$, $R_2 = 0.28\ \text{M}\Omega$, and $C_2 = 100\ \text{pF}$, the coaxial cable capacitance. Assume that the scope that is connected to the probe has an input impedance of $2\ \text{M}\Omega$ shunted by $10\ \text{pF}$. Calculate (a) the attenuation, (b) the C_1 value for the best response, and (c) the input impedance of the probe.

Answer. (a) $1/20$; (b) $5.75\ \text{pF}$; (c) $4.946\ \text{M}\Omega$ in parallel with $5.5\ \text{pF}$.

2-16. A current pulse of amplitude I is applied to a parallel RC circuit. Show that the capacitor current is given by

$$i_C = I\,e^{-t/RC}.$$

2-17. An LR' differentiator will be formed if the resistor R and capacitor C of an RC differentiator are replaced by an inductor L and resistor R', respectively. Prove that for the same input, the output from the two differentiating circuits will be the same if $RC = L/R'$. Assume that the initial conditions are those of rest (no voltage on C and no current in L).

2-18. (a) A narrow 16-V pulse is applied to the primary of the ideal pulse transformer in Figure p2-18. If $n = N_s/N_p = 2$ and $R_L = 1\ \text{k}\Omega$, determine v_o, i_s, and i_p. (b) From the equations used derive the following relationship:

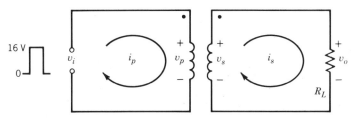

FIGURE P2-18

$$i_p = \frac{v_p}{R_L/n^2} = \frac{v_p}{R_{eq}}.$$

Answer. $v_o = 32\,\text{V}$; $i_s = 32\,\text{mA}$ peak; $i_p = 64\,\text{mA}$ peak.

2-19. A type RG-59/U coaxial cable has a capacitance of 20 pF/ft and a characteristic impedance of 73 Ω. Find the length required for a 0.4-μsec delay.

Answer. 274 ft.

2-20. (a) Derive the transfer function of the differentiating *RL* network shown in Figure p2-20. (b) Determine the condition necessary for the network to behave as a differentiator.

Answer. (a) $G(s) = \tau s/(\tau s + 1)$; $\tau = L/R$. (b) $v_o(t) = \tau(d/dt)v_i(t)$; how can this condition be satisfied?

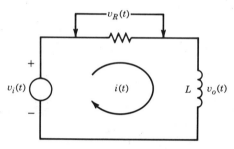

FIGURE P2-20

2-21. Determine the condition necessary for the output to be proportional to the derivative of the input for the high-pass *RC* network of Figure 2.14-6.

3

THEORY AND APPLICATION OF SEMICONDUCTOR-DIODE SWITCHING

3.1 STEADY-STATE SWITCHING CHARACTERISTICS OF *pn*-JUNCTION DIODES

Introduction

The semiconductor *pn*-junction diode is a nonlinear circuit element. The diode can be employed as a switch since it operates either in the OFF state (very high resistance) or in the ON state (very low resistance). If the input waveforms have transition times longer than the diode switching times, the diode in the switching circuits will have steady-state switching characteristics.

An ideal diode passes current in only one direction, that is, the forward direction. If a forward bias is applied to the diode, as in Figure 3.1-1*a*, the forward resistance is ideally zero, and the current flow is limited only by an external resistance. As the battery polarity is reversed, the diode is reverse biased, and, ideally, there is no current flow.

An actual diode only approximates the ideal. Typical characteristics of a low-current silicon diode are shown in Figure 3.1-1*b*. When it operates in the forward-bias region, the current increases exponentially with voltage and the forward resistance remains finite (a small value). Practically, an appreciable current does not flow until 0.6 V forward bias (silicon diode) is applied, but above this voltage, the current rises rapidly. In reality, there is a conduction threshold (cutin voltage) V_y at about 0.6 V. This fit to the actual exponential rise is termed the *piecewise approximation*, as shown in Figure 3.1-1*c*.

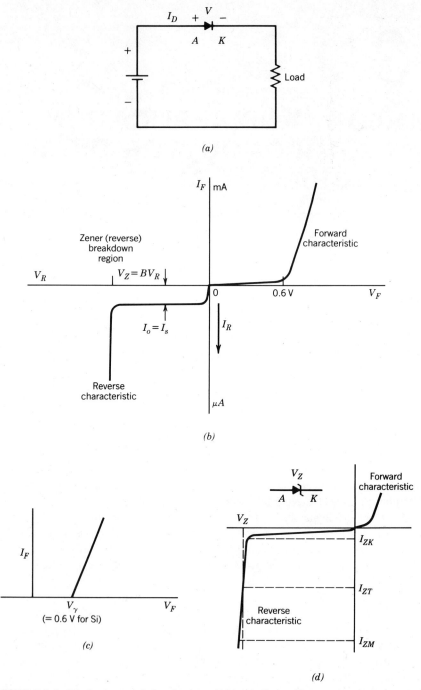

FIGURE 3.1-1 Diode characteristics: (*a*) forward-biased diode; (*b*) current–voltage curve of typical silicon diode; (*c*) piecewise-linear current–voltage curve; (*d*) Zener diode (Si) characteristics.

V–I Relationship

A physical analysis of a *pn*-junction diode (neglecting the Zener region) shows that the current I and voltage V are related by

$$I = I_o(e^{V/\eta V_T} - 1), \qquad (3.1\text{-}1)$$

where I = current through diode, A,
  V = voltage across diode, V,
  $I_o = I_s$ = reverse saturation current, A,
  V_T = volt equivalent of temperature = $kT/q = T/11{,}600$,
  k = Boltzmann's constant = 1.38×10^{-23} J/K,
  q = electron charge = 1.602×10^{-19} C,
  η = empirical constant between 1 (for Ge) and 2 (for Si).

We now assume $\eta = 1$ for simplicity. At room temperature (300 K),

$$V_T = \frac{T}{11{,}600} = 26 \text{ mV} \simeq 25 \text{ mV}. \qquad (3.1\text{-}2)$$

If V is negative and $|V| \gg V_T$, then $I \simeq -I_o$. If V is positive and $V \gg V_T$, then the forward current is

$$I \simeq I_o e^{V/V_T}. \qquad (3.1\text{-}3)$$

The actual characteristic of a typical diode shown in Figure 3.1-1*b* approximates the exponential curve represented by Equation (3.1-1). For large-signal applications, the actual characteristic can be approximated by a piecewise-linear characteristic (Figure 3.1-1*c*). For a forward bias ($V > V_\gamma$), the diode is said to be in its ON state. For a reverse bias ($V < V_\gamma$), the diode is said to be in its OFF state. The forward resistance is a constant dynamic resistance given by

$$R_f = \frac{dV}{dI} \quad \text{when } V > V_\gamma. \qquad (3.1\text{-}4)$$

For a current swing from cutoff to 10 mA with a typical silicon diode, reasonable values are $V_\gamma = 0.6$ V and $R_f = 15\,\Omega$, and for a typical germanium diode, $V_\gamma = 0.2$ V and $R_f = 20\,\Omega$. When conducting current, the typical voltage (V_F) at room temperature across a diode is 0.7 V for silicon and 0.3 V for germanium. A measure of the effect of reverse saturation current I_o is the value of the reverse resistance R_r defined as

$$R_r = \frac{BV_R}{I_o}, \qquad (3.1\text{-}5)$$

where BV_R is the peak reverse voltage (PRV). For large-signal applications the OFF state is indicated as a large reverse resistance R_r (larger than several hundred kilohms).

Imperfect Diode Switch

The *pn* diode is not a perfect switch, since its on-state resistance is not zero and its off-state resistance is not infinite. For example, typical values for the silicon 1N4153 diode are $V_F = 0.7$ V at $I_F = 10$ mA, corresponding to $R_F = R_{on} = 70\,\Omega$, and $I_R = 0.05\,\mu$A at $V_R = 50$ V, corresponding to $R_R = R_{off} = 1000$ MΩ.

Temperature Dependence of *V-I* Characteristic

The $V-I$ relationship [Equation (3.1-1)] contains the temperature implicitly in the two parameters V_T and I_o. From the experimental data we conclude that the reverse-saturation current I_o approximately doubles for every 10°C rise in temperature for both silicon and germanium. Since $(1.07)^{10} \simeq 2$, or $1.07 \simeq 2^{1/10}$, we may also conclude that I_o increases approximately 7%/°C. If $I_o = I_{o1}$ at $T = T_l$, then at a temperature T, I_o is expressed by

$$I_o(T) = I_{ol} \times 2^{(T - T_l)/10}. \tag{3.1-6}$$

The forward current increases at fixed voltage when the temperature is increased. However, if V is now reduced, then I may be brought back to its previous value. It is found that for either silicon or germanium (at room temperature)

$$\frac{dV}{dT} \simeq -2.5 \text{ mV/°C} \tag{3.1-7}$$

in order to maintain a constant value of I. Note that dV/dT decreases with increasing T.

Reverse (Zener Diode) Characteristics

A *Zener diode* is a semiconductor diode designed to operate in the reverse-breakdown region of its characteristics, as shown in Figure 3.1-1d. In a Zener diode, the *p*-type silicon and *n*-type silicon are heavily doped. This heavy doping results in a low value of the reverse breakdown voltage (BV_R) termed the Zener voltage V_z. The current that results in the reverse direction after breakdown is the Zener current I_z. If I_z is maintained within certain limits, the voltage drop across the diode is maintained at a reliable constant level V_z. In Figure 3.1-1d, V_z is measured at test current I_{ZT}. The knee current, I_{ZK}, is the minimum current that must pass through the device to maintain a constant V_z. The maximum permissible Zener current is $I_{z,\,\text{max}}$. ΔV_z (change of V_z) as a result of ΔT (change of temperature) is proportional to V_z as well as to ΔT. This variation is usually expressed as a temperature coefficient (TC), where

$$TC = \frac{\Delta V_z/V_z}{\Delta T} \times \frac{100\%}{°C}.$$ (3.1-8)

When $V_z \simeq 6$ V, $TC = 0$, and V_z is independent of temperature. If $V_z > 6$ V, TC is positive, and if $V_z < 6$ V, TC is negative. A typical $TC \simeq 0.1\%$; that is, the Zener voltage increases 0.001 V_z for each 1°C temperature rise. When the Zener diode is forward biased, it behaves as an ordinary diode.

3.2 DIODE CLIPPERS

Introduction

Clipper or *limiter circuits* are those that clip off a portion of an input waveform. The *pn* diode is a natural element for use in clipper circuits because of its ability to pass current in one direction and block it in the other. A clipper circuit requires at least a diode and a resistor. However a battery is also often used. If the positions of the various elements are interchanged and the magnitude of the battery is altered, then the output waveform from the limiter circuit can be clipped at different levels. For convenience, to illustrate the practical clipping circuits, the typical diode data are given in Tables 3.2-1 and 3.2-2.

Positive and Negative Series Clippers

In the positive series clipper of Figure 3.2-1, the diode is forward biased when the input becomes negative. Hence the output voltage at this time is the negative

TABLE 3.2-1
Data of High-Speed Silicon Switching Diodes[a]

	1N914A	1N914B	1N915	1N916B	1N917	Unit
V_R Reverse voltage at -65 to $+150°C$	75	75	75	75	30	V
$I_R = I_s$ Reverse current at V_R	5	5	5	5	5	μA
I_F Minimum forward current at $V_F = 1$ V	20	100	50	30	10	mA
P Power dissipation	250	250	250	250	250	mW
t_{rr} Maximum reverse-recovery time	4	4	10	4	3	ns

[a]Maximum ratings and characteristics at 25°C ambient temperature (unless otherwise noted).

TABLE 3.2-2
Data of Silicon Zener Diodes[a]

Parameter	V_Z, Zener Breakdown Voltage	α_z, Temperature Coefficient of Breakdown Voltage	Z_z, Small-Signal Breakdown Impedance	I_R, Static Reverse Current	
Test conditions	$I_{ZT} = 20\,\text{mA}$	$I_{ZT} = 20\,\text{mA}$	$I_{ZT} = 20\,\text{mA}$ $I_{zt} = 1\,\text{mA}$	$V_R = 1\,\text{V}$	$V_R = 1\,\text{V}$ $T_A = 150°C$
Limit →	Nominal	Typical	Maximum	Maximum	Maximum
Unit →	V	%/°C	Ω	µA	µA
1N746	3.3	−0.062	28	10	30
1N747	3.6	−0.055	24	10	30
1N748	3.9	−0.049	23	10	30
1N749	4.3	−0.036	22	2	30
1N750	4.7	−0.018	19	2	30
1N751	5.1	−0.008	17	1	20
1N752	5.6	+0.006	11	1	20
1N753	6.2	+0.022	7	0.1	20
1N754	6.8	+0.035	5	0.1	20
1N755	7.5	+0.045	6	0.1	20

[a]Characteristics at 25°C free-air temperature (unless otherwise noted).

peak input plus the diode voltage drop $(-V_m + V_F)$. When the input becomes positive, the diode is reverse biased, and the reverse saturation current I_s flows through resistor R_1. The output then is a very small positive voltage $I_s R_1$. The resultant output waveform is essentially the input with the positive portion clipped off.

The negative series clipper shown in Figure 3.2-2 operates in the same way

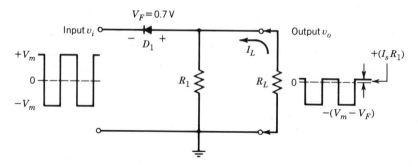

FIGURE 3.2-1 Positive series clipper.

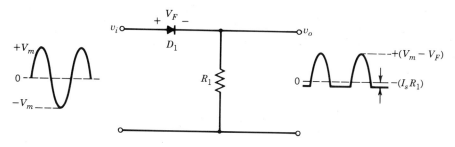

FIGURE 3.2-2 Negative series clipper.

as the positive clipper except that, in this case, the diode is reversed and the negative portion of the output waveform is clipped off.

Example 3.2-1. In the positive series clipper of Figure 3.2-1, input $v_i = \pm 60$ V, load current $I_L = 15$ mA, and output voltage $v_o \leqslant 0.5$ V. (a) Compute the value of R_1. (b) Specify the diode in terms of peak reverse voltage (PRV), forward current, and power dissipation. (c) Select a suitable diode from Table 3.2-1.

Solution.

(a) Since $V_i = \pm 60$ V, reverse voltage $V_R = +60$ V. From Table 3.2-1, $I_s = 5 \ \mu\text{A}$.

$$R_1 = \frac{V_o}{I_s} = \frac{0.5 \text{ V}}{5 \ \mu\text{A}} = 100 \text{ k}\Omega.$$

(b) Since diode reverse voltage $V_m = 60$ V, we need $PRV \geqslant 60$ V. For negative input,

$$I_{R1} = \frac{V_m}{R_l} = \frac{60 \text{ V}}{100 \text{ k}\Omega} = 0.60 \text{ mA}.$$

Power dissipation in R_l:

$$P_{R1} = \frac{V^2}{R_1} = \frac{(60 \text{ V})^2}{100 \text{ k}\Omega} = 36 \text{ mW}.$$

Diode forward current:

$$I_F = I_L + I_{R1} = 15 \text{ mA} + 0.60 \text{ mA} = 15.60 \text{ mA}.$$

Diode power dissipation:

$$P_{D1} = V_F \times I_F = 0.7 \text{ V} \times 15.60 \text{ mA}$$
$$= 10.92 \text{ mW}.$$

(c)   From parts (a) and (b): $I_s \leqslant 5\ \mu\text{A}$, PRV $\geqslant 60$ V, $I_F \geqslant 15.60$ mA.

Hence the diode selected should have

$$I_s \leqslant 5\ \mu\text{A},\ \text{PRV} \geqslant 60\ \text{V, and}\ I_F \geqslant 15.60\ \text{mA}.$$

The 1N914B, 1N915, and 1N916B diodes fulfill all the required conditions.

Series-Noise-Clipping Circuits

If an additional diode D_2 is shunted across the diode D_1 in the series clipper (Figure 3.2-1 or 3.2-2) with the D_1 (or D_2) anode connected to the D_2 (or D_1) cathode, then this new circuit, called the *series noise clipper*, can be used to eliminate the small noise ($< V_F$) contained in the input signal ($> V_F$). Because the peaks of noise voltage are not large enough to forward bias either D_1 or D_2, the output during the time between signals is zero. If the noise is too large for ordinary diodes, the series noise clipper can be made up using two Zener diodes and a resistor in series with the cathodes connected together, as shown in Figure 3.2-3. In this case, when the input signal goes positive, D_1 behaves as an ordinary forward-biased diode, while D_2 goes into breakdown; when the input signal goes negative, D_2 behaves as an ordinary forward-biased diode, while D_1 goes into breakdown. The dead zone is $\pm(V_F + V_Z)$, and only signals larger than this will be passed to the output. In the absence of a load current, R_1 must pass enough current to keep the diode conducting. To ensure that I_{R1} is larger than I_{ZK}, we can make $I_{R1} \simeq \frac{1}{4}I_{ZT}$, where I_{ZT} is obtained from the Zener-diode data sheet.

Example 3.2-2.   In the series Zener noise clipper of Figure 3.2-3, input peak $V_m = \pm 7$ V and noise amplitude $= \pm 2$ V. Find: (a) the required Zener voltage V_Z; (b) output amplitude; (c) R_1 value; (d) power dissipation in R_1.

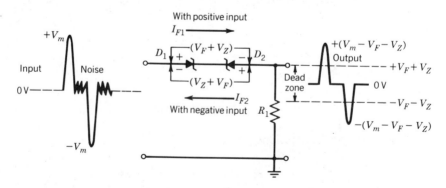

FIGURE 3.2-3   Series noise clipper using two Zener diodes.

Solution.

(a) $V_Z > 2$ V. From the Zener-diode data in Table 3.2-2, the 1N746 with $V_Z = 3.3$ V is a suitable device.

(b) $V_o = \pm(V_m - V_F - V_o) = \pm(7\text{ V} - 0.7\text{ V} - 3.3\text{ V}) = \pm 3$ V.

(c) From Table 3.2-2, $I_{ZT} = 20$ mA. To ensure $I_{R1} > I_{ZK}$, make

$$I_{R1} \simeq \tfrac{1}{4} I_{ZT} = 5\text{ mA},$$

when load current $I_L = 0$. Since $V_{R1} = V_o = \pm 3$ V, we obtain

$$R_1 = \frac{V_{R1}}{I_{R1}} = \frac{3\text{ V}}{5\text{ mA}} = 600\ \Omega.$$

(d) Power dissipation in R_1 is

$$P_{R1} = \frac{V_o^2}{R_1} = \frac{(3V)^2}{600\ \Omega} = 15\text{ mW}.$$

Biased Series Clippers

Figure 3.2-4 shows a biased negative series clipper, in which the diode is assumed to be ideal. With a 10-V reverse bias, the response to a sinusoidal input of 30-V amplitude is a positive 20-V peak. If the input signal is a square wave of $+20$-V peak to -30-V peak, the output waveform is of $+10$-V amplitude.

If the diode D_1 in Figure 3.2-4 is reversed, then the circuit becomes a positive series clipper.

Basic Shunt Clippers

A basic *shunt clipper* is a series combination of a resistor and a diode, across which the output voltage is developed. A shunt clipper is often used to protect the base-emitter junction of a transistor from excessive reverse bias. Most transistors cannot withstand more than 5 V applied in reverse direction across the base-emitter junction, and so the protective network of this type is usually employed when input signals are greater than 5 V.

In the positive shunt clipper of Figure 3.2-5, the positive portion of the input v_i is clipped off to protect the *pnp* transistor. When v_i becomes negative, D_1 is reverse biased and all the current I_o flows through the base. The clipper output is the base voltage given by

$$V_o = V_B = v_i - V_{R1} = -V_m - (-I_o R_1),$$

or

$$V_o = -(V_m - I_o R_1), \tag{3.2-1}$$

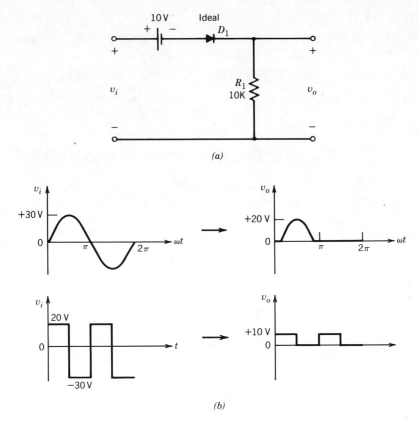

(a)

(b)

FIGURE 3.2-4   (*a*) A biased negative series clipper; (*b*) Output waveforms (v_o) for sinusoidal input and square-wave input.

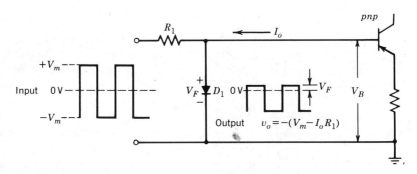

FIGURE 3.2-5   A positive shunt clipper used to protect the base-emitter junction of a transistor.

where V_m and I_o are the magnitude of the voltage and the current, respectively.

If in Figure 3.2-5, the diode is reversed, and the *pnp* transistor is replaced by an *npn* device, then a negative shunt clipper is formed and frequently used to protect the base-emitter junction from excessive reverse bias.

Example 3.2-3. In the positive shunt clipper of Figure 3.2-5, output voltage $V_o = -9.5\ \text{V}$ and output current $I_o \simeq 1\ \text{mA}$. If input voltage $v_i = \pm V_m = \pm 10.5\ \text{V}$, find the value of R_1 and the diode forward current I_F.

Solution. When $v_i = -10.5\ \text{V}$, $V_o = -9.5\ \text{V} = -(V_m - I_o R_1)$;

$$I_o R_1 = V_m - 9.5\ \text{V} = 10.5\ \text{V} - 9.5\ \text{V} = 1\ \text{V}.$$

Hence

$$R_1 = \frac{1\ \text{V}}{I_o} = \frac{1\ \text{V}}{1\ \text{mA}} = 1\ \text{k}\Omega$$

When $V_i = +10.5\ \text{V}$, D_1 is forward biased, and

$$V_F \simeq 0.7\ \text{V} = V_m - I_F R_1.$$

Hence

$$I_F = \frac{V_m - V_F}{R_1} = \frac{10.5\ \text{V} - 0.7\ \text{V}}{1\ \text{k}\Omega} = 9.8\ \text{mA}.$$

Biased Shunt Clippers

In the basic shunt clippers discussed above, the unwanted output is limited to a maximum of V_F above or below ground. In the biased shunt clipper of Figure 3.2-6, double diodes D_1 and D_2 are used, with D_1 cathode connected to $+V_B$ and D_2 anode connected to another bias $-V_B$. This circuit illustrates how the diode limits the signal on the top or on the bottom and how the double-diode clipper limits at two independent levels. Input signal v_i is a sine wave. Assume the magnitude of v_i peak greater than $(V_B + V_F)$. Then D_1 anode and D_2 cathode are both positive on each positive half-cycle; so D_1 conducts and D_2 does not, resulting in a positive output with a maximum of $+(V_B + V_F)$. Similarly a

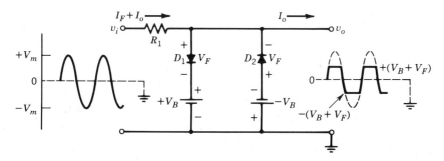

FIGURE 3.2-6 Biased double-diode shunt clipper circuit.

negative output with a maximum of $-(V_B + V_F)$ results on each negative half-cycle. The resultant output is then a square waveform, as shown in Figure 3.2-6. The biased double-diode shunt clipper is frequently employed to protect a circuit that has both positive and negative input signals. The bias voltages $+V_B$ and $-V_B$ are chosen to prevent the input from exceeding a maximum safe level.

A Zener-diode shunt clipper contains a resistor connected in series with two Zener diodes whose cathodes are joined together. This Zener clipper performs a function similar to that of the circuit of Figure 3.2-6.

Example 3.2-4. Referring to Figure 3.2-6, design a biased shunt clipper to protect a circuit that cannot accept voltages exceeding $V_o = \pm 2.9$ V. The input signal to the clipper is a ± 9-V square waveform, the maximum output current is $I_o = 1$ mA, and the minimum forward current I_F is taken as 10 mA. Find: (a) the value of R_1; and (b) specification of the silicon diodes in terms of I_F (at $V_F \simeq 0.7$ V) and PRV.

Solution.

(a) $V_o = V_B + V_F$;
   $V_B = V_o - V_F = 2.9$ V $- 0.7$ V $= 2.2$ V;
   $V_{R1} = (I_F + I_o)R_1 = V_m - V_B - V_F$;
$$R_1 = \frac{V_m - V_B - V_F}{I_F + I_o} = \frac{9 \text{ V} - 2.2 \text{ V} - 0.7 \text{ V}}{10 \text{ mA} + 1 \text{ mA}} = 555 \ \Omega;$$
   use 560 Ω standard value (see Appendix 1).

(b)   The diodes chosen must be low-current devices with $V_F \simeq 0.7$ V at $I_F = 10$ mA. They must have PRV > 10 V.

3.3   DIODE CLAMPERS

Introduction

A function that should often be performed with a periodic waveform called *clamping* is the establishing of the recurrent positive or negative extremity at some constant reference level V_R. A clamping circuit changes dc level rather than the shape of a waveform. The need to establish the extremity of the positive or negative signal excursion at V_R often appears in connection with a signal that has passed through a capacitive coupling network. Such a signal has lost its dc component, and the clamping circuit introduces a dc component. For this reason the clamping circuit is often called a *dc restorer* or *dc inserter*. Note that the dc component so introduced is not identical with the dc component lost in transmission.

Positive Clamping Circuit

Figure 3.3-1 shows a positive voltage clamper, in which $R_1 C_1 \geqslant 5t_p$, where t_p is the width of the pulse that forward biases the diode. When the input is negative, diode D_1 is forward biased, and capacitor C_1 charges with the polarity shown. During the negative input peak $(-V_m)$, the output cannot exceed the diode forward-bias voltage V_F. At this time, C_1 is charged to $V_m - V_F$, positive on the right side of C_1. When the input becomes positive, D_1 is reverse biased, and it has no further effect on the capacitor voltage. Since resistor R_1 has a large value, it cannot discharge the capacitor significantly during the positive input peak $(+V_m)$. While the input is positive, the output voltage is the sum of the input voltage and the capacitor voltage, or

$$\text{Positive output} = +V_m + (V_m - V_F) = +(2V_m - V_F). \qquad (3.3\text{-}1)$$

The peak-to-peak output $V_{o(p-p)}$ is the difference between the positive and negative peak voltages:

$$V_{o(p-p)} = (2V_m - V_F) - (-V_F) = 2V_m. \qquad (3.3\text{-}2)$$

Note that the amplitude of the output waveform is the same as that of the input. However the output negative peak is clamped to a level of V_F below ground.

The difference between clamper and clipper is that the clamper clamps the maximum positive or negative peak to a desired dc level, while the clipper clips off an unwanted portion of the input waveform.

To design a clamper circuit, C_1 must be chosen so that it becomes fully charged after about five cycles of the input waveform. Since it takes approximately five time constants for a capacitor to become completely charged,

$$5RC = 5t_p \quad \text{or} \quad RC = t_p,$$

where t_p is the width of the pulse that forward biases the diode, C equals C_1, and R is the sum of the source resistance R_s and the diode forward resistance R_F

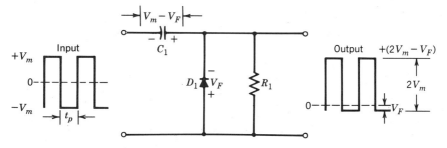

FIGURE 3.3-1 Positive voltage-clamper circuit.

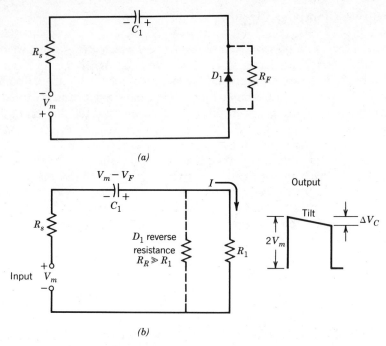

FIGURE 3.3-2 Positive clamping circuit: (*a*) charge of C_1 via R_s and R_F; (*b*) discharge of C_1.

(see Figure 3.3-2*a*). Since R_s is usually much larger than R_F,

$$(R_s + R_F)C_1 \simeq R_sC_1 = t_p. \tag{3.3-3}$$

When the capacitor in the positive clamper partially discharges during positive input peak, some tilt appears on the output, as shown in Figure 3.3-2*b*. The diode reverse resistance $R_R \gg R_1$, and so $R_R \| R_1 \simeq R_1$. The discharge current is

$$I \simeq \frac{2V_m}{R_1}.$$

Since the current I remains approximately constant during the discharge time, the change in capacitor voltage ΔV_c can be given by

$$\Delta V_c = \frac{It}{C_1} = \frac{2V_m}{R_1}\frac{t}{C_1},$$

from which we obtain

$$R_1 = \frac{2V_mt}{C_1\Delta V_c}. \tag{3.3-4}$$

Example 3.3-1.   Design a suitable circuit for a positive voltage clamper whose input is a 2-kHz square wave with an amplitude of ± 10 V. The signal source resistance R_s is 500 Ω, and the tilt on the output waveform is not to exceed 1%.

Solution.
    For the input square wave,

$$\text{period } T = \frac{1}{f} = \tfrac{1}{2} \text{ kHz} = 0.5 \text{ ms},$$

and

$$\text{pulse width } t_p = \tfrac{1}{2}T = 250 \text{ } \mu s.$$

From Equation (3.3-3),

$$C_1 = \frac{t_p}{R_s} = \frac{250 \text{ } \mu s}{500 \text{ } \Omega} = 0.5 \text{ } \mu F.$$

For 1% tilt on the output,

$$\Delta V_c = 0.01(2V_m) = 0.2 \text{ V}.$$

From Equation (3.3-4),

$$R_1 = \frac{2V_m t}{0.01(2V_m)C_1} = \frac{t}{0.01C_1},$$

and

$$t = t_p = 250 \text{ } \mu s;$$

$$R_1 = \frac{250 \text{ } \mu s}{0.01 \times 0.5 \text{ } \mu F} = 50 \text{ k}\Omega;$$

use 47 kΩ standard value.

Negative Clamping Circuit

Figure 3.3-3 shows a negative clamping circuit in which the resistance R_s of the signal source v_s is taken into account. During conduction the diode is assumed to be a forward-resistance R_f. The resistance R_f will lie in the range of tens to hundreds of ohms, depending on the type of diode used. The breakpoint

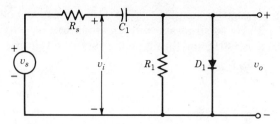

FIGURE 3.3-3 A clamping circuit with the source resistance R_s.

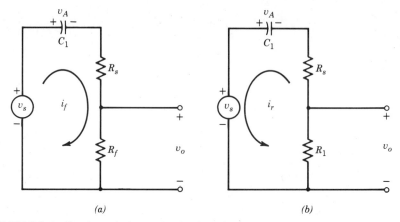

(a) (b)

FIGURE 3.3-4 Circuits equivalent to the circuit of Figure 3.3-3: (a) when the diode is conducting; (b) when the diode is not conducting. $R_R \gg R_1 \gg R_f$.

or threshold voltage V_γ in the diode characteristic is about 0.2 V for *Ge* and 0.6 V for *Si*. Beyond V_γ the forward current rises very rapidly. Now, for simplicity, we assume that the diode breakpoint V_γ occurs at zero voltage. The precision of operation of the circuit depends on the condition that $R_1 \gg R_f$. The equivalent circuit in Figure 3.3-4a applies when the diode is conducting, and the one in Figure 3.3-4b applies when the diode is not conducting.

Transient Approach to the Steady State.   In order to see how the transient waveform approaches the steady state, we shall make use of the equivalent circuits of Figure 3.3-4 and proceed as in Example 3.3-2.

Example 3.3-2.   In the circuit of Figure 3.3-3, $R_s = R_f = 100\,\Omega$, $R_1 = 10\,\text{K}$, and $C_1 = 1.0\,\mu\text{F}$. At $t = 0$ a symmetrical square-wave signal v_s of amplitude 10 V and frequency 5 kHz is applied, as indicated in Figure 3.3-5. Compute and draw the first several cycles of the output waveform.

Solution.   Assume that the capacitor C_1 is initially uncharged. At $t = 0$, the diode is forward biased and the equivalent circuit of Figure 3.3-4a is applicable. Thus, at the first 10-V jump of the input v_s, the output v_o jumps to +5 V. The

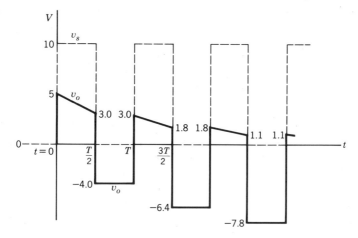

FIGURE 3.3-5   Example of the transient waveform in Figure 3.3-3.

capacitor now charges so that v_o delays toward zero exponentially, with a time constant

$$\tau = (R_s + R_f)C_1 = 200 \ \mu s.$$

Since the frequency is 5 kHz, the period $T = (5000)^{-1} \ s = 200 \ \mu s$; then at $t = T/2$, the output v_o, as indicated in Figure 3.3-5, has fallen to

$$v_o(t = T/2) = 5e^{-T/2} = 5e^{-1/2} = 5 \times 0.606 = 3.0 \ V.$$

At this time, the voltage across R_f or the voltage across R_s is 3.0 V, and so the capacitor voltage v_A is 4.0 V. At $t = (T/2)+$, the input drops back to zero and the diode does not conduct; we now use the equivalent circuit of Figure 3.3-4b. In this circuit, $v_A = 4.0$ V and $v_s = 0$, so that, neglecting R_s compared with R_1, $v_o = -4.0$ V, as in Figure 3.3-5. The output now starts to decay again toward zero. However the time constant is now $R_1C_1 = (10 \ K)(1.0 \ \mu F) = 10,000 \ \mu s$, which is 100 times larger than the time $T/2 = 100 \ \mu s$. Hence the decay is negligible.

Because in the interval $t = T/2$ to $t = T$, the capacitor voltage does not change ($v_A = 4.0$ V), then at $t = T+$, the diode is forward biased and the output returns to $+3.0$ V ($= 10 - 4 - 3$ V). Again the output decays toward zero exponentially with $\tau = 200 \ \mu s$, and the output at $t = 3T/2$ falls to $3e^{-1/2}$ or 1.8 V. The remaining calculations are repetitions of the preceding ones, and the results are shown in Figure 3.3-5. Cycle by cycle the output waveform approaches the steady-state case, at which the positive excursion of the waveform is clamped approximately to zero.

Steady-State Output Waveform.   Now consider a square-wave with an amplitude of $\pm V_m$ applied to the negative clamper, as shown in Figure 3.3-6.

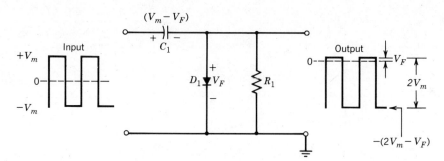

FIGURE 3.3-6 Negative clamping circuit.

When the input is positive, diode D_1 is forward biased and capacitor C_1 charges with the polarity shown. During the positive-input peak, output equals V_F and C_1 is charged to $V_m - V_F$, positive on the left side of C_1. During the negative-input peak, the reverse-biased diode has no further effect on the capacitor voltage, the high-value resistor R_1 cannot discharge the capacitor significantly, and so the output voltage is the sum of the input voltage and the capacitor voltage. Hence

$$\text{Negative output} = -(2V_m - V_F), \tag{3.3-5}$$

where $2V_m$ is the peak-to-peak output.

Biased Clampers

In the biased clamper of Figure 3.3-7, the diode (D_1) cathode is connected to a bias voltage V_B. During the positive-input peak, the output level is clamped to $V_B + V_F$, and so C_1 charges to $V_m - (V_B + V_F)$, positive on the left side of C_1. During the negative-input peak, the output is the sum of $-V_m$ and the capacitor voltage, or

$$\text{negative output} = -V_m - (V_m - V_B - V_F)$$

$$= -(2V_m - V_B - V_F), \tag{3.3-6}$$

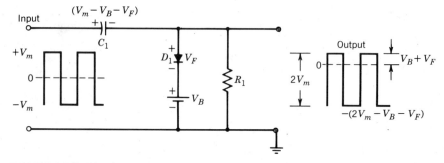

FIGURE 3.3-7 Biased clamping circuit to clamp output at approximately $+V_B$ maximum.

where $2V_m$ is the peak-to-peak output. The output is clamped to a maximum dc level of $V_B + V_F$. If the diode in Figure 3.3-7 is inverted, the circuit will clamp the output at $V_B - V_F$ minimum. If the bias voltage V_B in Figure 3.3-7 is replaced by V_Z of a Zener diode, the Zener-diode clamping circuit will perform the same function as the biased clamping circuit.

Example 3.3-3. Design a Zener-diode clamping circuit to clamp output to a maximum positive level of approximately 7.5 V. The input is a square wave having an amplitude of ± 12 V, a source resistance of 1 kΩ, and a frequency range of 500 Hz to 3 kHz. The output tilt is not to exceed 1%.

Solution. The Zener-diode clamping circuit is shown in Figure 3.3-8.

When frequency f is a minimum, the pulse width t_p is longest. In this case

$$T_{\max} = \frac{1}{f_{\min}} = \frac{1}{500 \text{ Hz}} = 2 \text{ ms}; \; t_p = \frac{T_{\max}}{2} = 1 \text{ ms}.$$

From Equation (3.3-3),

$$C_1 = \frac{t_p}{R_s} = \frac{1 \text{ ms}}{1 \text{ k}\Omega} = 1 \; \mu\text{F}.$$

For 1% tilt, $\Delta V_c = 0.01 \times 2V_m$.
From Equation (3.3-4)

$$R_1 = \frac{2V_m t_p}{\Delta V C_1} = \frac{2V_m \times 1 \text{ ms}}{0.01 \times 2V_m \times 1 \; \mu\text{F}} = 100 \text{ k}\Omega.$$

$$V_o = V_Z + V_F = 7.5 \text{ V}.$$

$$V_Z = V_o - V_F = 7.5 \text{ V} - 0.7 \text{ V} = 6.8 \text{ V}.$$

From the data in Table 3.2-2, the 1N754 Zener diode has $V_Z = 6.8$ V. Compared

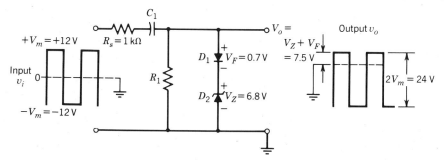

FIGURE 3.3-8 The Zener-diode clamping circuit.

to R_s, resistances R_f and R_z are very small. The charge current of C_1 is $V_m/(R_s + R_f + r_z) \simeq 12 \, \text{V}/1 \, \text{k}\Omega = 12 \, \text{mA}$. Hence, use a 1N754 Zener diode and a low-current diode such as 1N914A (see Table 3.3-1). The minimum capacitor voltage is $v_i + V_Z$ or 18.8 V.

3.4 *IC* DIODE ARRAY

Typical Diode Array

The diode array LM3019 shown in Figure 3.4-1 contains a diode quad and two isolated diodes, all fabricated at one time on the same silicon chip. The diodes in an IC chip have almost identical characteristics, while the characteristics of discrete diodes, selected at random, may differ widely. The diode array can be used as the analog switch, multiplier, phase detector, and so on.

Four-Diode Analog Switch

The analog switch shown in Figure 3.4-2 consists of the diode quad and the resistors. This circuit produces an output voltage v_o proportional to the input analog signal v_s when the control voltage $v_c = V_{\text{on}}$ and has $v_o = 0$ when $v_c = V_{\text{off}}$. When $v_c = V_{\text{on}}$ and $v_s = 0$, all four diodes are on with a voltage drop V_F across each diode, and thus the voltage drop from a to b is zero. The value V_{on} of the control voltage is adjusted so that the diodes remain on even if v_s is not zero. Therefore, when $v_c = V_{\text{on}}$, $v_o = v_s$ if $R_s = 0$ and v_o is proportional to v_s if R_s is not zero. The value V_{off} of the control voltage must be sufficiently small to ensure that when $v_c = V_{\text{off}}$, all four diodes are off and thus conduct no current; under these conditions, the current through R_L is zero and so the output voltage is also zero, independent of the input signal. It is reasonable to assume that $R_s = 0$ since the analog switch is practically driven from an op-amp whose output impedance is much less than 1 Ω.

Assume that ideal diodes are used in the analog switch. Then the circuit can

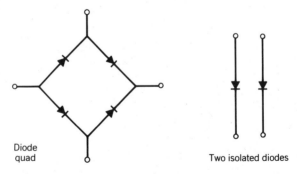

Diode quad

Two isolated diodes

FIGURE 3.4-1 The IC diode array LM3019.

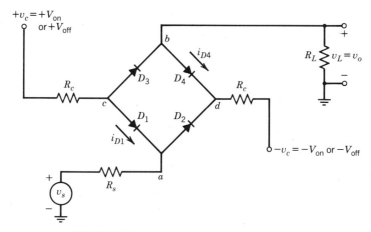

FIGURE 3.4-2 Four-diode-bridge analog switch.

be approximately analyzed. If $v_c = V_{on}$ now, V_F across each diode is zero, and thus

$$v_o = v_s. \tag{3.4-1}$$

Assume the maximum allowable input signal V_{sm}, which is a voltage sufficiently positive to cause D_1 and D_4 to turn off so that i_{D1} and i_{D4} are zero as $v_c = V_{on}$. When v_s increases to the point where $v_s = V_{sm}$, the four-diode bridge becomes completely opened, thereby disconnecting the input v_s from the load R_L, as shown in the equivalent circuit of Figure 3.4-3. Assume that the diodes D_1 and D_4 are just cut off so that v_{ac} and v_{bd} are both equal to zero. Then the

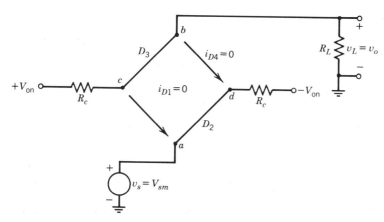

FIGURE 3.4-3 Equivalent switch circuit as $v_s = V_{sm}$ and diodes D_1 and D_4 are at the edge of cutoff ($v_{ac} = v_{bd} = 0$).

voltage drops with respect to ground are

$$v_a = v_d = V_{sm} \tag{3.4-2}$$

and

$$v_c = v_b = v_o = \frac{V_{on}R_L}{R_c + R_L}. \tag{3.4-3}$$

Setting $v_{ac} = v_a - v_c$ equal to zero, we obtain

$$V_{sm} = V_{on}\frac{R_L}{R_c + R_L}. \tag{3.4-4}$$

This equation indicates the value of V_{sm} at which diodes D_1 and D_4 are just cut off. If $R_c \gg R_L$, Equation (3.4-4) becomes

$$V_{sm} \simeq \frac{R_L}{R_c} V_{on}. \tag{3.4-5}$$

If the switch output is connected to the input of a noninverting op-amp whose input impedance R_L is much greater than R_c, then

$$V_{sm} \simeq V_{on}. \tag{3.4-6}$$

In either case, if input v_s is less than the value of V_{sm} given by Equation (3.4-4), the four-diode analog switch will operate to give $v_o = v_s$. If v_s exceeds V_{sm}, $v_o = V_{on}R_L/(R_c + R_L) =$ fixed [Equation (3.4-3)].

The value V_{off} of the control voltage is chosen so that the diodes are operating in the reverse region without suffering a Zener breakdown. Diodes in *IC* chips typically have a Zener breakdown when the reverse voltage is about $V_{ZB} = 6$ V; the symbol V_{ZB} represents the Zener breakdown voltage. To keep the diodes D_3 and D_4 off we must have

$$V_{cd} = V_c - V_d < 2(0.7) = 1.4 \text{ V},$$

so that no current flows in either resistor R_c; therefore

$$2V_{off} = V_{cd} < 1.4 \text{ V} \quad \text{and} \quad V_{off} < 0.7 \text{ V}.$$

There is a possibility that the diodes at cutoff may suffer a Zener breakdown if $v_s = V_{sm}$ is too large. Because no current flows in R_s when the diodes are cut off, $V_a = V_{sm}$, so that $V_{ac} = V_{sm} - V_{off}$. We set $V_{off} > V_{sm} - V_{ZB}$ to ensure that the diodes do not break down. Thus V_{off} should lie in the range

$$V_{sm} - V_{ZB} < V_{off} < 0.7 \text{ V}.$$

Because $V_{ZB} \simeq 6$ V, we might select $V_{off} = 0$ V if $V_{sm} = 5$ V. In this case, D_1 will turn on as $v_s < -0.7$ V and D_2 will turn on as $v_s > +0.7$ V. However, since D_3 and D_4 are off, v_o will remain equal to 0 V independent of v_s, and consequently the diode switch will remain cut off.

The circuit shown in Figure 3.4-2 is also called a *four-diode sampling gate*. A *sampling gate* is a switching circuit that is commonly used to sample the amplitude of dc or low-frequency signals. Sampling-gate circuits can also be constructed using bipolar transistors or FETs. A sampling gate is also known as a *transmission gate* or *linear gate*. It is distinct from a logic gate.

3.5 DIODE-SWITCHING COUNTERS

By means of the diode-switching characteristic, a diode-switching counter can be designed as shown in Figure 3.5-1. The circuit consists of ten stages of RC–bulb–diode combinations and an output inverter. The approximate characteristics of an NE-2 neon bulb are fire voltage = 50 V, operating voltage = 45 V, extinguishing voltage = 40 V, and operating current $\simeq 100 \, \mu$A. The PRV rating of the diode (e.g., 1N60) must be greater than 100 V. The transistor (e.g., 2N398) must have the magnitude of V_{CE} greater than 115 V.

The circuit in Figure 3.5-1 functions as a ring counter, in which the last stage is coupled back to the first. The trigger pulse has an amplitude of $+50$ V. The maximum count rate is 3000 pulses per second. The circuit action is described as follows.

When the reset button is pushed, only stage 0 conducts, and so only bulb 0 lights. C_1 then charges to $V_{c1} \simeq +5$ V.

When the first 50-V pulse is applied, only stage 1 conducts, and so only bulb 1 lights because of $+V_{c1}$ aid. C_2 then charges to $-V_{c2} \simeq -5$ V. Meanwhile C_1 discharges through R_1 toward the $+200$-V source so that the voltage across bulb 0 is decreased from 45 V to 40 V, causing bulb 0 to extinguish.

When the second pulse is applied, only stage 2 conducts, and so only bulb 2 lights because of $-V_{c2}$ aid. C_3 then charges to $+V_{c3} \simeq +5$ V. Meanwhile C_2 discharges through R_2 so that the voltage across bulb 1 is decreased from 45 to 40 V, causing bulb 1 to extinguish, and so on.

When the tenth pulse is applied, only stage 0 conducts, and so only bulb 0 lights due to $-V_{c10}$ aid. C_1 then charges to $V_{c1} \simeq +5$ V. Meanwhile C_{10} discharges through R_{10} so that the voltage across bulb 9 is decreased from 45 to 40 V, causing bulb 9 to extinguish. Note that a negative going voltage is developed across R_{10} as soon as the bulb 9 is extinguished; in this case, a positive pulse comes out from the inverter collector. The circuit of Figure 3.5-1 is a divide-by-10 counter; it will become a divide-by-6 counter if only six stages of the RC–bulb–diode combination are contained.

A timer can be made up by using a number of such counters (see Figure 3.5-1) connected in cascade. For example, a 1-min timer can be formed by using two divide-by-10 counters and two divide-by-6 counters connected in cascade with the trigger input of 60 pulses/second.

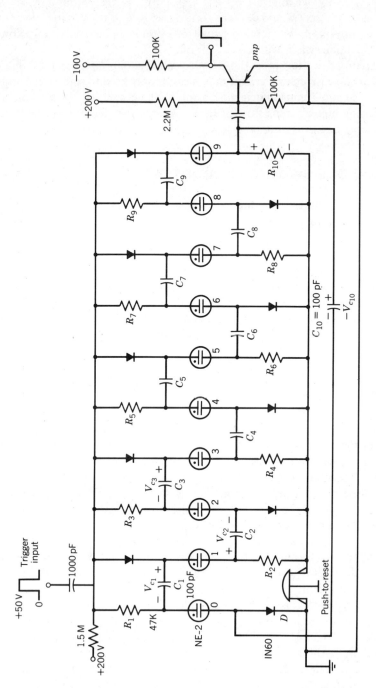

FIGURE 3.5-1 A diode-switching counter.

3.6   DUAL-DIODE SYNC DISCRIMINATORS FOR HORIZONTAL *AFC* IN TV RECEIVERS

Introduction

In order to make the synchronization more immune to noise, an AFC (automatic frequency control) circuit is employed for the horizontal-deflection oscillator in all television receivers. AFC is generally not used for the vertical-deflection oscillator, since the vertical RC integrator before the oscillator is a low-pass filter that removes high-frequency noise pulses. The block diagram of the AFC circuit for a horizontal oscillator is shown in Figure 3.6-1.

The dual-diode sync discriminator can detect a difference in frequency. The comparison sawtooth voltage can be taken from the horizontal-output circuit. The sync discriminator produces a dc-output voltage proportional to the difference in frequency between its two input voltages. The dc control voltage filtered by the RC integrator indicates whether the oscillator is on or off the sync frequency. The greater the difference between the oscillator and sync frequencies, the larger is the dc control voltage. The filtered dc control voltage changes the oscillator frequency by the amount necessary to make the oscillator frequency the same as the sync frequency. The horizontal oscillator generally uses either multivibrator or blocking-oscillator circuits. A rise of base forward bias increases the frequency for a common-collector transistor blocking oscillator (see Problem 6-13 and Table 4.1-3). At the synchronizing input of a commonly used emitter-coupled multivibrator, a decrease of base forward bias increases the oscillating frequency (Section 6.2).

Push-Pull Sync Discriminator

In the push-pull sync discriminator of Figure 3.6-2, opposite polarities of sync voltage are coupled to the two oppositely polarized diodes D_1 and D_2 so that D_1 is driven by negative sync voltage v_1 and D_2 is driven by positive sync voltage v_2. As a result, the sync input voltage drives both diodes into conduction. The comparison sawtooth voltage v_3 is applied at the junction of the two diodes.

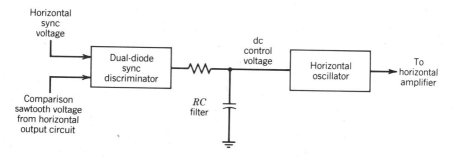

FIGURE 3.6-1   Block diagram of AFC circuit for horizontal oscillator in TV receivers.

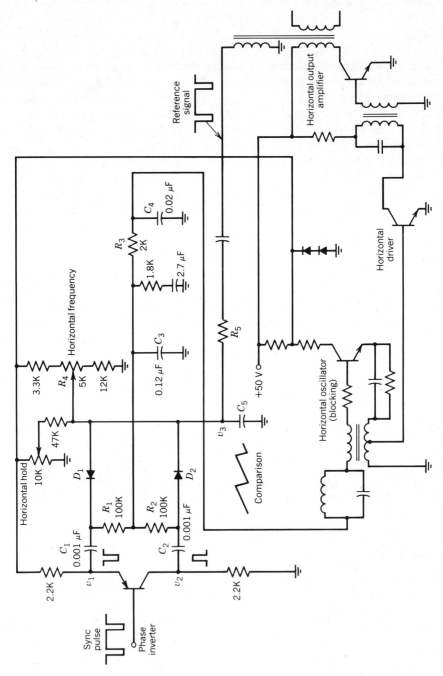

FIGURE 3.6-2 Push-pull sync discriminator connected to the input of the horizontal oscillator.

When v_3 is positive, it aids the D_1 current but opposes the D_2 current. The amount of diode conduction depends on how the sync input voltage is phased with respect to flyback on the sawtooth voltage.

The practical sync discriminator in Figure 3.6-2 can be simplified as in Figure 3.6-3. Note that v_1, v_2, and v_3 are applied to the simplified circuit. Since D_1 and D_2 are forward biased by v_1 and v_2, respectively, both diodes can be regarded as short circuits when the sync signal is applied. Thus the resultant voltages applied to D_1 and D_2 are $v_1 + v_3$ and $v_2 + v_3$, respectively. If the oscillator frequency is correct, then $|v_1 + v_3| = |v_2 + v_3|$. In this case, both diode currents are equal in magnitude but flow in opposite directions; thus no control-voltage output develops across C_4, as shown in Figure 3.6-4.

If $|v_1 + v_3| \neq |v_2 + v_3|$ (i.e., the oscillator frequency is incorrect), then the currents through both diodes cannot cancel each other, and thus C_4 develops the output control voltage (v_c); its magnitude and polarity are determined by the current difference and direction. If the oscillator frequency is greater than the sync frequency (i.e., the oscillator signal has a leading phase), then the sync signal applied to each diode is at the negative flyback end of the sawtooth wave, and so both diode currents cannot cancel each other. As indicated in Figure 3.6-5, the resultant output control voltage v_c is less than zero. Then the *npn* transistor used for the horizontal blocking oscillator has a decreased base bias, and thus the oscillator frequency decreases with the result of returning to synchronization. On the other hand, the output control voltage v_c is greater than zero as the oscillator frequency becomes less than the sync frequency, thus increasing the oscillator frequency with the result of returning to synchronization.

Single-End Sync Discriminator

In the single-end sync discriminator of Figure 3.6-6, negative sync pulses are coupled by C_1 to the common cathode of the two diodes D_1 and D_2. The diodes

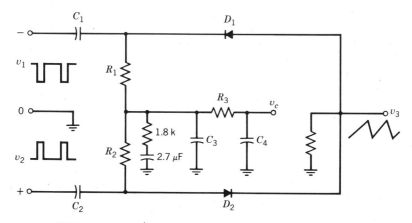

FIGURE 3.6-3 Simplified push-pull sync-discriminator circuit.

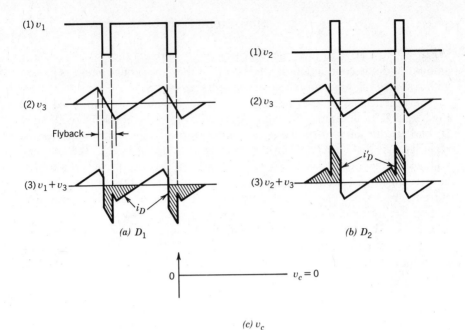

FIGURE 3.6-4 Waveforms for push-pull sync discriminator in Figure 3.6-3. $v_c = 0$ since $|v_1 + v_3| = |v_2 + v_3|$ at correct frequency.

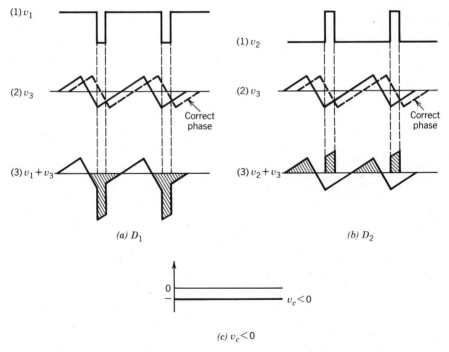

FIGURE 3.6-5 Waveforms for push-pull sync discriminator in Figure 3.6-3. $v_c < 0$ since $|v_1 + v_3| > |v_2 + v_3|$ when the oscillator frequency is greater than sync frequency.

110

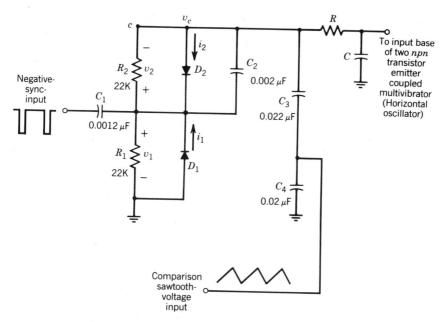

FIGURE 3.6-6 Single-ended discriminator for horizontal *AFC*.

are effectively in parallel for the sync input. The comparison sawtooth voltage coupled via C_3 is applied across the two diodes in series. Hence each diode has one-half the sawtooth input voltage.

The sawtooth voltage is applied anode to cathode for D_2 but cathode to anode for D_1. For convenience, let us consider the sawtooth as cathode-to-anode voltage for D_2 the same way sync voltage is applied to both diodes. Then the sawtooth voltage for D_2 can be shown inverted for the discriminator wave-shapes in Figure 3.6-7. In fact, the two diodes have equal and opposite sawtooth input voltages operated as a push-pull sawtooth input.

Both diode currents i_1 and i_2 flow due to the negative sync voltage applied at the cathode combined with a sawtooth-voltage input. When D_1 conducts, the result is dc voltage v_1 shown across R_1. Similarly conduction in D_2 results in the dc voltage v_2 shown across R_2. Note that R_1 and R_2 are the diode load resistors. The polarities of v_1 and v_2 mainly result from the stored minority charges when the diode forward voltages fall to zero (see Section 3.8). The dc control voltage v_c taken from point c with respect to ground consists of v_1 and v_2 in series opposition. Suppose the discriminator output voltage is applied to the input base of a two *npn*-transistor emitter-coupled multivibrator that serves as the horizontal oscillator (see Figure 6.2-1 or 12.7-1). *AFC* action is explained referring to the waveshapes in Figure 3.6-7: (*a*) at the correct oscil-lator frequency, the sync pulse is in the middle of flyback, $|v_2| = |v_1|$, and so $v_c = 0$ V; (*b*) when the oscillator is too fast with a shorter sawtooth cycle, the sync pulse is at the end of flyback, $|v_2| < |v_1|$, and so v_c is positive—this increases

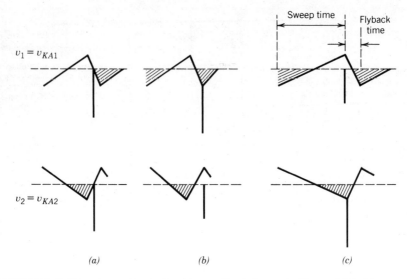

FIGURE 3.6-7 Waveshapes for single-end sync discriminator in Figure 3.6-6: (*a*) at correct oscillator frequency: sync pulse in the middle of flyback ($|v_1| = |v_2|$); (*b*) oscillator too fast with shorter sawtooth cycle: sync pulse at the end of flyback ($|v_1| > |v_2|$); (*c*) oscillator too slow with longer sawtooth cycle: sync pulse at start of flyback ($|v_1| < |v_2|$).

the input base forward bias and the collector current of the first transistor in the oscillator, thus reducing its frequency due to more time needed for the coupling capacitor (C_c in Figure 6.2-1) to discharge down to cutin (V_y) for conduction in the second transistor; (*c*) when the oscillator is too slow with a longer sawtooth cycle, sync pulse is at the start of flyback, $|v_2| > |v_1|$, and so v_c is negative—this reduces the input base forward bias, thus increasing the oscillator frequency. v_c is filtered by the RC network for the horizontal oscillator. The anode of D_2 is returned to ground through C_3 and C_4 in series. C_2 equalizes the amount of sawtooth voltage across the diodes. In order to help balance the input to the diodes, C_2 is larger than C_1 but smaller than the C_3-C_4 combination. Usually the peak-to-peak value of the sawtooth voltage is smaller than that of the sync pulse.

3.7   TWO BASIC LOGIC GATES

Introduction

The two basic diode logic gates are the OR gate and the AND gate. Each of these has two or more inputs and one output. The output will be either a high voltage or a low voltage depending upon the voltages on the various inputs. Digital circuits and systems contain voltages that exist at either HIGH level (e.g., 5–6 V) or LOW level (e.g., 0–1 V).

The Diode–Resistor OR Gate

The output of a positive-logic OR gate goes HIGH if any one or more of its inputs go HIGH and goes LOW only if all the inputs are LOW. Figure 3.7-1 shows the two-input diode–resistor OR gate with its function table and truth table. Both switches at inputs A and B control logic LOW (L) and HIGH (H). As shown in row 1 of the truth table, the output voltage at Y is LOW only when both inputs are LOW. With both inputs LOW, there is no source voltage, and the voltage at Y must be 0 V or logic LOW. When input A is LOW while input B is HIGH, the resulting output Y is HIGH, as shown in row 2 of the truth table; in this case, input V_i is passed directly to output Y via the conducting diode D_2 while diode D_1 is reverse biased by V_i via the conducting diode D_2. When input A is HIGH while input B is LOW, the output Y is also HIGH, as shown in row 3 of the truth table; in this case, D_1 is forward biased and conduct-

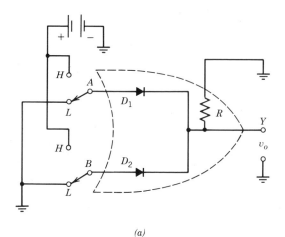

(a)

Row	Inputs		Output
	A	B	Y
1	L	L	L
2	L	H	H
3	H	L	H
4	H	H	H

(b)

A	B	Y
0	0	0
0	1	1
1	0	1
1	1	1

(c)

FIGURE 3.7-1   *(a)* Two-input positive-logic OR gate; *(b)* function table; *(c)* truth table. [Circuit *(a)* will become an AND gate if negative logic is used.] The way logic gate works can be itemized on a truth table. The function table can be used as a truth table to determine the output signal whether positive or negative logic is applied.

ing while D_2 is reverse biased and nonconducting. When both inputs are HIGH, as in row 4 of the truth table, both diodes conduct and so both of the HIGH inputs are directly connected to the output Y through their respective conducting diodes.

An OR gate with an inverter (a NOT circuit; see Section 4.3) at its output performs the same logic function as a NOR gate does.

The Diode–Resistor AND Gate

The output of a positive-logic AND gate is HIGH only when all its inputs are HIGH; it is LOW if any one or more of its inputs are LOW. Figure 3.7-2 shows the two-input diode–resistor AND gate with its function table and truth table. When both inputs are LOW, as shown in row 1 of the table, the resulting output

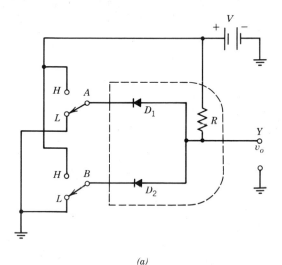

(a)

	Inputs		Output
Row	A	B	Y
1	L	L	L
2	L	H	L
3	H	L	L
4	H	H	H

(b)

A	B	Y
0	0	0
0	1	0
1	0	0
1	1	1

(c)

FIGURE 3.7-2   (a) Two-input positive-logic AND gate; (b) function table; (c) truth table. [Circuit (a) will become an OR gate if negative logic is used.]

Y is LOW; two LOW inputs are caused if both switches are down in the L positions; in this case, grounding both inputs forward biases both diodes, and since the output terminal Y is effectively grounded through these conducting diodes, the resulting output voltage V_o is 0 V or logic LOW. If the switch at input B is placed up in the H position while switch A remains in the L position, row 2 of the table indicates the resulting input and output signals; in this case, the HIGH logic signal at input B reverse biases diode D_2 through the conducting diode D_1 and the forward-biased D_1 (behaving like a short circuit) holds the output Y at ground potential or logic LOW. When input A is HIGH and input B is LOW, as shown in row 3 of the truth table, the output Y is at logic LOW; in this case, diode D_1 is reverse biased, diode D_2 is forward biased, and so the output is at ground potential. When both inputs are HIGH, both diodes are nonconducting, and hence the output Y is HIGH, as shown in row 4 of the truth table.

If we change from positive logic to negative logic, the AND gate becomes an OR gate and the OR gate becomes an AND gate. We can see this in the function table of Figure 3.7-2b for the circuit of Figure 3.7-2a. If the low-level voltage is defined as binary 1 while the high level is binary 0, then the negative-logic truth table of Figure 3.7-3 results.

An AND gate followed by an inverter (a NOT circuit) performs the same function as a NAND gate does.

3.8 JUNCTION-DIODE TRANSIENT CHARACTERISTICS

The preceding analyses are not valid if the input waveform transition times are shorter than the diode switching (transient) times. As a diode is driven from the reversed condition to the forward state or in the opposite direction, the diode response is accompanied by a transient, and an interval of time elapses before the diode recovers to its steady state.

A practical *pn*-junction diode possesses a certain amount of capacitance

A	B	Y
L	L	L
L	H	L
H	L	L
H	H	H

A	B	Y
1	1	1
1	0	1
0	1	1
0	0	0

(a)   (b)

FIGURE 3.7-3 Truth table for the negative-logic OR gate of Figure 3.7-2a: (a) function table; (b) truth table for negative logic.

across its reverse-biased *pn* junction. This junction capacitance C_j is caused by the majority ions on either side of the junction that have been uncovered by the removal of majority carriers from the junction region due to reverse bias. The diode-resistor circuit of Figure 3.8-1 includes the junction capacitance C_j. As the input is at $-V$, the diode is OFF and C_j is charged to $-V$. At $t = 0$ when the input switches to $+V$, the voltage across C_j cannot change instantaneously and so C_j will begin charging toward $+V$. As the voltage of C_j reaches $+0.7$ V, the diode turns ON and holds its voltage drop at $+0.7$ V. Because of C_j, the output voltage V_o at $t = 0$ will jump to 2 V and then drop to $V - 0.7$ when C_j charges to 0.7 V. Therefore the presence of C_j causes an overshoot on the v_o waveform and a deterioration of the rise time on the v_{cj} waveform.

A diode's junction capacitance increases as the reverse bias is decreased. A fast-switching diode might have the C_j value of 2 to 4 pF as the reverse bias varies from 25 V to 0 V. If the diode turn-ON time is much faster than the rise time of the input signal, there will be no overshoot on the resistor voltage since the diode turns ON before the input completes its transition.

Assume that the voltage in Figure 3.8-2a is applied to the diode–resistor circuit in Figure 3.8-2b. Consider the situation in this circuit where the diode is ON and it is being turned OFF. There are the effects of junction capacitance C_j and storage time t_s on the circuit operation. The t_s effect takes place due to the charge storage in a forward-biased diode. In the forward-bias condition, majority carriers flow across the junction (holes from *p* to *n* and electrons from *n* to *p*), whereupon they become minority carriers. Upon the application of reverse bias, these stored minority carriers should return to their respective majority regions. This removal of stored minority charges takes time, which is referred to as *storage time* t_s.

The effect of minority-charge storage is essentially to keep the diode ON for a time t_s after the input v_i has switched negatively, and thus the diode voltage

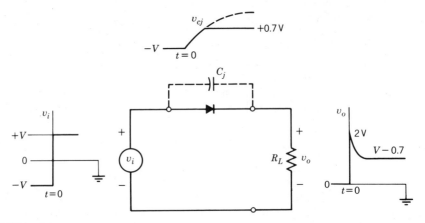

FIGURE 3.8-1 Effect of the junction capacitance C_j on the v_o waveform when switching diode from OFF to ON.

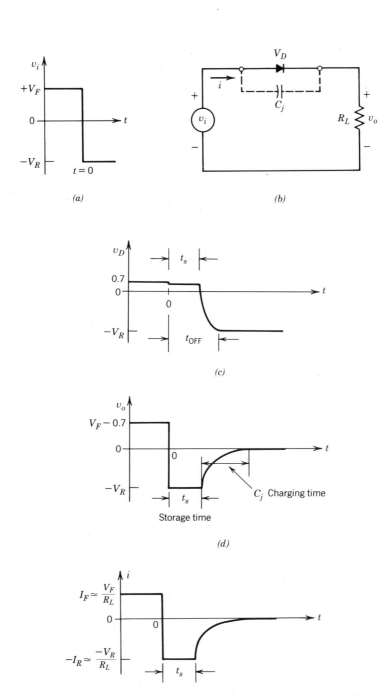

FIGURE 3.8-2 The waveform in (a) applied to the diode circuit in (b). Effects of storage time t_s and junction capacitance C_j on switching diode from ON to OFF: (c) diode waveform v_D; (d) output waveform v_o; (e) current waveform i.

v_D remains at about $+0.7$ V for the storage time t_s after $t = 0$, as indicated by the waveform v_D in Figure 3.8-2c. In the t_s interval, the negative input $-V_R$ pushes current $-I_R$ in the reverse direction through the diode, as indicated by the i waveform in Figure 3.8-2e. Current $-I_R \simeq -V_R/R_L$ is essentially removing the diode's minority stored charge, and so the larger this current, the shorter t_s will be. At the end of the t_s interval, the diode will start to turn OFF. Since C_j must charge to $-V_R$, the turn-OFF will be gradual, as shown in Figure 3.8-2d. Hence the total time for the diode to switch from ON to OFF, represented by t_{OFF}, is the sum of the storage time t_s and the C_j charging time t_c. t_{OFF} is also called the *reverse recovery time* t_{rr}. Typical t_{OFF} values for fast-switching diodes can be 5 ns or less. Both a smaller positive amplitude and a larger negative amplitude on input v_i will serve to decrease storage time t_s. By decreasing R_L, the t_c value will reduce, resulting in a shorter t_{OFF}.

REFERENCES

1. Boylestad, R. and L. Nashelsky, *Electronic Devices and Circuit Theory*, 3rd ed., Prentice-Hall, Englewood Cliffs, New Jersey, 1982, Chaps. 1 and 2.

2. Bell, D. A., *Solid State Pulse Circuits*, 2nd ed., Reston Publishing Company, Reston, Virginia, 1981. Chap. 3.

3. Millman, J. and H. Taub, *Pulse, Digital, and Switching Waveforms*, McGraw-Hill, New York, 1965, Chaps. 6–8.

4. *The Transistor and Diode Data Book for Design Engineers*, Diode Data Sheets, Texas Instruments, Canton, Massachusetts, 1973.

QUESTIONS

3-1. Under what condition will diodes in switching circuits have steady-state switching characteristics?

3-2. Describe the temperature dependence of the diode characteristic.

3-3. What is a diode clipper?

3-4. Describe the operation of a biased double-diode shunt clipper.

3-5. Define the term *clamping*.

3-6. Why is the clamping circuit often called a dc restorer?

3-7. What is a difference between clipping and clamping circuits?

3-8. How should the control voltage be selected for a four-diode-bridge analog switch?

3-9. Why does only one bulb light when the diode-switching counter in Figure 3.5-1 is operating?

3-10. Bulb 1 lights while bulb 0 automatically extinguishes as the first trigger pulse is applied to the circuit of Figure 3.5-1. Why?

3-11. How does the push-pull sync discriminator work?

3-12. How does the single-ended sync discriminator work?

3-13. How does the diode-resistor OR gate operate?

3-14. How does the diode-resistor AND gate operate?

3-15. How does the junction capacitance C_j affect the output-voltage waveform of the diode-resistor circuit when switching diode from OFF to ON?

3-16. What is the storage time t_s of a switching diode?

3-17. How do the junction capacitance and the storage time affect switching the diode from ON to OFF?

3-18. How can the storage time be reduced when switching the diode from ON to OFF?

PROBLEMS

3-1. (a) A silicon diode at 300 K conducts 7 mA at 0.7 V. Find the diode current at 0.8 V. (b) Find the reverse saturation current.

Answer. (a) 47.89 mA; (b) 9.97 nA.

3-2. Two silicon diodes are connected in opposing series. A 6-V dc source is applied across this series arrangement. Find the voltages across the forward-biased junction and the reverse-biased junction.

Answer. 0.036 V; 5.964 V.

3-3. A silicon diode operates at a forward bias of 0.7 V. Find the factor by which the current will be multiplied to indicate its reduction when the temperature is decreased from 25 to $-45°C$.

Answer. 0.579.

3-4. A Zener diode has the following specifications:

$$V_Z = 7.5 \text{ V at } I_{ZT} = 20 \text{ mA}; r_Z = Z_z = 6\,\Omega; P_{z,\max} = 0.4 \text{ W}; I_o \simeq 0 \simeq I_{z,\min}.$$

Find: (a) $V_{z,\max}$ and $V_{z,\min}$; (b) the percentage change in V_Z over this range.

Answer. (a) 7.70 V and 7.38 V; (b) 4.27%.

3-5. Design a circuit to clip the negative peaks off a ± 10-V square wave. A silicon diode is available with a maximum reverse saturation current of 10 μA. The negative-output voltage is not to exceed 0.5 V. Calculate the amplitude of the positive-output peak.

Answer. $R_1 = 50 \text{ k}\Omega$; +9.3 V.

3-6. A positive shunt clipper circuit has a square-wave input of ± 20 V. The output voltage is to be -18 V and $+0.7$ V, and the diode reverse saturation current is to be 240 μA. Calculate the required resistance and the diode forward current.

Answer. 8.2 kΩ; 2.35 mA.

3-7. Draw the circuit of a series diode noise clipper, showing typical input and output waveforms. Briefly explain how the circuit operates.

3-8. A positive-voltage clamper has a 5-kHz square-wave input with an amplitude of ± 8 V. The signal source resistance is 1 kΩ, and the tilt on the output waveform is not to exceed 1%. Design a suitable circuit for the required clamping.

Answer. 0.1 μF; 100 kΩ.

3-9. Design a biased clamping circuit to clamp a ± 20-V square wave to a minimum level of 4 V. The input-signal frequency range is from 2 to 8 kHz, and the signal-source resistance is 1 kΩ. The tilt on the output waveform does not exceed 1%.

Answer. 0.25 μF; 100 kΩ.

4

TRANSISTOR SWITCHING

4.1 THREE OPERATION REGIONS OF BIPOLAR TRANSISTOR CHARACTERISTICS

Introduction

Basic transistor (i.e., the bipolar junction transistor; BJT) operation includes forward biasing the emitter-base junction so that the emitter injects majority carriers (holes in *pnp* or free electrons in *npn*) into the base region and reverse biasing the collector-base junction so that most of these injected carriers are collected by the collector. Under these conditions, the collector current will be approximately equal to the emitter current, while the base current will be much smaller than either. The base current is so small because of the following facts: (1) the ratio of the total transistor width to that of the center layer (the base) is typically 150:1; and (2) the doping of the center layer is also considerably less than that of the outer layers; this lower doping level decreases the conductivity of this material by limiting the number of "free" carriers.

The transistor data sheet includes maximum ratings and electrical characteristics. The chief maximum ratings of the general-purpose silicon *npn* transistors 2N3903 and 2N3904 at $T_A = 25°C$ (ambient temperature) are the following:

Collector-base voltage $V_{CB} = 60$ Vdc.
Collector-emitter voltage $V_{CEO} = 40$ Vdc.
Emitter-base voltage $V_{EB} = 6$ *Vdc*.
Collector current $I_C = 200$ mAdc.
Total device dissipation $P_D = 310$ mW
$\qquad\qquad\qquad (P_D = 210$ mW at $T_A = 80°C)$.
Junction operating temperature $T_j = 135°C$.
Storage temperature range $T_{stg} = -55$ to $+135°C$.

CE Output Characteristics

Figure 4.1-1 shows the common-emitter (CE) output characteristics ($I_C - V_{CE}$ curves) of an *npn* transistor connected in a CE configuration. For a fixed value of base current I_B, the collector current is not a very sensitive value of V_{CE}.

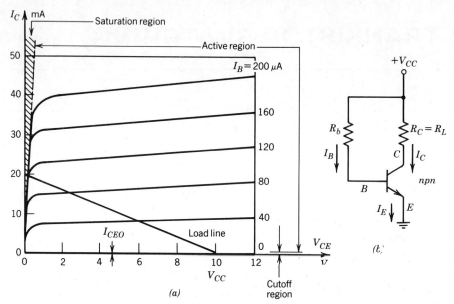

(a)

FIGURE 4.1-1   (*a*) Common-emitter (CE) output characteristics of an *npn* transistor connected in (*b*) the CE configuration. The dc load line defines all corresponding current and voltage conditions that can exist in the circuit. In (*a*) a load line corresponding to $V_{CC} = 10$ V and $R_L = 500\ \Omega$ is superimposed.

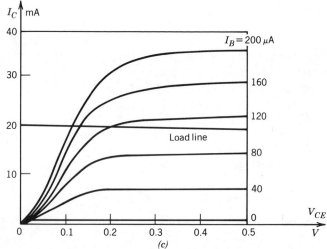

(c)

FIGURE 4.1-1c   Saturation-region *CE* characteristics of the same transistor as in Figure 4.1-1*b*. The load line corresponds to $V_{CC} = 10$ V and $R_L = 500\ \Omega$.

The load line superimposed on the CE output characteristics is determined by the chosen value of a load $R_L = 500\,\Omega$ and a supply $V_{CC} = 10$ V. This load line passes through the point $I_C = 0$, $V_{CE}(= V_{CC}) = 10$ V, and $I_C\ (= V_{CC}/R_L) = 20$ mA, $V_{CE} = 0$. Its slope equals $-1/R_L$ independently of the output characteristics. The CE output characteristics have three operation regions of interest: the active, cutoff, and saturation regions.

Example 4.1-1. In the circuit of Figure 4.1-1b, $V_{CC} = 10$ V, $R_C = R_L = 500\,\Omega$, and $R_b = 250$ kΩ. The load line is drawn in Figure 4.1-1a. (a) Obtain the Q-point (= quiescent or operating point). (b) Find V_{CEQ}, I_{CQ}, $I_{CQ}R_C$, and I_{EQ} from the graph.

Solution.

(a) The two points for the load line are (0 mA, 10 V) and (20 mA, 0 V).

$$\text{Base current } I_B = \frac{V_{CC}}{R_b} = \frac{10\text{ V}}{250\text{ k}\Omega} = 0.04\text{ mA} = 40\ \mu\text{A}.$$

The intersection of the load line and the transistor curve for $I_B = 40\ \mu$A defines the operating point.

(b) From the curve $V_{CEQ} = 6$ V, $I_{CQ}R_C = 10 - 6 = 4$ V $(= 8$ mA $\times\ 500$ $\Omega)$, $I_{CQ} = 8$ mA, and $I_{EQ} \simeq 8$ mA.

CE Active Region

In the active region the emitter-base junction is forward biased and the collector-base junction is reverse biased. In the CE characteristics of Figure 4.1-1, the active region is the area to the right of the ordinate $V_{CE} = $ a few tenths of a volt and above $I_B = 0$. Since the transistor output current in this region responds most sensitivity, the transistor amplifier stage must be restricted to operate in this region for minimum distortion. Applying Kirchhoff's current law to the transistor of Figure 4.1-1b as if it were a single node, we obtain

$$I_E = I_C + I_B. \tag{4.1-1}$$

For common-base configuration, the short-circuit amplification factor is

$$\alpha = \frac{I_C - I_{CO}}{I_E}, \tag{4.1-2}$$

where I_{CO} is the reverse saturation current (leakage). Including the effects of I_{CO}, the collector current in the CE circuit is

$$I_C = \alpha I_E + I_{CO}. \tag{4.1-3}$$

Then

$$I_E = \frac{I_C - I_{C0}}{\alpha},$$ (4.1-4)

and

$$I_B = I_E - I_C = \frac{I_C - I_{C0}}{\alpha} - I_C.$$

Hence

$$I_C - \alpha I_C = I_{C0} + \alpha I_B,$$

or

$$I_C = \frac{I_{C0}}{1 - \alpha} + \frac{\alpha I_B}{1 - \alpha}.$$ (4.1-5)

Let

$$\beta \equiv \frac{\alpha}{1 - \alpha}; \quad \text{then } \alpha = \frac{\beta}{1 + \beta}$$ (4.1-6)

and Equation (4.1-5) becomes

$$I_C = (1 + \beta)I_{C0} + \beta I_B.$$ (4.1-7)

Usually $I_B \gg I_{C0}$ and thus in the active region

$$I_C \simeq \beta I_B \quad \text{or } \beta \simeq \frac{I_C}{I_B},$$ (4.1-8)

where β is the CE forward current amplification factor. Note that α is not a constant. From Equation (4.1-6) we see that a slight change in α (e.g., $\Delta\alpha = 0.997 - 0.996$) has a large effect on β ($\Delta\beta = 332 - 249$), and also on the CE curves. This is why these curves are not actually horizontal lines. The small signal characteristics of the npn silicon transistors 2N2903 and 2N2904 are listed in Table 4.1-1.

CE Cutoff Currents

When $I_E = 0$ and $I_C = I_{C0}$, the transistor is at cutoff. But when $I_B = 0$ or the base is open circuited, the transistor is not at cutoff. From Equations (4.1-1) and

TABLE 4.1-1
Small Signal Characteristics of 2N3903 and 2N3904 Silicon *npn* Transistors at $T_a = 25°C$

Characteristic		Symbol	Minimum	Maximum	Unit
Current gain (I_C = 10 mA,	2N3903	h_{fe}	2.5	—	—
V_{CE} = 20 V, f = 100 MHz)	2N3904		3.0	—	—
Current gain-bandwidth product	2N3903	f_T	250	—	MHz
(I_C = 10 mA, V_{CE} = 20 V,	2N3904		300	—	
f = 100 MHz)					
Output capacitance (V_{CB} = 5 Vdc,		C_{ob}	—	4	pF
I_E = 0, f = 100 kHz)					
Input capacitance ($V_{BE(rev)}$ = 0.5 Vdc,		C_{ib}	—	8	pF
I_C = 0, f = 100 kHz)					
Small signal-current gain	2N3903	h_{fe}	50	200	—
(I_C = 1.0 mA, V_{CE} = 10 V,					
f = 1 kHz)	2N3904		100	400	—
Voltage feedback ratio					
(I_C = 1.0 mA, V_{CE} = 10 V,	2N3903	h_{re}	0.1	5.0	$\times 10^{-4}$
f = 1 kHz)	2N3904		0.5	8.0	
Input impedance (I_C = 1.0 mA,	2N3903	h_{ie}	0.5	8	kΩ
V_{CE} = 10 V, f = 1 kHz)	2N3904		1.0	10	
Output admittance					
(I_C = 1.0 mA, V_{CE} = 10 V, f = 1 kHz)		h_{oe}	1.0	40	μmhos
Noise figure (I_C = 100 μA,	2N3903	NF	—	6	dB
V_{CE} = 5 V, R_g = 1 kΩ,	2N3904		—	5	
Noise bandwidth = 10 Hz to 15.7 kHz)					

(4.1-5), if $I_B = 0$, then $I_E = I_C$ and

$$I_C = I_E = \left. \frac{I_{C0}}{1 - \alpha} \right|_{I_B = 0} = I_{CE0}. \qquad (4.1\text{-}9)$$

The symbol I_{CE0} designates the actual collector current with the collector junction reverse biased and the base open circuited. Substituting Equation (4.1-6) into Equation (4.1-9) we find

$$I_{CE0} = (\beta + 1)I_{C0} \simeq \beta I_{C0} \simeq \beta I_{CB0}. \qquad (4.1\text{-}10)$$

$|I_{CB0}|$ is slightly greater than $|I_{C0}|$ due to the surface leakage and some other factors. Typically $|I_{CE0}|$ for Si (a few μA) is much smaller than for Ge (a few hundred μA). For $\alpha = 0.996$,

$$I_C = \frac{I_{C0}}{1 - 0.996} = \frac{I_{C0}}{0.004},$$

TABLE 4.1-2

Off Characteristics of 2N3903 and 2N3904 Silicon *npn* Transistors at $T_A = 25°C$

Off Characteristics	Symbol	Minimum	Maximum	Unit
Collector-base breakdown voltage ($I_C = 10 \mu Adc$, $I_E = 0$)	BV_{CBO}	60	—	Vdc
Collector-emitter breakdown voltage[a] ($I_C = 1$ mAdc)	$BV_{CEO}{}^a$	40	—	Vdc
Emitter-base breakdown voltage $I_E = 10 \mu Adc$, $I_C = 0$)	BV_{EBO}	6	—	Vdc
Collector cutoff current ($V_{CE} = 40$ Vdc, $V_{BE(rev)} = 3$ Vdc)	I_{CEX}	—	50	nAdc
Base cutoff current ($V_{CE} = 40$ Vdc, $V_{BE(rev)} = 3$ Vdc)	I_{BL}	—	50	nAdc

[a]Pulse test: Pulse width = 300 μs; duty cycle = 2%.

or

$$I_{CEO} = 250 I_{CO}.$$

In silicon, cutoff occurs at $V_{BE} \simeq 0$ V, corresponding to a base short circuited to the emitter. However, in order to cut off the germanium transistor, a reverse-biasing voltage of approximately 0.1 V is ordinarily needed to apply across the emitter junction.

The temperature sensitivity of I_{CBO} is the same as that of the reverse saturation current I_o of a *pn* diode (Section 3.1). Silicon transistors may be used up to junction temperature of about 200°C, whereas germanium transistors are limited to about 100°C.

The off characteristics of silicon *npn* transistor 2N3903 and 2N3904 are listed in Table 4.1-2.

CE Saturation Region

In the saturation region, the collector and emitter junctions are forward biased by at least the cutin voltage (= a few tenths of a volt), resulting in the exponential change in collector current with a small change in voltage V_{CB}. Then $V_{CE} = V_{BE} - V_{BC}$ is only a few tenths of a volt at saturation. Thus the saturation region is very close to the zero-voltage axis, as indicated in Figure 4.1-1c, where all the curves merge and fall rapidly toward the origin. In the saturation region the collector current is approximately independent of base current. The beginning of saturation occurs at the knee of the characteristic curves. We observe from Figure 4.1-1c that the load line corresponding to $R_L = 500 \Omega$ intersects with the $I_B = 0.12$ mA curve at the point $I_C = 20$ mA, $V_{CE} \simeq 0.19$ V, and find that V_{CE} and I_C no longer respond appreciably to base current I_B after the base current has reached 0.12 mA. At $I_B = 0.12$ mA the transistor goes into saturation. At $I_B = 0.20$ mA, V_{CE} has dropped to $V_{CE} \simeq 0.12$ V. Large magnitudes of I_B will reduce $|V_{CE}|$ slightly further. $V_{CE(sat)}$ depends somewhat on the values of I_B and I_C. For a transistor operating in the saturation region at a given operating point, a quantity of interest is the *CE* saturation resistance R_{CS} $(= R_{CES}) = V_{CE(sat)}/I_C$. The current that enters the very thin base region across the emitter junction must flow through a long narrow path to reach the base terminal. The cross-sectional area for current flow in the base is very much smaller than in the collector or emitter, and so the ohmic resistance of the base is usually very much larger than that of the collector or emitter. The dc ohmic base resistance is called the *base-spreading resistance* $(r_{bb'})$ and is of the order of 100 Ω. In saturation the transistor consists of two forward-biased diodes back to back in opposing series. A reasonable value for the temperature coefficient of $V_{BE(active)}$, $V_{BE(sat)}$, or $V_{BC(sat)}$ is -2.5 mV/°C, and that of $V_{CE(sat)}$ is about -0.25 mV/°C.

The forward bias at the collector-base junction when a transistor is saturated results in a decrease of the dc current gain (β or h_{FE}). This effect is true because to draw the maximum number of charge carriers from emitter to collector, the collector-base junction should be reverse biased.

The ON characteristics of transistors 2N3903 and 2N3904 are given in Table 4.1-3.

npn Transistor-Junction Voltages

The volt-ampere characteristic between base and emitter at constant collector-to-emitter voltage is similar to the volt-ampere characteristic of a simple *pn* junction diode. The typical *npn* transistor-junction voltages at 25°C are listed in Table 4.1-4.

Example 4.1-1. Determine whether or not the silicon transistor in Figure 4.1-2 is in saturation and calculate I_B and I_C.

Solution. Applying KVL to the base circuit yields

$$60I_B + 0.8 = 6$$

or

$$I_B = \frac{5.2}{60} = 0.0867 \text{ mA}.$$

Applying KVL to the collector circuit gives

$$3I_C + 0.2 = 12$$

TABLE 4.1-3
ON Characteristics of 2N3903 and 2N3904 Silicon *npn* Transistors at T_A=25°C

Characteristic		Symbol	Minimum	Maximum	Unit
dc current gain[a]		h_{FE}[a]			
(I_C = 0.1 mAdc, V_{CE} = 1 Vdc)	2N3903		20	—	
	2N3904		40	—	
(I_C = 1.0 mAdc, V_{CE} = 1 Vdc)	2N3903		35	—	
	2N3904		70	—	
(I_C = 10 mAdc, V_{CE} = 1 Vdc)	2N3903		50	150	
	2N3904		100	300	
(I_C = 50 mAdc, V_{CE} = 1 Vdc)	2N3903		30	—	
	2N3904		60	—	
(I_C = 100 mAdc, V_{CE} = 1 Vdc)	2N9303		15	—	
	2N9304		30	—	
Collector-emitter saturation voltage[a]		$V_{CE(sat)}$[a]			Vdc
(I_C = 10 mAdc, I_B = 1 mAdc)			—	0.2	
(I_C = 50 mAdc, I_B = 5 mAdc)			—	0.3	
Base-emitter saturation voltage[a]		$V_{BE(sat)}$[a]			Vdc
(I_C = 10 mAdc, I_B = 1 mAdc)			0.65	0.85	
(I_C = 50 mAdc, I_B = 5 mAdc)			—	0.95	

[a]Pulse test: Pulse width = 300 μs; duty cycle = 2%.

TABLE 4.1-4
Typical *npn* Transistor-Junction Voltages (*V*) at T_A=25°C

Material	$V_{CE(sat)}$	$V_{BE(sat)} \equiv V_\sigma$	$V_{BE(active)}$	$V_{BE(cutin)} \equiv V_\gamma$	$V_{BE(cutoff)}$
Si	0.2	0.8	0.7	0.5	0.0
Ge	0.1	0.3	0.2	0.1	−0.1

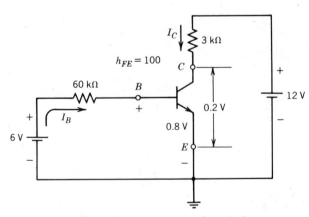

FIGURE 4.1-2 An example illustrating how to determine whether or not a transistor is in saturation.

or

$$I_C = \frac{11.8}{3} = 3.933 \text{ mA}.$$

The minimum value of base current required for saturation is

$$I_{B(min)} = \frac{I_C}{h_{FE}} = \frac{3.933}{100} = 0.0393 \text{ mA}.$$

Since $I_B = 0.0867$ mA $> I_{B(min)} = 0.0393$ mA, the transistor is indeed in saturation.

Example 4.1-2. The *npn* silicon transistor in Figure 4.1-1*b* has the following characteristics: I_{CEX} (= collector cutoff current = I_{CO}) = 50 nA, $V_{CE(sat)} = 0.2$ V at $I_C = 10$ mA, and $I_B = 1$ mA. If $V_{CC} = 15$ V and $R_L = 1.5$ kΩ, calculate the transistor power dissipation (a) at cutoff, (b) at saturation, and (c) when $V_{CE} = 2$ V.

Solution.

(a) At cutoff, $P_D \simeq I_{CEX}V_{CC} = 50 \text{ nA} \times 15 \text{ V} = 0.75 \text{ } \mu\text{W}.$

(b) At saturation, $I_C \simeq V_{CC}/R_L = 15 \text{ V}/1.5 \text{ kΩ} = 10 \text{ mA}$,

$$P_D = I_C V_{CE(sat)} = 10 \text{ mA} \times 0.2 \text{ V} = 2 \text{ mW}.$$

(c) When $V_{CE} = 2$ V, $I_C = (V_{CC} - V_{CE})/R_L = (15 \text{ V} - 2 \text{ V})/1.5 \text{ kΩ} = 8.667$ mA;

$$P_D = I_C V_{CE} = 8.667 \text{ mA} \times 2 \text{ V} = 17.3 \text{ mW}.$$

4.2 APPLICATION OF EBERS–MOLL EQUATIONS TO BIPOLAR-TRANSISTOR CHARACTERISTICS

Assume an *npn* transistor as shown in Figure 4.2-1 with positive directions for the currents as indicated. When the *npn* transistor junctions are identified as the base-collector and the base-emitter junctions, the corresponding junction currents I_C and I_E can be written as the Ebers–Moll equations:

$$I_C = \alpha_N I_E - I_{CO}(e^{V_{BC}/V_T} - 1) \qquad (4.2\text{-}1)$$

and

$$I_E = \alpha_R I_C + I_{EO}(e^{V_{BE}/V_T} - 1). \qquad (4.2\text{-}2)$$

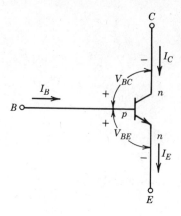

FIGURE 4.2-1 Voltage drops across the junctions (V_{BC} and V_{BE}) and positive directions for the currents (I_C, I_E, and I_B) in an *npn* transistor.

Here the first terms on the right sides of Equations (4.2-1) and (4.2-2) represent the transistor component of current due to the minority carriers that have crossed the other junction and have diffused across the base; the second terms represent the diode component [see Equation (3.1-1)]. When transistors are used as linear devices in the active region, carriers that diffuse from emitter to collector are viewed as diffusing in the normal (N) direction, while carriers diffusing in the other direction are judged to be diffusing in the reverse (R) direction. Thus the subscripts N and R are on the α's. The currents I_{C0} and I_{E0} are the reverse saturation currents for the collector junction and emitter junction, respectively, corresponding to I_0 in Equation (3.1-1). Collector and base currents I_C and I_B are positive when these currents flow into the *npn* transistor, and the emitter current I_E is positive when I_E flows out of the transistor. V_{BC} (V_{BE}) represents the voltage drop across the collector (emitter) junction from the *p*-type base to the *n*-type collector (emitter). The collector (emitter) junction is forward biased when V_{BC} (V_{BE}) is positive.

Assume that the *npn* transistor is a linear device. Equations (4.2-1) and (4.2-2) can be solved for I_E and I_C simultaneously.

$$I_C - \alpha_N I_E = -I_{C0}(e^{V_{BC}/V_T} - 1). \tag{4.2-1a}$$

$$-\alpha_R I_C + I_E = I_{E0}(e^{V_{BE}/V_T} - 1). \tag{4.2-2a}$$

$$I_E = \frac{\begin{vmatrix} 1 & -I_{C0}(e^{V_{BC}/V_T} - 1) \\ -\alpha_R & I_{E0}(e^{V_{BE}/V_T} - 1) \end{vmatrix}}{\begin{vmatrix} 1 & -\alpha_N \\ -\alpha_R & 1 \end{vmatrix}};$$

$$I_E = \frac{I_{E0}}{1 - \alpha_N\alpha_R}(e^{V_{BE}/V_T} - 1) - \frac{\alpha_R I_{C0}}{1 - \alpha_N\alpha_R}(e^{V_{BC}/V_T} - 1). \tag{4.2-3}$$

$$I_C = \frac{\begin{vmatrix} -I_{C0}(e^{V_{BC}/V_T} - 1) & -\alpha_N \\ I_{E0}(e^{V_{BE}/V_T} - 1) & 1 \end{vmatrix}}{1 - \alpha_N \alpha_R};$$

$$I_C = \frac{\alpha_N I_{E0}}{1 - \alpha_N \alpha_R}(e^{V_{BE}/V_T} - 1) - \frac{I_{C0}}{1 - \alpha_N \alpha_R}(e^{V_{BC}/V_T} - 1). \qquad (4.2\text{-}4)$$

The difference of I_E and I_C is the third current I_B, or

$$I_B = I_E - I_C. \qquad (4.2\text{-}5)$$

The junction voltages in terms of the currents from Equations (4.2-2) and (4.2-1) are expressed as follows:

$$V_{BE} = V_T \ln\left(1 + \frac{I_E - \alpha_R I_C}{I_{E0}}\right); \qquad (4.2\text{-}6)$$

$$V_{BC} = V_T \ln\left(1 - \frac{I_C - \alpha_N I_E}{I_{C0}}\right). \qquad (4.2\text{-}7)$$

The reverse saturation currents are related to the current gains by Einstein's relation

$$\alpha_N I_{E0} = \alpha_R I_{C0}. \qquad (4.2\text{-}8)$$

BJT in Active Region

For a transistor operating in its active region, the CE dc current gain is

$$\beta = h_{FE} = \frac{I_C}{I_B}, \qquad (4.1\text{-}8a)$$

and the CE incremental current gain is

$$h_{fe} = \frac{\Delta I_C}{\Delta I_B}. \qquad (4.2\text{-}9)$$

These current-gain parameters can be related to the parameters that appear in the Ebers–Moll equations. In the active region, the base-collector junction is reverse biased and the magnitude of this bias is usually large in comparison with $V_T (\simeq 25 \text{ mV})$. Thus Equation (4.2-1) can be written as

$$I_C = \alpha_N I_E + I_{C0}. \qquad (4.1\text{-}3a)$$

From Equations (4.2-5) and (4.1-3a) we find

$$I_C = \frac{\alpha_N}{1 - \alpha_N} I_B + \frac{I_{C0}}{1 - \alpha_N}. \tag{4.1-5a}$$

From Equation (4.1-5a) we find

$$h_{fe} = \frac{\Delta I_C}{\Delta I_B} = \frac{\alpha_N}{1 - \alpha_N}. \tag{4.2-10}$$

Since I_{C0} is usually very small in comparison with the transistor currents, we may ignore the last term in Equation (4.1-5a) and so have the expression

$$h_{FE} = \frac{\alpha_N}{1 - \alpha_N}. \tag{4.1-6a}$$

Since the right sides of Equations (4.1-6a) and (4.2-10) are the same, we shall assume no distinction between h_{FE} and h_{fe}. Typically $h_{FE} \simeq 50$ while $\alpha_N = \alpha = 0.98$.

The parameter h_{FE} varies widely over the operating range of a transistor. h_{FE} is also temperature sensitive and may also show a variation by a factor of 3 or more over a temperature range from -50 to $+150°$C. The subscripts on the parameter h_{FE} are derived from the fact that h_{FE} is the hybrid parameter characterized as the forward common-emitter (CE) current gain. In the reverse direction, the collector becomes the common terminal, and we have the parameter h_{FC}, corresponding to f_{FE} in the normal direction. Like h_{FE} given by Equation (4.1-6a), h_{FC} is related to α_R by

$$h_{FC} = \frac{\alpha_R}{1 - \alpha_R}. \tag{4.2-11}$$

h_{FC} is found in the range 0.01 to 0.25, much smaller than h_{FE}. While α_N is normally quite close to unity, α_R is in the range 0.01 to 0.2; this difference results from the geometry and doping of transistors. In the normal mode of operation, $V_{BC} \leq 0$, $V_{BE} \gg V_T$ ($\simeq 25$ mV), and $\alpha_R \ll 1$; under these conditions, Equation (4.2-3) becomes

$$I_E \simeq I_{E0} e^{V_{BE}/V_T}. \tag{4.2-12}$$

BJT at Cutoff

Consider an *npn* transistor switch with an R_L load and a V_{CC} supply connected between the collector and emitter. If the voltage between base and emitter is not large enough to forward bias the emitter junction adequately, the emitter current will be decreased nominally to zero and the transistor will be cut off. The Ebers–Moll equations can yield some information about the cut-in voltage of the transistor. Near the cut-in voltage, the base-collector junction will be substantially reverse biased, and thus $V_{BC}/V_T \ll 0$. Then Equation (4.2-3) becomes

$$I_E = \frac{I_{E0}}{1 - \alpha_N \alpha_R}(e^{V_{BE}/V_T} - 1) + \frac{\alpha_R I_{C0}}{1 - \alpha_N \alpha_R}. \tag{4.2-13}$$

From Equation (4.2-8), $\alpha_R I_{C0} = \alpha_N I_{E0}$. Substituting,

$$I_E = \frac{I_{E0}}{1 - \alpha_N \alpha_R}(e^{V_{BE}/V_T} - 1 + \alpha_N). \tag{4.2-14}$$

Since usually $\alpha_N \simeq 1$, Equation (4.2-14) can be simplified as

$$I_E \simeq \frac{I_{E0}}{1 - \alpha_R} e^{V_{BE}/V_T}. \tag{4.2-15}$$

Generally $\alpha_R = 0.01$ to 0.2, and

$$\frac{I_{E0}}{1 - \alpha_R} \simeq I_{E0} \text{ to } 1.25\, I_{E0}.$$

If $I_{E0}/(1 - \alpha_R) \simeq I_{E0}$, then Equation (4.2-15) becomes identical to the diode equation, Equation (3.1-3), I_{E0} being the reverse saturation current of the base-emitter junction. Thus we can use the same method to determine the cut-in point V_γ of a diode and the base-to-emitter cut-in voltage. For a silicon diode and a silicon transistor, V_γ usually lies in the range 0.5 to 0.65 V.

BJT in Saturation

In the saturation region, both the base-emitter junction and the base-collector junction are forward biased, so that the normal-direction emitter current and the reverse-direction collector current are flowing simultaneously. Assume that $V_{BE}/V_T \gg 1$ and $V_{BC}/V_T \gg 1$; then the -1 terms in Equations (4.2-3) and (4.2-4) can be neglected. Since $I_B = I_E - I_C$, we have

$$I_B = \frac{I_{E0}}{1 - \alpha_N \alpha_R}(1 - \alpha_N)e^{V_{BE}/V_T} + \frac{I_{C0}}{1 - \alpha_N \alpha_R}(1 - \alpha_R)e^{V_{BC}/V_T}. \tag{4.2-16}$$

Dividing Equation (4.2-4) by Equation (4.2-16) yields

$$\frac{I_C}{I_B} = \frac{\alpha_N I_{E0}e^{V_{BE}/V_T} - I_{C0}e^{V_{BC}/V_T}}{(1 - \alpha_N)I_{E0}e^{V_{BE}/V_T} + (1 - \alpha_R)I_{C0}e^{V_{BC}/V_T}}. \tag{4.2-17}$$

Note the following expressions: $\alpha_N I_{E0} = \alpha_R I_{C0}$ [Equation (4.2-8)],

$$V_{CE} = V_{BE} - V_{BC}, \quad h_{FE} = \alpha_N/(1 - \alpha_N), \quad h_{FC} = \alpha_R/(1 - \alpha_R),$$

and $h_{FE} \gg h_{FC}$. Then

$$\frac{I_C}{I_B} = h_{FE} \frac{(\alpha_R/\alpha_N)I_{C0} e^{V_{BE}/V_T} - (1/\alpha_N)I_{C0} e^{V_{BC}/V_T}}{(\alpha_R/\alpha_N)I_{C0} e^{V_{BE}/V_T} + [(1 - \alpha_R)/(1 - \alpha_N)]I_{C0} e^{V_{BC}/V_T}}$$

$$= h_{FE} \frac{(\alpha_R/\alpha_N)e^{(V_{BE}-V_{BC})/V_T} - 1/\alpha_N}{(\alpha_R/\alpha_N)e^{(V_{BE}-V_{BC})/V_T} + (1 - \alpha_R)/(1 - \alpha_N)}$$

$$= h_{FE} \frac{e^{V_{CE}/V_T} - 1/\alpha_R}{e^{V_{CE}/V_T} + (\alpha_N/\alpha_R)[(1 - \alpha_R)/(1 - \alpha_N)]},$$

or

$$\frac{I_C}{I_B} = h_{FE} \frac{e^{V_{CE}/V_T} - 1/\alpha_R}{e^{V_{CE}/V_T} + h_{FE}/h_{FC}}. \tag{4.2-18}$$

We rearrange Equation (4.2-18) to express the collector-emitter voltage in terms of I_C/I_B:

$$\frac{I_C}{I_B h_{FE}}\left(e^{V_{CE}/V_T} + \frac{h_{FE}}{h_{FC}}\right) = e^{V_{CE}/V_T} - \frac{1}{\alpha_R};$$

$$e^{V_{CE}/V_T}\left(1 - \frac{I_C}{I_B h_{FE}}\right) = \frac{1}{\alpha_R} + \frac{I_C}{I_B h_{FC}}.$$

Hence

$$V_{CE} = V_T \ln\left[\frac{1/\alpha_R + (I_C/I_B)/h_{FC}}{1 - (I_C/I_B)/h_{FE}}\right]. \tag{4.2-19}$$

The ratio I_C/I_B given by Equation (4.2-18) is plotted against the ratio V_{CE}/V_T for a typical transistor, as in Figure 4.2-2. The edge of saturation is defined as the point where $I_C/I_B = 0.9\, h_{FE}$ ($\gg 1 + h_{FC}$). Substituting this value to Equation (4.2-19), we find V_{CE} at this point as follows:

$$V_{CE} = V_{CE(\text{sat})} = V_T \ln\left[\frac{(1 + h_{FC})/h_{FC} + 0.9 h_{FE}/h_{FC}}{1 - 0.9}\right]$$

$$= V_T \ln\left[10 \frac{1 + h_{FC} + 0.9 h_{FE}}{h_{FC}}\right]$$

$$\simeq V_T \ln\left[\frac{9 h_{FE}}{h_{FC}}\right] = V_T\left\{2.2 + \ln\left[\frac{h_{FE}}{h_{FC}}\right]\right\}.$$

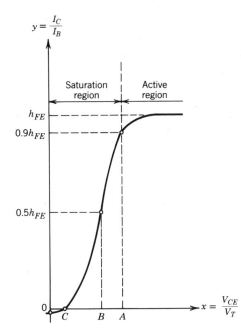

FIGURE 4.2-2 Modified collector characteristics of a typical transistor ($h_{FE} \gg h_{FC}$) with the ratio $I_C/I_B = y$ plotted against the ratio $V_{CE}/V_T = x$. Corresponding to the I_C/I_B points of 0.9 h_{FE}, 0.5 h_{FE}, and 0, the V_{CE}/V_T points on the x-axis are $A = 2.2 + \ln(h_{FE}/h_{FC})$, $B = \ln(h_{FE}/h_{FC})$, and $C = \ln(1/\alpha_R)$, respectively.

For example, at the 0.9 h_{FE} point, $V_{CE(\text{sat})} \simeq 285$ mV when $h_{FE} = 100$ and $h_{FC} = 0.01$. At the point where $I_C/I_B = 0.5h_{FE}$,

$$V_{CE} = V_T \ln \left[\frac{(1 + h_{FC})/h_{FC} + 0.5h_{FE}/h_{FC}}{1 - 0.5} \right]$$

$$= V_T \ln \left[2 \frac{1 + h_{FC} + 0.5h_{FE}}{h_{FC}} \right] \simeq V_T \frac{h_{FE}}{h_{FC}}.$$

From the preceding analysis, we know that the Ebers–Moll equations are useful for determining circuit conditions when a transistor is operating in the saturation region.

4.3   BIPOLAR TRANSISTOR INVERTER AND ITS LOAD

Basic Transistor-Inverter Circuit

The most common transistor switching circuit is essentially an overdriven common-emitter circuit used as an inverter in digital logic circuits. The basic inverter circuit is shown in Figure 4.3-1. When input v_i is at 0 V, the transistor is OFF, and so output $v_o = V_o$. When v_i is at V_i, base current I_B flows, producing a flow of collector current I_C. If I_B is large enough, it will saturate the transistor; thus output $v_o = V_{CE(\text{sat})} \simeq 0$. Therefore the transistor operates in either the

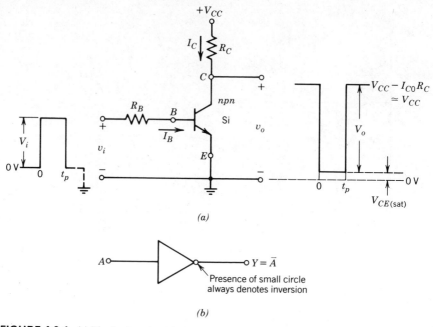

FIGURE 4.3-1 (a) Bipolar transistor inverter with pulse input; (b) symbol for the inverter (NOT circuit).

OFF state or the ON state, depending on the state of input v_i. Since the circuit essentially inverts the input pulse, the positive-going input pulse produces a negative-going output pulse, or vice versa.

Since the values of R_B and R_C determine the values of I_B and I_C, respectively, both the resistor values must be selected so that the transistor will saturate when v_i goes positive. If V_{CC} and I_C are known, R_C is computed simply as

$$R_C = \frac{V_{CC} - V_{CE(sat)}}{I_C}. \tag{4.3-1}$$

A value of R_C that is larger than calculated gives saturation with a lower I_C level. The transistor current gain h_{FE} determines the minimum value of I_B needed to produce I_C for saturation. Thus the minimum value of I_B required for saturation is

$$I_{B(min)} = \frac{I_C}{h_{FE(min)}} = \frac{V_{CC} - V_{CE(sat)}}{h_{FE(min)}R_C} \simeq \frac{V_{CC}}{h_{FE(min)}R_C}. \tag{4.3-2}$$

The value of R_B is computed as

$$R_B = \frac{V_i - V_{BE}}{I_{B(min)}}. \tag{4.3-3}$$

If R_B is smaller than calculated, I_B is greater than $I_{B(\text{min})}$ and the transistor saturation will occur. The actual base current produced is given by

$$I_B = \frac{V_i - V_{BE}}{R_B}. \tag{4.3-4}$$

For saturation to occur, we need

$$\frac{V_i - V_{BE}}{R_B} \geq \frac{V_{CC}}{h_{FE(\text{min})}R_C}. \tag{4.3-5}$$

Example 4.3-1. In the inverter circuit of Figure 4.3-1, $V_{CC} = 10\,\text{V}$, $I_C = 10\,\text{mA}$, $V_i = 10\,\text{V}$, and $h_{FE} = 50$ to 100. Find the values of (a) R_C and (b) R_B needed to ensure saturation.

Solution.

$$\text{(a)} \quad R_C \simeq \frac{V_{CC}}{I_C} = \frac{10\,\text{V}}{10\,\text{mA}} = 1\,\text{k}\Omega.$$

(b) Since the transistor can have an h_{FE} as low as 50, the value of R_B must be selected so that it will produce $I_C = 10\,\text{mA}$ for saturation even at the lowest h_{FE}. Therefore

$$I_B = I_{B(\text{min})} = \frac{I_C}{h_{FE(\text{min})}} = \frac{10\,\text{mA}}{50} = 0.2\,\text{mA},$$

and

$$R_B = \frac{V_i - V_{BE}}{I_B} = \frac{10\,\text{V} - 0.7\,\text{V}}{0.2\,\text{mA}} = 46.5\,\text{k}\Omega.$$

Effect of Load on Inverter Output

Resistive Loading.   Figure 4.3-2 shows an inverter circuit with its output driving a load R_L. The transistor is OFF during the 0-V portion of v_i and ON during the $+V_i$ portion. When the transistor is saturated, the load has no effect on the output. In this situation, all the current furnished by the V_{CC} collector supply flows through the transistor collector and none through the load since the ON transistor will have a very low resistance R_{ON} ($\ll R_L$). When the transistor is OFF, R_C and R_L are in series, and so v_o will be determined by their voltage divider ratio:

$$v_o \simeq \left(\frac{R_L}{R_C + R_L}\right)V_{CC}.$$

Clearly any R_L will cause v_o to be less than V_{CC}.

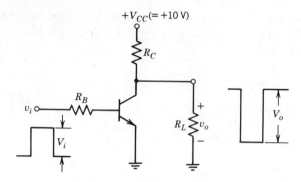

FIGURE 4.3-2 An inverter circuit with its output driving a load R_L.

Capacitive Loading.   Figure 4.3-3 shows an inverter circuit with its output driving a capacitive load C_L. When v_i drops from $+V_i$ to zero, the transistor turns OFF and so C_L is exponentially charged up to $\approx V_{CC}$ with a rise time of $t_r = 2.2R_CC_L$. When v_i rises from 0 V to V_i, the transistor turns on, and thus C_L is discharged rapidly through the low resistance of the on transistor. The complete v_i and v_o waveforms appear in Figure 4.3-3.

If a very small rise time were required, a very small value of R_C would be used. However the resultant I_C could be too large to be handled. In this case, we usually employ additional circuitry, such as the emitter follower Q_2 shown in Figure 4.3-4. Due to the operation of the emitter follower Q_2, the actual load impedance seen at the output (Y) of the inverter Q_1 will be $(h_{FE} + 1)$ times larger than the impedance at the emitter of Q_2. Thus R_E and C_L will appear as $(h_{FE} + 1)R_E$ and $C_L/(h_{FE} + 1)$ to the inverter Q_1 output, respectively. The Q_1 output actually sees a load consisting of $(h_{FE} + 1)R_E$ in parallel with $C_L/(h_{FE} + 1)$; this results in a value of t_r that is much smaller than if the inverter were driving C_L directly.

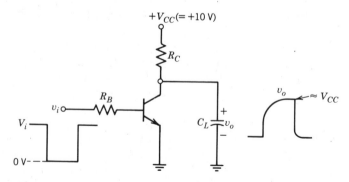

FIGURE 4.3-3 An inverter circuit with its output driving a capacitive load C_L.

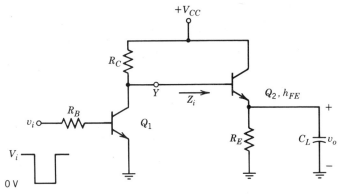

FIGURE 4.3-4 Emitter follower Q_2 connected to the inverter Q_1 output (Y) for the reduction of capacitive loading effect when Q_1 is turned OFF ($v_i = 0$).

4.4 BIPOLAR-TRANSISTOR SWITCHING TIMES AND SWITCHING SPEED IMPROVEMENT

Response of Transistor to Pulse Waveform

The transistor switch in Figure 4.4-1a is driven by the pulse waveform in Figure 4.4-1b. The response of the collector current i_C to the input waveform v_i is shown in Figure 4.4-1c; waveform v_{CE} is shown in Figure 4.4-1d. The waveform v_i operating between levels V_2 and V_1 drives the transistor from cutoff to saturation and back to cutoff. Current i_C does not immediately respond to input v_i. Instead there is a delay, and the delay time t_d is the time required to bring i_C to $0.1 I_{CS}$, where I_{CS} ($\simeq V_{CC}/R_C$) is the saturation collector current. The time to carry i_C from $0.1 I_{CS}$ to $0.9 I_{CS}$ is termed the *rise time* t_r. The sum of the delay and rise time is the total *turn-on time* t_{ON} ($= t_d + t_r$). When input v_i returns to its initial state at $t = T$, current i_C again does not respond immediately. The time that elapses after the reversal of base current i_B (in Figure 4.4-1e) before i_C falls again to $0.9 I_{CS}$ is termed the *storage time* t_s. The storage time is followed by the *fall time* t_f, which is the time required for i_C to fall from 90 to 10% of I_{CS}. The sum of the storage and fall times is termed the *turn-off time* t_{OFF} ($= t_s + t_f$).

Delay Time

The delay time results from the following three factors. First, when the input signal drives the base, a time interval is required to charge up the emitter-junction capacitance so that the transistor can be brought from cutoff to the active region. Second, a time is required before the minority carriers in the base can cross the base region to the collector junction; the minority carriers are those carriers that have crossed the emitter junction into the base. Third, a time is required for current i_C to rise to $0.1 I_{CS}$.

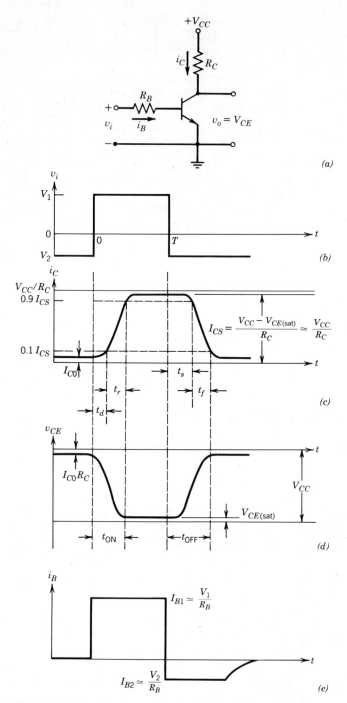

FIGURE 4.4-1 Time relationship of the input and output voltages and of the i_B and i_C waveforms for the transistor switching circuit.

Rise Time and Fall Time

The rise time and fall time result from the following fact. When a base-current step is used to saturate the transistor or return it from saturation to cutoff, the collector current must traverse the active region. The collector current increases or decreases along an exponential curve whose time constant can be given by

$$\tau_r = h_{FE}\left(C_c R_C + \frac{1}{2\pi f_T}\right), \tag{4.4-1}$$

where C_c is the collector transition capacitance and f_T is the gain-bandwidth product. The 3-dB frequency f_β for the short-circuit CE current gain is the frequency at which the current gain equals 0.707 of its low-frequency value h_{fe}. The frequency range up to f_β is referred to as the *bandwidth* of the circuit. The parameter f_T is defined as the frequency at which the short-circuit common-emitter current gain attains unit magnitude, or

$$\frac{h_{fe}}{\sqrt{1 + (f_T/f_\beta)^2}} \simeq \frac{h_{fe}f_\beta}{f_T} \simeq 1.$$

Since $f_T \simeq h_{fe}f_\beta$, this parameter represents the short-circuit current gain-bandwidth product.

Storage Time

The storage time results from the fact that a transistor in saturation has a saturation charge of excess minority carriers stored in the base. The transistor cannot respond until this saturation excess charge has been removed.

Typical Transistor-Switching-Time Data

The transistor switching characteristics listed in Table 4.4-1 are useful data for designers.

Example 4.4-1. In the switching circuit of Figure 4.4-1, $V_{CC} = 15$ V, $R_C = 2.7$ kΩ, a 2N3903 transistor is used, and the input pulse has a 5-μs width. Determine the time from the beginning of the input pulse until the transistor switches off. Also calculate the level of output voltage V_{CE} (a) before the input pulse is applied, (b) at the end of the delay time, and (c) at the end of the turn-on time.

Solution. From Table 4.4-1, $t_{OFF} = t_s + t_f = 175 + 50 = 225$ ns. The time from beginning of input to transistor switching off is

$$T + t_{OFF} = 5 \text{ } \mu\text{s} + 225 \text{ ns} = 5.225 \text{ } \mu\text{s}.$$

TABLE 4.4-1
Transistor Switching Characteristics

Type (Si)	t_d (ns)[a]	t_r (ns)[a]	t_s (ns)[b]	t_f (ns)[b]
npn 2N3903	35	35	175	50
npn 2N3904	35	35	200	50
pnp 2N3905	35	35	200	60
pnp 2N3906	35	35	225	75

Remark: 2N3903 and 2N3904 are complementary with 2N3905 and
2N3906, respectively.
[a] $\pm V_{CC} = 3$ V; $V_{BE(rev)} = 0.5$ V; $I_C = 10$ mA; $I_B = 1$ mA.
[b] $\pm V_{CC} = 3$ V; $I_C = 10$ mA; $I_B = 1$ mA.

From Table 4.1-2, $I_{CEX} = 50$ nA. Before the transistor switches on,

$$V_{CE} = V_{CC} - I_{CEX}R_C = 15 \text{ V} - (50 \text{ nA} \times 2.7 \text{ k}\Omega) = 15 \text{ V} - 0.000135$$

$$= 14.9999 \text{ V}.$$

At the end of t_d,

$$V_{CE} = V_{CC} - 0.1 I_{cs}R_C = 15 \text{ V} - 0.1 \times 15 \text{ V} = 13.5 \text{ V}.$$

At the end of t_{ON},

$$V_{CE} \simeq V_{CC} - 0.9 \times \frac{V_{CC}}{R_C} \times R_C = 15 \text{ V} - 0.9 \times 15 \text{ V} = 1.5 \text{ V}.$$

Switching-Speed Improvement

In the transistor switching circuit, the switching speed will be affected by the
E-B junction capacitance $C_{b'e}$ and the C-B junction capacitance $C_{b'e}$ ($=$ transition capacitance C_c) due to their charge and discharge action. To minimize the
turn-on time t_{ON}, V_{VE} must be zero or have a very small reverse bias before
switch-on. If I_B is made larger than the minimum required for saturation, the
junction capacitances will be charged faster, thus reducing the turn-on time.
However, as the transistor is overdriven with the larger I_B, the storage time t_s is
extended by the larger current flow across the forward-biased C-B junction
when the transistor is in saturation.

In order to increase the switching speed, the following two requirements
must be satisfied. First, v_{BE} must begin at 0 V, and i_B must initially be large at
switch-on but must rapidly settle down to the minimum required for saturation.
Second, switch-off must be achieved by a large reverse bias voltage that quickly
returns to zero. These conditions can be accomplished if the base resistor R_B
is shunted by a speed-up capacitor C_1, as shown in Figure 4.4-2. C_1 is initially

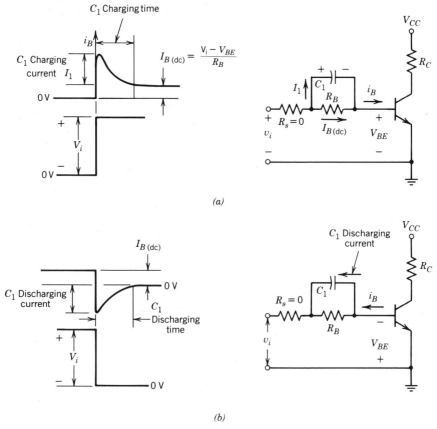

FIGURE 4.4-2 Improvement in transistor switching speed by means of the speed-up capacitor C_1 in parallel with R_B: (a) effect of C_1 charge at switch-on; (b) effect of C_1 discharge at switch-off.

uncharged before v_i is applied. When v_i rises, C_1 begins to charge to $(V_i - V_{BE})$. The charging current I_1 flows into the base terminal. Hence i_B is initially large, but it quickly settles down to its minimum dc level as C_1 becomes charged. The settled base current is given by

$$I_B = \frac{V_i - V_{BE}}{R_s + R_B} \ll I_1 = C_1 \text{ charging current.} \qquad (4.4\text{-}2)$$

When the transistor is switched off, the capacitor discharge produces a reverse base current that rapidly returns to zero. If the value of C_1 is properly selected, it tends to decrease t_d, t_r, t_s, and t_f.

For the best possible improvement in switching speed, a speed-up capacitor must be chosen that is large enough to maintain the charging current (I_B) nearly constant at its maximum level during the turn-on time. The charging current will drop by only 10% from its maximum level if the capacitor is allowed to charge by 10% during the turn-on time. C_1 charges by 10% during a time of $0.1\,R_sC_1$ (Section 2.2). Hence

$$t_{\text{ON}} = 0.1\,R_sC_1$$

or

$$C_1 = \frac{t_{\text{ON}}}{0.1\,R_s}. \qquad (4.4\text{-}3)$$

For $t_{\text{ON}} = 200$ ns, and $R_s = 1$ kΩ,

$$C_1 = \frac{200\ \text{ns}}{0.1 \times 1\ \text{k}\Omega} = 2000\ \text{pF}.$$

A several-to-ten-times improvement in switching time may be achieved by reducing C_1 to a proper value under the following conditions: (1) the transistor initially operates well below its maximum switching speed; and (2) the input pulse has a rise time much less than the minimum switching time sought.

When the transistor is switched off, C_1 discharges through R_B. For correct switching, C_1 must be at least 90% discharged during the time interval between transistor switch-off and switch-on. C_1 discharges by 90% in a recovery or resolving time $t_{re} = 2.3R_BC_1$ (Section 2.2). Hence

$$C_{1(\text{max})} = \frac{t_{re}}{2.3\,R_B}, \qquad (4.4\text{-}4)$$

where t_{re} is between switch-off and switch-on.

Example 4.4-2. In the circuit of Figure 4.4-2, $R_B = 10$ kΩ, $R_s = 1$ k$\Omega \neq 0$, and v_i is a 100-kHz square wave. Find the maximum value of C_1 that can be used.

Solution. $T = 1/f = 1/100$ kHz $= 10\ \mu$s. $t_{re} = T/2 = 5\ \mu$s.

$$C_{1(\text{max})} = \frac{t_{re}}{2.3\,R_B} = \frac{5\ \mu\text{s}}{2.3 \times 10\ \text{k}\Omega} = 217\ \text{pF}.$$

4.5   JUNCTION FIELD-EFFECT TRANSISTORS (JFETs)

Field-effect transistors (FETs) are available in two types, the junction field-effect transistor (JFET) and the metal-oxide-semiconductor field-effect transistor (MOSFET). The JFET is shown schematically and with its characteristics in Figure 4.5-1. It consists of an n-type silicon bar with two ohmic contacts—the source S and the drain D—along with two rectifying contacts—the p-type

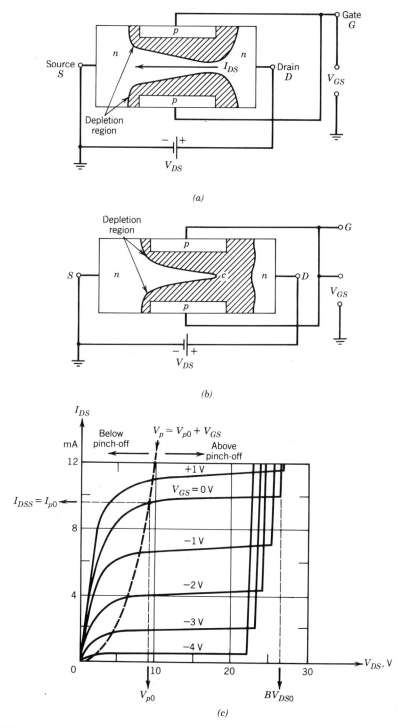

FIGURE 4.5-1 An *n*-channel junction field-effect transistor: (*a*) below pinch-off; (*b*) above pinch-off; (*c*) common-source drain characteristics.

gates G; the conducting path between S and D is the n-channel. Also available are p-channel JFETs with n-type gates.

In Figure 4.5-1a, a supply voltage V_{DS} connected across the silicon bar causes current I_{DS} to flow. At the junction between the p- and n-type semiconductor, there develops a depletion region, which is devoid of free current carriers. Between depletion regions, there is a channel of n-type silicon connecting the source to the drain. The depletion region can be made to spread further into the channel by further reverse biasing the pn junction between gate and channel. With sufficient reverse bias, the depletion regions will spread completely across the JFET and the channel will thereby have been pinched off. The drain-source voltage V_{DS} that causes the channel to pinch off is termed the *pinch-off voltage* V_p. For $V_{DS} > V_p$, the JFET is above pinch-off and the depletion region expands as shown in Figure 4.5-1b. However we find that the potential between point c and source S remains at the pinch-off voltage V_p, and so the current I_{DS} remains almost constant. This is called *saturation* in the FET, and it is shown in the common-source drain characteristics of Figure 4.5-1c. The drain-to-source voltage at which saturation begins when $V_{GS} = 0$ is called the *pinch-off voltage* V_{p0}. The dashed line in the plot of Figure 4.5-1c is the locus of V_p for various gate-to-source voltages. Typically

$$V_p \simeq V_{p0} + V_{GS}. \tag{4.5-1}$$

Corresponding to V_{p0}, the symbol I_{p0} represents the pinch-off current equal to I_{DS} when $V_{GS} = 0$. The current I_{p0} is also referred to as I_{DSS}, the drain-to-source current with the gate-source shorted ($V_{GS} = 0$ V).

Consider the plot of Figure 4.5-1c for $V_{GS} = 0$ V. When V_{DS} is low, the uniform channel between S and D is open and the transistor behaves like a resistor. The volt-ampere characteristic between drain and source is linear. When V_{DS} is increased, the depletion region spreads and constricts the channel. The drain current I_{DS} increases rapidly as V_{DS} increases toward V_{p0}. Above V_{p0}, the transistor is in saturation and I_{DS} tends to level off at I_{p0} and then rise slowly. The drain current increases hardly at all with further increase in drain-to-source voltage. However, when V_{DS} equals breakdown voltage BV_{DS0}, an avalanche breakdown occurs and I_{DS} again rises rapidly. Breakdown occurs at a fixed junction voltage BV_{DS}, and when the gate supplies a part of this voltage, breakdown occurs at a lower value of V_{DS}. Thus

$$BV_{DS} = BV_{DS0} + V_{GS}, \tag{4.5-2}$$

where BV_{DS0} is the breakdown voltage when $V_{GS} = 0$ V.

Again, consider keeping V_{DS} fixed and varying V_{GS}. When V_{GS} is made negative, the pn junction is reverse biased, increasing the depletion region between the gate and the source. This decreases the channel width, increasing the channel resistance. The current I_{DS} consequently decreases. As V_{GS} is made positive, the depletion region decreases until, for large positive gate voltages, the channel opens. Thus the *pn* junction between gate and source becomes forward biased, and the current flows from gate to source. The *n*-channel JFET is usually operated so that the voltage V_{GS} is either negative or slightly positive to avoid gate-to-source current.

At drain-to-source voltages between 0 V and pinch-off, the FET is said to be operating in the linear (nonsaturation) region since the drain-to-source current I_{DS} is approximately proportional to V_{DS} as in a linear resistor. At drain-to-source voltages between pinch-off and breakdown, the FET is said to be operating in the saturation region, since the current has saturated and does not change appreciably as a function of V_{DS}. When a reverse bias is applied to the gates, the supply of carriers is depleted progressively as the bias is increased. The range of operation of the JFET is consequently a range over which the available carriers are depleted to a greater or lesser extent. Thus the JFET operates in the depletion mode. The JFET finds extensive application in linear circuits and in a number of nonlinear circuits used for processing analog signals.

4.6   METAL-OXIDE-SEMICONDUCTOR FETs (MOSFETs)

The MOSFET finds extensive application in digital logic circuitry. In comparison with the bipolar transistor (BJT), one immediate advantage of the MOSFET is that the silicon-chip area it requires is only about 15% as large as that required for a BJT. MOSFETs are not yet capable of operating at the speeds of BJTs. There are basic differences that result in MOSFETs having lower capacitance and higher input impedance than JFETs. The operation of the MOSFET is similar to the operation of the JFET.

The basic construction of the *n*-channel MOSFET is shown in Figure 4.6-1*a*. It consists of a *p*-type substrate into which two n^+ regions have been diffused. These two regions form the source and the drain. No channel is actually fabricated in this transistor. The gate is formed by covering the semiconductor region between the drain and the source with a silicon dioxide layer, on top of which is deposited a metal plate. This device is operated in an enhancement mode as a positive voltage is applied between the gate and the source. When the gate is positive, an *n*-type channel is induced between the source and the drain as a result of electrons leaving the source and the drain and being attracted toward the gate by its positive voltage. The enhancement-mode MOSFET is different from the JFET, which operates in the depletion mode.

With the positive gate voltage fixed, an increase in drain voltage (V_{DS}) will result in an increase in drain current (I_{DS}), producing a resistor-type operation. When V_{DS} continues to increase, the electric field produced beneath the gate varies, being largest at the source and smallest at the drain, since V_{GS} is larger than $V_{GD} = V_{GS} - V_{DS}$. Pinch-off occurs as V_{DS} is large enough to decrease the field near the drain to zero. Above pinch-off the drain current I_{DS} remains almost constant. The value of V_{DS} required for pinch-off is

$$V_{DS} = V_{GS} - V_T, \tag{4.6-1}$$

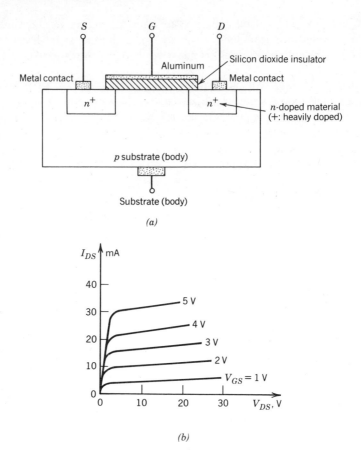

FIGURE 4.6-1 Enhancement n-channel MOSFET: (a) basic construction; (b) volt-ampere characteristics.

where V_T is a threshold voltage. Typical values of V_T for an enhancement NMOSFET lie between 2 and 4 V. The value of V_T is given by the manufacturer. No current can flow in an enhancement n-channel MOSFET until the channel between the source and the drain is formed; the current I_{DS} can flow only when V_{GS} exceeds the threshold voltage V_T. Gate current of this device is normally of the order of picoamperes.

A depletion-mode NMOSFET can be built by arranging an additional diffusion of n-type impurities into the surface of the substrate between the two n-type regions (see Figure 4.6-1a). For a depletion NMOSFET a channel is present even when the gate-source voltage is zero. In this case V_{GS} should become negative in order to cut the transistor off so that $I_{DS} = 0$ for all V_{DS}. Similarly a PMOSFET should have V_{SG} negative to achieve cutoff. Therefore, in a depletion-mode MOSFET, the threshold voltage V_T is negative. Typical values of V_T for a depletion MOSFET lie between -4 and -10 V. The value of V_{DS} required for pinch-off in a depletion-mode NMOSFET is given by Equation (4.6-1), where V_T is negative. For a PMOSFET, the value of V_{SD} needed for pinch-off is

$$V_{SD} = V_{SG} - V_T. \tag{4.6-2}$$

For an enhancement-mode PMOSFET, V_T is again a positive number; it is a negative number for a depletion-mode PMOSFET. The enhancement MOSFET is usually preferred in digital circuitry since it is a great convenience that the device be cut off at zero gate voltage.

The depletion NMOSFET can be operated in both the depletion mode and the enhancement mode, since the drain current would decrease with increasingly negative gate voltage when the diffused n channel is depleted, and the drain current would increase with increasingly positive gate voltage when the n channel is enhanced. Of course, the depletion PMOSFET can also be operated in both modes.

An enhancement NMOSFET is cut off when the gate voltage is decreased to a value less than the threshold voltage. In this case the channel is depleted of all the induced free charges. Thus, for a positive drain-source voltage, an npn transistor exists that is operating at cutoff. Therefore drain current flows that is equivalent to I_{CB0} in an npn transistor. Typical drain currents are of the order of picoamperes. A PMOSFET is cut off when the source-to-gate voltage is decreased to a value less than the threshold voltage.

Consider that an n channel has been formed as a result of the application of a positive voltage to the insulated gate. Then a pn junction is formed between this n channel and the rest of the p substrate. If a bias is applied to the substrate, then at this junction there is a depletion region that is devoid of carriers. Thus the width of the depletion region (and therefore the width of the conducting channel) will be affected by the bias voltage on the substrate. When the reverse bias between the substrate and source is increased, the drain current for a fixed voltage on the insulated gate will be decreased. Therefore pinch-off is affected, and the threshold voltage is consequently made larger by such reverse bias. The change in threshold voltage can be approximated by the formula

$$\Delta V_T \simeq C\sqrt{V_{SB}}, \tag{4.6-3}$$

where V_{SB} = source-to-substrate voltage,
$\quad\quad \Delta V_T$ = corresponding change in threshold voltage,
$\quad\quad\quad C$ = parameter depending on the doping of the substrate; value of C lies in the range 0.5–20.

4.7   JFET AND MOSFET SWITCHES

JFET Switch

Figure 4.7-1 shows the n-channel JFET common-source circuit operated as a switch. The output voltage is given by

$$v_o = v_{DS} = V_{DD} - I_D R_L. \tag{4.7-1}$$

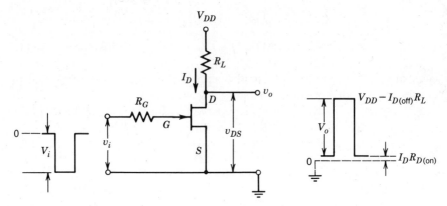

FIGURE 4.7-1 The JFET common-source circuit operated as a switch.

When the JFET is biased on,

$$v_o = v_{DS(on)} = I_D R_{D(on)} = I_D r_{ds(on)}, \tag{4.7-2}$$

or typically

$$v_{DS(on)} = I_D R_{D(on)} = 100 \ \mu A \times 40 \ \Omega = 4 \ mV.$$

When the JFET is off biased, a small drain-gate leakage-current $I_{D(off)}$ flows across the reverse-biased gate-channel junctions. Typically $I_{D(off)} = 0.25$ nA at 25°C. Comparing it with a BJT switch, $V_{DS(on)} \ll V_{CE(sat)}$, and the JFET has a much higher input resistance than a BJT. Therefore a signal having large source resistance can easily switch the JFET.

MOSFET Switch

Figure 4.7-2 shows the enhancement n-channel metal-oxide-semiconductor field-effect transistor (MOSFET) switching circuit. The enhancement device requires no external bias voltage to turn it off since its channel does not exist while the gate is at the same potential as the source. When v_i ($= v_{GS}$) becomes positive, v_o ($= v_{DS}$) drops from V_{DD} to $I_D R_{D(on)}$ due to I_D flow. The MOSFET has no drain-gate leakage current since the gate terminal is insulated from the channel. This is why the MOSFET has a higher input resistance than a JFET. When the MOSFET is off, some voltage drop develops across R_L due to the small drain-source leakage current. The p-channel MOSFET operates in exactly the same manner as the n-channel MOSFET except that it uses voltages of the opposite polarity.

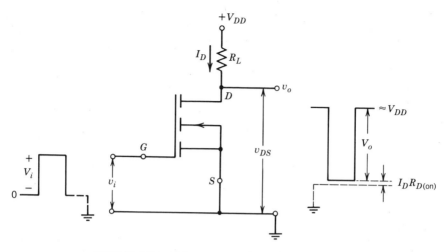

FIGURE 4.7-2 Enhancement n-channel MOSFET switch.

4.8 CMOS SWITCHES

Basic CMOS Switch

Complementary devices are two devices that are identical in every way except for their supply voltage polarities. Thus complementary MOS or CMOS is the circuitry consisting of enhancement p-channel and n-channel MOSFETs, as shown in Figure 4.8-1. In the enhancement devices, no channel exists until one of them is switched on. As v_i is 0 V at the common gate terminal, the PMOSFET is biased on and the NMOSFET is off. Under this condition, there is only a very small v_{DS} for the PMOSFET and v_o is very close to V_{DD}. As v_i becomes positive, the NMOSFET is biased on and the PMOSFET is off. There is now only a very

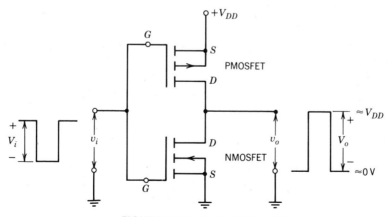

FIGURE 4.8-1 Basic CMOS switch.

small v_{DS} for NMOSFET, and therefore v_o is very close to ground potential. The main advantages of CMOS circuits are small power dissipation, small v_{DS} drop across the ON transistor, and high input impedance. The main disadvantage of FETs is their relatively small gain-bandwidth product in comparison with that which can be obtained with BJTs. Also BJTs operate at higher speeds than are possible with FETs.

CMOS Bilateral Switch

The CMOS bilateral switch is a transmission gate, as shown in Figure 4.8-2a. It acts essentially as a single-pole, single-throw switch controlled by an input logic level. This switch will pass current in both directions, and it is useful for digital and analog applications. The bilateral switch consists of a PMOSFET and an NMOSFET in parallel so that both polarities of input voltage can be switched. The fabrication process of the IC chip for this gate starts with a p-type substrate (body) into which an n-type well is diffused. The PMOS is formed in this n-type well, and NMOS is formed in the p-type body.

    The n-type well (W) of the PMOS is tied to the most positive potential $V(1)(= +V_{DD})$ in the circuit for reverse biasing the pn diodes formed between the n-type well and the drain, source, and channel. For the same purpose, the p substrate (B) of the NMOS is tied to $V(0)(= -V_{SS})$, the most negative voltage. In Figure 4.8-2a, the control voltage C is applied to G_1 and the inverter. Assume $C = 1$ so that $v_{G1} = V(1)$ $(= +5 \text{ V}$, typically) and $V_{G2} = V(0)$ $(= -5 \text{ V}$, typically). If $v_i = V(1)$, then $v_{GS1} = V(1) - V(1) = 0$, and NMOS is OFF. However $V_{GS2} = |V(1) - V(0)| >$ threshold voltage $|V_T|$, and v_{GS2} is negative,

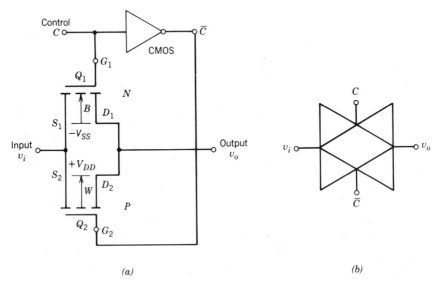

(a)   (b)

FIGURE 4.8-2 CMOS bilateral switch (transmission gate) consisting of a PMOSFET and NMOSFET in parallel: (a) circuit; (b) symbol.

causing PMOS to conduct. Because there is no applied drain voltage, PMOS operates in the nonsaturated region where $v_{DS} \simeq 0$. Thus $v_o = V(1) = v_i$. If $v_i = V(0)$, then PMOS is OFF, whereas NMOS conducts and $v_o = V(0) = v_i$.

Assume $C = 0$ so that $v_{G1} = V(0)$ and $v_{G2} = V(1)$. If $v_i = V(1)$, then v_{GS1} is negative and NMOS is OFF, whereas $v_{GS2} = 0$ and PMOS is also OFF. In this situation, the resistance between input and output is of the order of $10^9 \, \Omega$.

The symbol for the transmission gate is shown in Figure 4.8-2b. The control C is binary, but the input v_i can be either digital or an analog signal whose instantaneous value should lie between $V(0)$ and $V(1)$ (see Problems 4-8 and 4-9).

CMOS Single-Pole Double-Throw Switch

The CMOS single-pole double-throw (SPDT) switch configuration is shown in Figure 4.8-3. It consists of a CMOS inverter feeding an op-amp follower that drives R_1 from a very low output resistance. A positive logic system is indicated with $V(1) = V_R = +5$ V and $V(0) = 0$ V. The complement $\bar{Q}$ of the bit $Q = a_n$ under consideration is applied to the input. Thus, if $a_n = 1$, then $\bar{Q} = 0$, the output of the inverter is logic 1, and 5 V is applied to R_1. On the other hand, if the nth is a binary 0, $\bar{Q} = 1$, and the output of the CMOS inverter is 0 V, so that R_1 is connected to ground. This confirms that the circuit of Figure 4.8-3 operates as a SPDT switch.

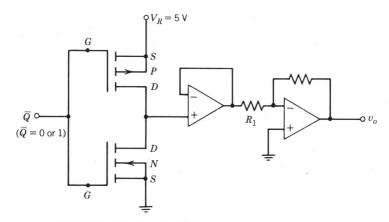

FIGURE 4.8-3 The CMOS single-pole double-throw switch.

REFERENCES

1. Bell, D. A., *Solid State Pulse Circuits*, 2nd ed., Reston Publishing Company, Reston, Virginia, 1981, Chaps. 4 and 5.

2. Tocci, R. J., *Fundamentals of Pulse and Digital Circuits*, 2nd ed., Charles E. Merrill, Columbus, Ohio, 1977, Chap. 7.

3. Pettit, J. M. and M. M. McWhorter, *Electronic Switching, Timing, and Pulse Circuits*, 2nd ed., McGraw-Hill, New York, 1970, Chap. 3.

4. Taub, H. and D. Schilling, *Digital Integrated Electronics*, McGraw-Hill, New York, 1977, Chap. 1.

5. *The Transistor and Diode Data Book for Design Engineers*, (Transistor Sections), Texas Instruments, Canton, Massachusetts, 1973.

QUESTIONS

4-1. Define the three operation regions of the bipolar transistor characteristics.

4-2. How does the reverse collector saturation current I_{CBO} vary with temperature?

4-3. Briefly explain the resistive loading effect on the transistor inverter output.

4-4. Briefly explain the capacitive loading effect on the transistor inverter output.

4-5. The emitter follower connected to the transistor inverter output can reduce the capacitive loading effect. Why?

4-6. What times do the bipolar-transistor switching times contain?

4-7. Describe the factors from which the switching transistor delay time results.

4-8. Describe the fact from which the switching transistor rise time and fall time result.

4-9. Describe the fact from which the switching transistor storage time results.

4-10. Explain how the speed-up capacitor improves the switching speed of a bipolar transistor.

4-11. Why can a junction field-effect transistor be used as a switch? Compare the JFET switch to a BJT switch.

4-12. Compare the JFET switch to a MOSFET switch.

4-13. What are the major advantages of a CMOS circuit?

4-14. Briefly describe the operation of a CMOS bilateral switch.

4-15. How does the CMOS SPDT switch work?

PROBLEMS

4-1. In a common-emitter silicon transistor configuration, $V_{CC} = 24$ V, $R_L = 2.7$ kΩ, and $I_B = 0.3$ mA. Find $h_{FE(min)}$ if the transistor is to be saturated.
Answer. 29.4.

4-2. A common-emitter circuit, using a silicon transistor, has $V_{CC} = 20$ V, $R_L = 3.3$ kΩ, and $h_{FE(min)} = 40$. Determine the minimum base current needed to achieve saturation.

Answer.   0.15 mA.

4-3. Design a transistor inverter circuit in which $V_{CC} = 15$ V, $V_{CE(sat)} = 0.2$ V, and $h_{FE(min)} = 35$ at $I_C = 1$ mA. The input is a ± 5-V square wave.

Answer. $R_L = 15$ kΩ; $R_B = 150$ kΩ.

4-4. A 10-V pulse is applied to the inverter circuit in Figure 4.3-2, which uses a silicon transistor with $h_{FE(min)} = 40$. The inverter output should drive a 1.2 kΩ load without its output dropping below 9 V in the transistor OFF state. Determine the proper values for R_C and R_B.

Answer. $R_C = 100$ Ω selected; $R_B = 3.3$ kΩ selected.

4-5. Assume $C_L = 100$ pF in Figure 4.3-3. Determine the values for R_C and R_B needed to provide a rise time of at most 0.1 μs in the output. Use $h_{FE(min)} = 60$.

Answer. $R_C = 390$ Ω selected; $R_B = 18$ kΩ selected.

4-6. In the circuit of Figure 4.4-2a, $R_B = 10$ kΩ, $C_1 = 200$ pF, and $R_s = 1$ kΩ $\neq 0$. Determine the input frequency.

Answer.   108.7 kHz.

4-7. In the circuit of Figure 4.7-1, $V_{DD} = 15$ V and $R_L = 3.3$ kΩ. The transistor is a JFET 2N4856, which has $I_{D(off)} = 0.25$ nA (maximum) and $R_{D(on)} = r_{ds(on)} = 25$ Ω (from Texas Instruments FET data sheet). Determine the output voltage (a) when the device is cut off and (b) when it is switched on.

Answer.   (a) ≈ 15 V; (b) 113.6 mV.

4-8. Consider the CMOS bilateral switch in Figure 4.8-2 with the control voltages $V(1) = +5$ V and $V(0) = -5$ V and a 5-V peak-input sinusoid. Assume that the threshold voltage V_T is zero. (a) Verify that the entire sinusoid appears at the output if the binary control C corresponds to $V(1) = +5$ V. (b) Show that transmission is inhibited if the control voltage is $V(0) = -5$ V.

4-9. Repeat Problem 4-8 if $|V_T| = 2$ V. Indicate the range of v_i over which both transistors conduct.

5

TRANSISTOR AND IC
COMPARATOR CIRCUITS

5.1 INTRODUCTION TO COMPARATORS

A *comparator* is a circuit used to detect the instant at which the amplitude of an arbitrary input waveform (v_i) attains a reference level (V_R). The comparator output usually consists of a voltage that changes abruptly when the input signal reaches the reference level. In this chapter we shall discuss the following three categories of comparator circuits.

1. Transformer-coupled regenerative comparator circuits, in which the output pulse is produced precisely at the moment an input ramp voltage reaches a reference level.

2. Schmitt trigger circuits. A basic Schmitt circuit uses an emitter-coupled binary with only one input. The input is dc coupled, and its amplitude determines the instant at which the circuit switches from one stable state to the other. When the input voltage reaches the upper or lower triggering levels (UTL or LTL), the output voltage rapidly changes. The difference between the UTL and LTL is the *Schmitt trigger hysteresis voltage V_H*. In normal operation V_H can be varied all the way down to zero if desired. The Schmitt trigger operates from almost any input waveform and always produces a pulse-type output. Transistor Schmitt trigger circuits are usually designed to trigger at specified upper and lower levels of the input voltage. Several types of Schmitt trigger integrated circuits are available. IC differential comparator and operational amplifiers (op-amps) can also be employed as Schmitt trigger circuits. The main applications of the Schmitt trigger are its uses as a voltage comparator and a pulse shaper.

3. Differential comparators and op-amp voltage comparators. This class of circuits produces an output pulse when the input voltage level becomes exactly equal to a reference voltage. The differential comparator compares two voltages and provides an indication of which one is greater. The op-amp is ideally suited for this application because of its differential input and single-ended output.

5.2 A TRANSFORMER-COUPLED REGENERATIVE COMPARATOR

Figure 5.2-1 shows the circuit of a transformer-coupled regenerative comparator with zero volts as a reference level V_R. An output pulse is produced precisely at

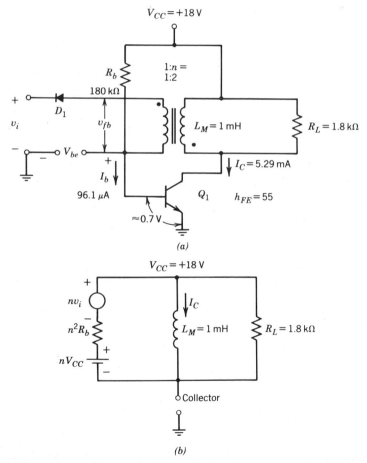

FIGURE 5.2-1 (*a*) Transformer-coupled regenerative comparator that produces an output when v_i crosses zero volts; (*b*) equivalent circuit for the transformer and circuit of Figure 5.2-1*a* when the transistor is off; the left branch represents quantities reflected from the base winding.

the instant (t_0) an input negative-going ramp voltage v_i attains $V_R = 0$ V, as shown in Figure 5.2-2. For $t < t_0$ the transistor Q_1 is in active region with $I_B = 96.1$ μA, $I_C \simeq 5.29$ mA, and $V_{CE} \simeq 18$ V. The Q_1 output across the transformer is coupled back to its input so as to give regenerative or positive feedback. However, for $t < t_0$, there is no feedback and the circuit is stable since the positive value of v_i keeps the diode D_1 turned off. As v_i attains zero, D_1 is forward biased by $V_{BE} \simeq 0.7$ V and the negative-going input waveform is coupled through the transformer to the base of Q_1. This begins to turn off Q_1, causing

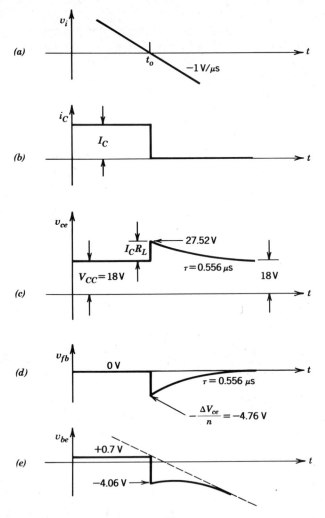

FIGURE 5.2-2 Waveforms in the transformer-coupled regenerative comparator circuit of Figure 5.2-1a: (a) negative-going input ramp (v_i); (b) collector current i_C waveform; (c) v_{ce} waveform; (d) base-winding voltage (feedback) v_{fb} waveform; (e) v_{be} waveform resulting from the sum of v_i and v_{fb}.

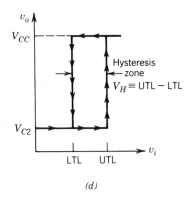

(d)

FIGURE 5.3-1   Schmitt trigger: (a) basic circuit; (b) input and output waveforms; (c) an approximate plot of v_o versus v_i for the illustrative example; (d) the general hysteresis zone.

and

$$V_{B2} = V_E + V_{BE(sat)} \simeq 2 \text{ V} + 0.8 \text{ V} = 2.8 \text{ V}.$$

Since the base of Q_1 is reverse biased by $V_E \simeq 2$ V, any voltage less than 2.5 V $(= V_E + V_\gamma)$ applied to the input will not produce the base current I_{B1}. When $v_i = 2.5$ V, the current I_{B1} begins to flow, and so the collector voltage V_{C1} is decreased. This reduced voltage coupled to the base of Q_2 decreases the collector current I_{C2} from saturation, so that the emitter voltage V_E is reduced. This regeneration continues until the UTL $(= V_E + V_{BE(sat)} = V_{B2})$ of 2.8 V is reached. When v_i is at the UTL, Q_2 is off and Q_1 is on; thus $v_o \simeq 10$ V, and $V_E = 10$ V $\times 1 \text{ k}\Omega/(9 \text{ k}\Omega + 1 \text{ K}\Omega) = 1$ V. The switching occurs very fast, so that output v_o rapidly changes from about 2 V to 10 V. In order to reduce v_o from 10 V to about 2 V, input v_i must be decreased. As v_i is decreased to the LTL $(= V_E + V_{BE(sat)} \simeq 1.8$ V), current I_{C1} reduces from saturation and I_{C2} rises from zero, resulting in an increase of V_E. The rapid regeneration causes Q_1 to cut off and Q_2 to saturate, so that $v_o \simeq 2$ V is finally reached. The relationship of output v_o and input v_i is shown as the plot of Figure 5.3-1c. Output v_o is either about 2 V at the lower state or 10 V at the upper state. When v_o is at the lower state, v_i must be increased to the UTL $(\simeq 2.8$ V), and then v_o can be switched to the upper state. Once the output is at the upper state, v_o maintains about 10 V until v_i is reduced to the LTL $(\simeq 1.8$ V), at which v_o rapidly returns to about 2 V.

Figure 5.3-1b illustrates the input and output waveforms when the Schmitt trigger is used as a pulse shaper or squaring circuit. The arbitrary input signal has a large enough excursion to carry the input beyond the limits of the hysteresis range $V_H \equiv$ UTL $-$ LTL. The output is a square wave, as shown, whose amplitude is independent of the amplitude of the input waveform.

5.4   SCHMITT TRIGGER CIRCUIT ANALYSIS

In Figure 5.3-1a transistors Q_1 and Q_2 share a common emitter voltage V_E. For either transistor to be in a conducting state, the base voltage is the sum of the emitter voltage and the base-emitter voltage, or

$$V_B = V_E + V_{BE}. \tag{5.4-1}$$

With input $v_i = 0$, Q_1 is off and Q_2 is on (saturated). The base current I_{B2} is given by

$$I_{B2} = I_1 - I_2. \tag{5.4-2}$$

The emitter voltage is given by

$$V_E = I_{E2}R_E, \tag{5.4-3}$$

where

$$I_{E2} = (h_{FE} + 1)I_{B2} \tag{5.4-4}$$

if Q_2 is in the active region of operation. If Q_2 is in saturation,

$$I_{E2(max)} = \frac{V_{CC} - V_{CE(sat)}}{R_{C2} + R_E}. \tag{5.4-5}$$

The equivalent circuits for the condition $v_i = 0$ are illustrated in Figure 5.4-1. Voltage V_{TH} is the Thevenin equivalent voltage at the base of Q_2. From Figure 5.4-1a

$$V_{TH} = \frac{R_2 V_{CC}}{R_{C1} + R_1 + R_2}. \tag{5.4-6}$$

The Thevenin resistance is given by

$$R_{TH} = \frac{(R_{C1} + R_1)R_2}{R_{C1} + R_1 + R_2}. \tag{5.4-7}$$

From Figure 5.4-1b, $V_{TH} - I_{B2}R_{TH} - V_{BE2} - (h_{BE} + 1)I_{B2}R_E = 0$. Hence the base current of Q_2 is

$$I_{B2} = \frac{V_{TH} - V_{BE(sat)}}{R_{TH} + (h_{FE} + 1)R_E}. \tag{5.4-8}$$

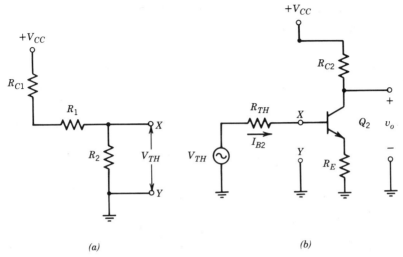

FIGURE 5.4-1 Equivalent circuits for the Schmitt trigger when input $v_i = 0$: (a) determining the Thevenin voltage at the base of Q_2; (b) the Thevenin equivalent circuit.

Substitution of Equations (5.4-4) and (5.4-8) in Equation (5.4-3) yields

$$V_E = (h_{FE} + 1)\frac{[V_{TH} - V_{BE(\text{sat})}]R_E}{R_{TH} + (h_{FE} + 1)R_E} \qquad (5.4\text{-}9)$$

for Q_2 in the active region, or

$$V_E = I_{E2(\text{max})}R_E = \frac{R_E}{R_{C2} + R_E}[V_{CC} - V_{CE(\text{sat})}] \qquad (5.4\text{-}10)$$

for Q_2 in the saturation region.

For Q_1 to turn on (saturate), the required minimum input voltage is the UTL:

$$\text{UTL} = V_E + V_{BE(\text{sat})}, \qquad (5.4\text{-}11)$$

where $V_{BE(\text{sat})} \simeq 0.8$ V(si). When Q_1 begins to turn on, voltage V_{C1} begins to fall. This results in a decrease in I_{B2}, causing voltage V_{C2} to rise. During these changes, current $I_E = I_{E1} + I_{E2} = $ constant. Finally, when Q_1 saturates, Q_2 turns off. Now voltage $V_{CE(\text{sat})} \simeq +0.2$ V(si) is not sufficient to turn on Q_2. The transition time is very short; it is indicated as the interval between t_1 and t_2 in Figure 5.3-1b. The output level remains at about V_{CC} as long as Q_2 is off.

The input potential level required to turn off Q_1 and turn on Q_2 is the LTL. When input v_i drops to a point where Q_1 begins to turn off, voltage V_{C1} increases until it is high enough to supply current I_{B2}. Then Q_2 turns on. Again a transition period exists, with both transistors conducting until Q_1 turns off. The LTL is calculated with the aid of the equivalent circuit of Figure 5.4-2. From Figure 5.4-2a, the Thevenin voltage and resistance at the collector of Q_1 are

$$V_{TH} = \frac{(R_1 + R_2)V_{CC}}{R_{C1} + R_1 + R_2} \tag{5.4-12}$$

and

$$R_{TH} = \frac{R_{C1}(R_1 + R_2)}{R_{C1} + R_1 + R_2}. \tag{5.4-13}$$

For Q_2 to turn on,

$$V_{B2} = V_E + V_{BE(sat)},$$

where

$$V_E = \frac{R_2 V_{C1}}{R_1 + R_2} - V_{BE(sat)}, \tag{5.4-14}$$

and

$$V_{C1} = V_{TH} - I_{C1}R_{TH}, \tag{5.4-15}$$

where Q_1 is at the end of the transition period.

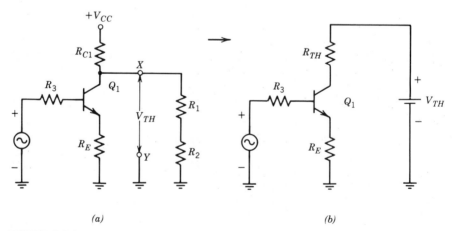

(a)   (b)

FIGURE 5.4-2   Determining the LTL of the Schmitt trigger: (a) equivalent circuit; (b) simplified circuit.

Substitution of Equations (5.4-12) and (5.4-13) in Equation (5.4-15) yields

$$V_{C1} = \frac{(V_{CC} - I_{C1}R_{C1})(R_1 + R_2)}{R_{C1} + R_1 + R_2}. \qquad (5.4\text{-}16)$$

Thus

$$V_E = \frac{(V_{CC} - I_{C1}R_{C1})R_2}{R_{C1} + R_1 + R_2} - V_{BE(sat)}. \qquad (5.4\text{-}17)$$

Substituting V_E/R_E for I_{C1} and solving for V_E, we obtain

$$V_E = \frac{V_{CC}R_E R_3 - V_{BE(sat)}R_E(R_{C1} + R_1 + R_2)}{R_E(R_{C1} + R_1 + R_2) + R_{C1}R_2}. \qquad (5.4\text{-}18)$$

The LTL is

$$\text{LTL} = V_E + V_{BE(sat)}, \qquad (5.4\text{-}19)$$

where the value of V_E is less than that in the UTL expression in Equation (5.4-11).
   A significant difference between the Schmitt trigger and a flip-flop is that the Schmitt circuit has no memory. The hysteresis zone ($V_H \equiv \text{UTL} - \text{LTL}$) is shown in the plot of Figure 5.3-1d. Arrows in the plot indicate the switching sequence. The hysteresis zone may be reduced to zero by adjusting the loop gain to unity ($A\beta = 1$). A loop gain $A\beta \geqslant 1$ must be maintained for switching to occur. To ensure proper switching, the following relationships must hold true:

$$R_{C2} \leqslant R_{C1} \geqslant \frac{R_1}{h_{FE} - 1}, \qquad (5.4\text{-}20)$$

and

$$R_3 \ll h_{FE}R_E. \qquad (5.4\text{-}21)$$

The value of R_3 equals the sum of the source and external base resistances. In practice, a loop gain greater than unity is maintained.

Example 5.4-1. The Schmitt trigger of Figure 5.3-1a uses npn silicon transistors with values of $V_{CE(sat)} = 0.2$ V, $V_{BE(sat)} = 0.8$ V, and $h_{FE1} = h_{FE2} = 100$. The circuit parameters are $V_{CC} = 10$ V, $R_{C1} = 1.3$ kΩ, $R_{C2} = R_3 = 1$ kΩ, $R_1 = 6.8$ kΩ, $R_2 = 8.2$ kΩ, and $R_E = 240\,\Omega$. Determine (a) if switching occurs, (b) the UTL, (c) the LTL, and (d) the output levels.

Solution.

(a) For switching to occur, the loop gain must be greater than unity. By Equation (5.4-20),

$$\frac{R_1}{h_{FE} - 1} = \frac{6.8 \text{ k}\Omega}{100 - 1} \simeq 68\Omega \ll R_{C1} = 1.3 \text{ k}\Omega.$$

By Equation (5.4-21), $h_{FE}R_E = 100 \times 240 = 24 \text{ k}\Omega \gg R_3 = 1 \text{ k}\Omega$. Hence we conclude that switching will occur.

(b) By Equations (5.4-6) and (5.4-7),

$$V_{TH} = 8.2 \times \frac{10}{1.3 + 6.8 + 8.2} = 5.03 \text{ V}$$

and

$$R_{TH} = \frac{(1.3 + 6.8)8.2}{1.3 + 6.8 + 8.2} \text{ k}\Omega = 4.07 \text{ k}\Omega.$$

By Equation (5.4-8),

$$I_{B2} = \frac{V_{TH} - V_{BE2}}{R_{TH} + (h_{FE} + 1)R_E} = \frac{5.03 - 0.8}{4.07 + (100 + 1)0.24} = 149 \, \mu\text{A}.$$

For saturation,

$$I_{B2(min)} = \frac{I_{C2(max)}}{h_{FE}} = \frac{V_{CC} - V_{CE(sat)}}{(R_{C2} + R_E)h_{FE}} = \frac{10 - 0.2}{(1 + 0.24)100} \simeq 80 \, \mu\text{A}.$$

Since $I_{B2} > I_{B2(min)}$, Q_2 is in saturation. Hence $V_{BE2} = V_{BE(sat)} = 0.8$ V. The value of V_E is

$$V_E = \left(\frac{V_{CC} - V_{CE(sat)}}{R_{C2} + R_E}\right) R_E = \left(\frac{10 - 0.2}{1 + 0.24}\right) 0.24 = 2 \text{ V}.$$

Thus UTL $= V_E + V_{BE(sat)} = 2 + 0.8 = 2.8$ V.

When v_i reaches UTL, Q_1 begins conducting and goes into saturation. The emitter voltage is

$$V_{E1} = I_{E1(max)}R_E = \frac{(10 - 0.2)0.24}{1.3 + 0.24} = 1.53 \text{ V}.$$

Assuming that Q_1 ultimately becomes saturated,

$$V_{B2} = \frac{(1.53 + 0.2)8.2}{6.8 + 8.2} = 0.95 \text{ V.}$$

Since $V_{B2} < V_{E1}$, Q_2 is indeed off.

(c) By Equation (5.4-18),

$$V_E = \frac{10 \times 0.24 \times 8.2 - 0.8 \times 0.24(1.3 + 6.8 + 8.2)}{0.24(1.3 + 6.8 + 8.2) + 1.3 \times 8.2} = 1.14 \text{ V.}$$

By Equation (5.4-19),

$$\text{LTL} = V_E + V_{BE(\text{sat})} = 1.14 + 0.8 = 1.94 \text{ V.}$$

(d) When Q_1 saturates, $v_o \simeq V_{CC} = 10$ V. When Q_2 saturates, $v_o \simeq V_{CC} - I_{C2(\text{max})}R_{C2} = 10 - (100 \times 80 \ \mu\text{A})(1 \text{ k}\Omega) = 2$ V.

5.5 DESIGN OF TRANSISTOR SCHMITT TRIGGER CIRCUITS

Designing for a Specified UTL

The most important parameters for the Schmitt trigger are the upper trigger level UTL and the lower trigger level LTL. The circuit can be simply designed to meet a specified UTL and to completely ignore the LTL. In this case, the LTL is normally located between the UTL and ground. With the pulse or sawtooth input waveform, the LTL can be anywhere between UTL and ground. It should be noted that the UTL equals Q_2 base voltage V_{B2} when Q_2 is on (Section 5.3). Hence the circuit is designed to have V_{B2} equal to the specified UTL. The circuit used to illustrate the design procedure is shown in Figure 5.5-1. The $R_1 - R_2$ divider and R_{L1} resistor must provide a stable bias voltage V_{B2} for Q_2. R_1 and R_2 must be several times larger than R_{L1}. A good rule of thumb here is to take the value of $I_2 \simeq I_{E2}/10$ for the calculation of $R_2 = V_{B2}/I_2$. A potentiometer may be connected between R_1 and R_2 with Q_2 base joined to its moving contact, thus providing adjustment for a precise UTL.

Example 5.5-1. Design a Schmitt trigger circuit with UTL = 6 V. Use a silicon-*npn* transistor with $h_{FE(\text{min})} = 50$ and $I_C = 1$ mA. The available supply is $V_{CC} = 15$ V.

Solution. UTL $= V_{B2} = 6$ V.

$V_E = V_{B2} - V_{BE(\text{sat})} = 6$ V $- 0.8$ V $= 5.2$ V.

$I_E \simeq I_C = 1$ mA.

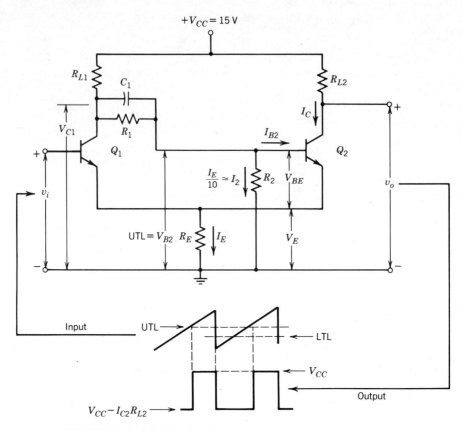

FIGURE 5.5-1   The Schmitt trigger circuit for illustrating design procedure.

$R_E = V_E/I_E = 5.2 \text{ V}/1 \text{ mA} = 5.2 \text{ k}\Omega$; use $5.6 \text{ k}\Omega$ standard value (Appendix 1).
$I_C R_{L2} = V_{CC} - V_E - V_{CE(sat)} = 15 - 5.2 - 0.2 = 9.6 \text{ V}$.
$R_{L2} = 9.6 \text{ V}/1 \text{ mA} = 9.6 \text{ k}\Omega$; use $10 \text{ k}\Omega$ standard value.
$I_2 \simeq I_E/10 = 1 \text{ mA}/10 = 0.1 \text{ mA}$.
$R_2 = V_{B2}/I_2 = 6 \text{ V}/0.1 \text{ mA} = 60 \text{ k}\Omega$; use $56 \text{ k}\Omega$ standard value.
Now I_2 becomes $6 \text{ V}/56 \text{ k}\Omega = 0.107 \text{ mA}$.
$I_{B2} = I_{C2}/h_{FE(min)} = 1 \text{ mA}/50 = 0.02 \text{ mA} = 20 \text{ }\mu\text{A}$.
$I_2 + I_{B2} = 0.107 \text{ mA} + 0.02 \text{ mA} = 0.127 \text{ mA}$.

$$R_{L1} + R_1 = \frac{V_{CC} - V_{B2}}{I_2 + I_{B2}} = \frac{15 \text{ V} - 6 \text{ V}}{0.127 \text{ mA}} = 70.87 \text{ k}\Omega.$$

As the LTL is not specified, we may select

$$R_{L1} = R_{L2} = 10 \text{ k}\Omega.$$

Hence $R_1 = 70.87 \text{ k}\Omega - 10 \text{ k}\Omega = 60.87 \text{ k}\Omega$; use $56 \text{ k}\Omega$ standard value.

Determination of *LTL*

As explained in Section 5.3, the input v_i must be decreased in order to reduce the output v_o from a high state to a low state. When v_i decreases to LTL, the circuit is about to change state but Q_1 is still on and Q_2 is off. Thus LTL should be determined referring to the situation illustrated in Figure 5.5-2. When v_i approaches LTL, v_{B1} is decreasing and v_{B2} is increasing. LTL occurs as $v_{B2} = v_{B1} = v_i$. These are the circuit conditions when Q_1 is on and v_i is exactly at LTL.

Example 5.5-2.   Design a Schmitt trigger circuit with UTL = 6 V and LTL = 5 V. Use a silicon *npn* transistor with $h_{FE(min)} = 50$ and $I_C = 1$ mA. The available supply is $V_{CC} = 15$ V. (With the exception of the LTL, the circuit is exactly as specified for Example 5.5-1.)

Solution.   From Example 5.5-1, $R_E = 5.6\text{ k}\Omega$, $R_{L2} = 10\text{ k}\Omega$, $R_2 = 56\text{ k}\Omega$, and $R_{L1} + R_1 = 70.87\text{ k}\Omega$. Since v_i is exactly at LTL as Q_1 is on, $v_i = v_{B1} = v_{B2} = \text{LTL} = 5$ V.
Referring to Figure 5.5-2,

$$I_1 = \frac{V_{B2}}{R_2} = \frac{5\text{ V}}{56\text{ k}\Omega} = 0.0893\text{ mA} = 89.3\ \mu\text{A};$$

$$I_{C1} \simeq I_E = \frac{V_{B1} - V_{BE}}{R_E} = \frac{5\text{ V} - 0.8\text{ V}}{5.6\text{ k}\Omega} = 0.75\text{ mA}.$$

$$V_{CC} = (I_{C1} + I_1)R_{L1} + I_1(R_1 + R_2);$$

$$15\text{ V} = (0.75\text{ mA} + 0.0893\text{ mA})R_{L1} + 0.0893\text{ mA}(R_1 + 56\text{ k}\Omega).$$

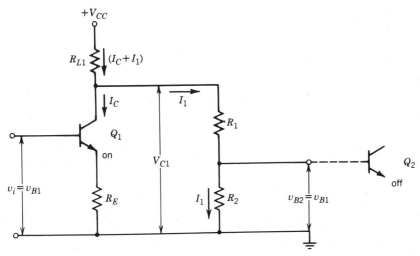

FIGURE 5.5-2   The circuit illustrating the conditions when Q_1 is on and v_i is exactly at LTL.

From Example 5.5-1, $R_1 = 70.87 \text{ k}\Omega - R_{L1}$; substituting,

$$15 \text{ V} = (0.75 \text{ mA} + 0.0893 \text{ mA})R_{L1}$$
$$+ 0.0893 \text{ mA}(70.87 \text{ k}\Omega - R_{L1} + 56 \text{ k}\Omega)$$
$$= (0.8393 \text{ mA})R_{L1} + 6.33 \text{ V} - (0.0893 \text{ mA})R_{L1} + 5 \text{ V};$$
$$15 \text{ V} - 6.33 \text{ V} - 5 \text{ V} = (0.75 \text{ mA})R_{L1};$$

$$R_{L1} = \frac{3.67 \text{ V}}{0.75 \text{ mA}} = 4.89 \text{ k}\Omega.$$

Use 4.7 kΩ standard value.

$$R_1 = 70.87 \text{ k}\Omega - R_{L1} = 70.87 \text{ k}\Omega - 4.7 \text{ k}\Omega = 66.17 \text{ k}\Omega.$$

Use 68 kΩ standard value.

Determination of the Largest Speed-Up Capacitor

Consider the situation in Figure 5.5-1 when the speed-up capacitor is not used. During switching, the voltage change at the collector of Q_1 is potentially divided across R_1 and R_2 before being applied to Q_2 base. If R_1 is now shunted by a speed-up capacitor C_1, then the potential division is eliminated, or $\Delta V_{B2} = \Delta V_{C1}$. Therefore the switching process is speeded up. For C_1 to charge through 90% of its total voltage charge, the value of the largest speed-up capacitor C_1 is given by

$$C_{1(\text{max})} = \frac{t_{re}}{2.3R}, \tag{4.3-4a}$$

where the resolving time $t_{re} = 1/(\text{triggering frequency})$ and R is the resistance in parallel with the C_1 terminals when Q_1 is off.

$$R = R_1 \| (R_{L1} + R_2). \tag{5.5-2}$$

For a Schmitt circuit to trigger at UTL and LTL as designed, the capacitor voltage with C_1 determined by Equation (4.3-4a) will settle to the dc level across R_1 in the time interval between triggering.

5.6 SCHMITT TRIGGER INTEGRATED CIRCUITS

Schmitt trigger integrated circuits are commercially available. For example, the 7413 chip is a dual four-input Schmitt trigger. Each 7413 chip contains two complete Schmitt trigger circuits, each of which has four input terminals. With $V_{CC} = +5 \text{ V}$, the typical triggering levels are UTL $= V_{T+} = 1.7 \text{ V}$ and LTL $=$

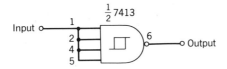

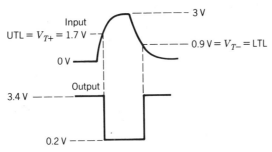

FIGURE 5.6-1 IC Schmitt trigger, illustrating the operation of one of 7413 NAND gates: V_{T+} = positive-going threshold; V_{T-} = negative-going threshold.

V_{T-} = 0.9 V. Typically the circuit output switches between a low level of 0.2 V and a high level of 3.4 V. The switching time between levels is 15 to 18 ns. Input and output currents might be as high as 1 mA and 55 mA, respectively.

The 7413 Schmitt trigger is a TTL (transistor–transistor logic) device. Its input terminals are connected in a way that permits the circuit to function as a NAND gate. Hence the 7413 is actually a dual four-input NAND gate. Figure 5.6-1 illustrates the operation of $\frac{1}{2}$7413. The slow-changing input will cause the NAND gate output to abruptly change states when it goes through the UTL of 1.7 V and the LTL of 0.9 V, thereby producing an output with transition times that can reliably drive other TTL devices. Compared with an IC op-amp or transistor Schmitt trigger circuit, the 7413 switches very much faster and is much less expensive; however its trigger levels are fixed, and so we cannot obtain a different UTL and LTL from the 7413.

5.7 IC Op-AMPS AS SCHMITT TRIGGER CIRCUITS

Operational Amplifiers (Op-Amps)

The op-amp shown in Figure 5.7-1 is a very high gain dc amplifier with two input terminals: (1) the inverting input ($-$), at which a positive-going signal produces a negative-going output voltage or vice versa; and (2) the noninverting input ($+$), at which a positive-going signal produces a positive-going output or vice versa. The input impedance is extremely high, and the output impedance is very low.

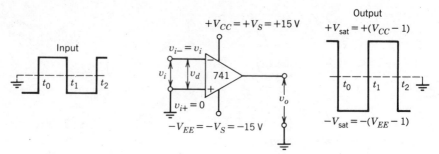

FIGURE 5.7-1   IC 741 op-amp circuit. At $v_i < 70\ \mu$V below ground, $v_o \simeq +(V_{CC} - 1)$; at $v_i > 70\ \mu$V above ground, $v_o \simeq -(V_{EE} - 1)$.

For a supply voltage $V_S = \pm 15$ V, the IC 741 op-amp is specified in terms of the following typical quantities:

1. Open-loop voltage gain $A_v = 200{,}000$, output $V_o = 10$ V, and thus $V_i = 10$ V/$200{,}000 = 50\ \mu$V. This is the difference voltage v_d between inverting and noninverting input terminals. v_d will cause v_o to go to ± 10 V.

2. Output impedance $Z_o = 75\ \Omega$. Input impedance $Z_i = 2$ MΩ.

3. Input bias current $I_B = 80$ nA: I_B is the current flowing into each of the two input terminals when they are biased to the same voltage level.

4. Input offset current $I_{IOS} = 20$ nA: I_{IOS} is the difference between the two input bias currents.

5. Input offset voltage $V_{IOS} = 1$ mV: V_{IOS} is the voltage difference that may have to be applied between the two input terminals in order to adjust the output level to exactly zero.

In switching applications, the following typical quantities for 741 should be noted:

1. Maximum output voltage $V_{o(max)} = \pm 14$ V when a ± 15-V supply is used. $V_{o(max)} = \pm 10$ V (minimum) if a 2-kΩ load is connected to the output terminals. $V_i \simeq V_o/A_v$ is required to drive the output to its extreme levels. For a ± 15-V supply, $V_i \simeq 14$V/$200{,}000 = 70\ \mu$V (min) $= V_d$. The maximum output voltage for an op-amp is called its *saturation voltage* V_{sat}.

2. The *slew rate* is the rate of change of output voltage, or the speed with which the output changes. For the 741, the slew rate $= 0.5$ V/μs. Thus the output moves from -10 V to $+10$ V in a time of 20 V/0.5 V $= 40\ \mu$s. If the slew rate of the 741 is too slow in an application, another op-amp with a faster slew rate must be employed.

3. The 741 has a maximum input bias current $I_{B(max)} = 500$ nA.

The 715 op-amp has a slew rate of 100 V/μs. The 710 is described as a high-speed differential comparator; its response time is typically 40 ns.

A differential voltage comparator is a high-gain, differential input, single-ended output amplifier. Clearly an op-amp can be used as a voltage comparator; however its response time is often too slow for many applications, and the large output-voltage swing desired for op-amps is often a disadvantage when the op-amp comparator is used to drive low-level logic circuits. Tables 5.7-1 and 5.7-2 list major characteristics of commonly used typical IC operational amplifiers and differential comparators, respectively.

IC 741 Op-Amp as Inverting Schmitt Trigger Circuit

In the inverting Schmitt trigger circuit of Figure 5.7-2, the input triggering voltage is applied to the inverting input terminal and the noninverting input terminal is connected to the junction of the R_1-R_2 voltage divider. The voltage across R_2 is V_2 at the noninverting terminal. As $v_i < V_2$ or $V_2 > v_i$, the output V_o is positive. In this case $V_o \simeq V_{CC} - 1$ if $R_L \geqslant 10$ kΩ. The voltage V_2 is given by

$$V_2 \simeq \frac{R_2}{R_1 + R_2} (V_{CC} - 1). \qquad (5.7\text{-}1a)$$

When v_i is increased to V_2, output v_o begins to go negative, causing V_2 to reduce; then the noninverting terminal quickly becomes negative with respect to the inverting terminal. At this time, v_o changes over rapidly from about $+(V_{CC} - 1)$ to about $-(V_{EE} + 1)$. When v_o is positive, the UTL equals V_2 given by Equation (5.7-1a). On the other hand, when v_o is negative and equal to $-(V_{EE} - 1)$, V_2 becomes

$$V_2 \simeq -\frac{R_2}{R_1 + R_2} (V_{EE} - 1) = LTL.$$

Hence we have

$$\text{UTL} \simeq \frac{R_2}{R_1 + R_2} (V_{CC} - 1) = \frac{R_2}{R_1 + R_2} V_{\text{sat}} \qquad (5.7\text{-}1b)$$

and

$$\text{LTL} \simeq -\frac{R_2}{R_1 + R_2} (V_{EE} - 1) = -\frac{R_2}{R_1 + R_2} V_{\text{sat}}, \qquad (5.7\text{-}2)$$

where $V_{CC} = +V_S$ = positive supply voltage and $-V_{EE} = -V_S$ = negative supply voltage.

TABLE 5.7-1
Major Characteristics of Several Typical IC Operational Amplifiers ($T_A = 25°C$)

Op-Amp	V_{IDS} (mV)	I_{IOS} (nA)	I_B	A_V	R_i typ	V_i range (V)	R_o Ω typ	I_S mA max	V_{OSW} (V) typ	P_{dis} (mW) max	Slew Rate (V/μs)	CMRR (dB) typ
715	5 max ($R_S \leqslant 10$ K)	250 max	750 μA max	15 K ($R_L \geqslant 2$ K) ($V_o = \pm10$ V)	1 M	±12 typ	75	7	±13 ($R_L \geqslant 2$ K) ($V_S = \pm15$ V)	210	100 ($A_V = 1$, invert.)	92
741	5 max, 1 typ	200 max, 20 typ	80 typ, 500 max nA	200 K typ ($V_o = 10$ V) ($R_L \geqslant 2$ K)	2 M	±12 min	75	2.8	±14 ($R_L \geqslant 10$ K)	85	0.5 ($R_L \geqslant 2$ K)	90 ($R_S \leqslant 10$K)
747 dual	5 max ($R_S \leqslant 10$ K)	200 max	80 typ, 500 max nA	200 K typ ($V_o = \pm10$ V) ($R_L \geqslant 2$ K)	2 M	±13 typ	75	5.6	±13 ($V_S = \pm15$ V)	170	0.5 ($R_L \geqslant 2$ K)	90 ($R_S \leqslant 10$K)

TABLE 5.7-2
Major Characteristics of Several Typical IC Differential Comparators (Diff Comp) ($T_A = 25°C$)

Diff Comp	V_{IOS} (mV)	I_{IOS} (μA)	I_B (μA)	A_v	R_i	$\pm V_i$ (V)	R_o (Ω)	$\pm V_{di}$ (V)	$+I_S$ $-I_S$ mA	Δv_o (V)	P_{dis} (mW)	T_{resp} nA	CMRR dB
710	0.6 typ ($R_S \leqslant 200\,\Omega$)	0.75 typ	20 max	1 K min	—	±5 min ($V- = -7\,V$)	200 typ	±5 min	max: +9 ($V_o \leqslant 0$); −7 (V_o = gnd, V_i- = +5 mV)	4 − 0 max $V+$ = 12 V $V-$ = −6 V	150 max (V_o = gnd, V_i- = +10 mV)	40 ns typ	80 min ($R_S \leqslant 200\,\Omega$)
711 dual	5 max ($R_S \leqslant 200\,\Omega$, V_o = +1.4 V)	10 max	75 max	1.5 K typ	—	±5 min ($V-$ = −7 V)	200	±5 min	+8.6 −3.9 typ	5 − 0 max ($V+$ = 12 V, $V-$ = −6 V), ($V_i \geqslant$ 10 mV)	200 max	40 ns typ	—
760	6 max ($R_S \leqslant 200\,\Omega$)	7.5 max	60 max	12 K typ (1 MHz)	—	±4.5 typ	100	±5 typ	+32 −16 max (V_S = ±6.5 V)	3 − .25 typ (V_S = ±4.5 to ±6.5 V)	—	30 ns	—

Notes for Tables 5.7-1 and 5.7-2:

V_{IOS} = input offset voltage;
I_B = input bias current;
R_i = input resistance;
R_o = output resistance;
V_S = supply voltage;
V_{OSW} ($=\Delta v_o$) = output voltage swing;
T_{resp} = response time;

I_{IOS} = input offset current;
A_v = voltage gain;
$\pm V_i$ = input voltage range;
$\pm V_{di}$ = differential input voltage range;
I_S = supply current;
P_{dis} = power consumption;
CMRR = common mode rejection ratio.

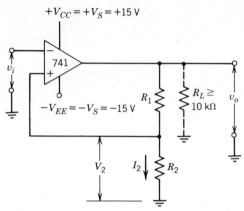

FIGURE 5.7-2 IC 741 op-amp as inverting Schmitt trigger circuit.

Example 5.7-1. The inverting Schmitt trigger circuit in Figure 5.7-2 has its trigger levels of ± 2 V. The 741 has a maximum input bias current ($I_{B,\,max}$) of 500 nA. Calculate the actual trigger levels when the standard values of R_1 and R_2 are selected.

Solution. For a stable level of V_2, $I_2 \gg I_{B(max)}$.
Let $I_2 = 100 \times 500\,\text{nA} = 50\,\mu\text{A}$.
Then

$$R_2 = \frac{\text{UTL}}{I_2} = \frac{2\,\text{V}}{50\,\mu\text{A}} = 40\,\text{k}\Omega.$$

Use 39 kΩ standard value. Now, I_2 becomes $\dfrac{2\,\text{V}}{39\,\text{k}\Omega} = 51.3\,\mu\text{A}$.

$V_{R1} = (V_{CC} - 1) - V_2 = 15 - 1 - 2 = 12$ V.

$R_1 = \dfrac{V_{R1}}{I_2} = \dfrac{12\,\text{V}}{51.3\,\mu\text{A}} = 234\,\text{k}\Omega.$

Use 220 kΩ standard value.

Actual UTL $\simeq \dfrac{V_o R_2}{R_1 + R_2} = \dfrac{14\,\text{V} \times 39\,\text{k}\Omega}{220\,\text{k}\Omega + 39\,\text{k}\Omega} = 2.1$ V.

Actual LTL $\simeq -2.1$ V.

Inverting Op-Amp Schmitt Trigger with Reference Voltage

The inverting op-amp Schmitt trigger of Figure 5.7-3a has a reference voltage V_R in series with the noninverting terminal resistor R_1. The hysteresis zone of the

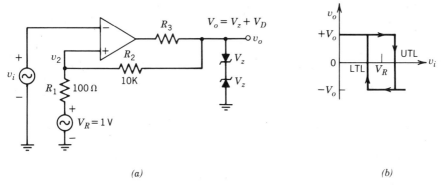

(a)   *(b)*

FIGURE 5.7-3 (*a*) The inverting op-amp Schmitt trigger with reference voltage V_R; (*b*) its hysteresis zone.

circuit is a function of the feedback resistors R_1 and R_2 and is given by

$$V_H = \text{UTL} - \text{LTL} = \frac{2R_1 V_O}{R_1 + R_2}, \qquad (5.7\text{-}3)$$

where

$$V_o = V_z + V_D, \qquad (5.7\text{-}4)$$

$$\text{UTL} = V_R + \frac{R_1}{R_1 + R_2}(V_o - V_R) = v_2 \quad \text{for } v_i < v_2, \qquad (5.7\text{-}5)$$

and

$$\text{LTL} = V_R - \frac{R_1}{R_1 + R_2}(V_o + V_R) = v_2 \quad \text{for } v_i > \text{UTL}. \qquad (5.7\text{-}6)$$

If v_i is less than v_2 and is now increased, then v_o remains constant at $+V_o$ and $v_2 = \text{UTL} = \text{constant}$ until $v_i = \text{UTL}$. At this triggering potential, the output regeneratively switches to $v_o = -V_o$ and remains at this value as long as $v_i > \text{UTL}$. This transfer characteristic is indicated in Figure 5.7-3*b*. If v_i is now decreased, then the output remains at $-V_o$ until $v_i = \text{LTL}$. At this triggering potential, a regenerative transition takes place and the output returns to $+V_o$ almost instantaneously. For the parameter values given in Figure 5.7-3*a* and with $V_o = 10\,\text{V}$, $\text{UTL} = 1 + (0.1 \times 9)/10.1 = 1.089\,\text{V}$, and $\text{LTL} = 1 - (0.1 \times 11)/10.1 = 0.891\,\text{V}$. Hence $V_H = \text{UTL} - \text{LTL} = 0.198\,\text{V}$.

IC Op-Amp as Noninverting Schmitt Trigger Circuit

In the noninverting Schmitt circuit of Figure 5.7-4, the output v_o goes positive when input v_i is increased to UTL and negative when v_i is decreased to LTL. Assume that v_o is negative at a level of about $-(V_{EE} - 1) = -V_o$. If $v_i = 0$, the voltage across R_1 is

$$V_{R1} = -\frac{R_1}{R_1 + R_2}(V_{EE} - 1) = -\frac{R_1}{R_1 + R_2}V_{sat}.$$

This negative voltage at the noninverting input (+input) keeps the output v_o at its negative saturation level ($= -V_{sat} = -V_o$). For v_o to go positive, v_i must be increased until v_{i+} goes slightly above ground level ($=v_{i-} = 0$). When this occurs, $v_d \simeq 0$ and the voltage across R_2 is $0 \text{ V} - (-V_o) = V_o$. Hence the current through R_2 is

$$I_2 = \frac{V_o}{R_2}, \tag{5.7-7}$$

the current flowing into the + input is almost zero, and the current through R_1 is also I_2. Therefore

$$v_i = V_{R1} = I_2 R_1 = \text{UTL.} \tag{5.7-8}$$

The LTL is numerically equal to the UTL but with reversed polarity. There are two steps in designing for a given UTL and LTL: first, select I_2 very much larger than the maximum input bias current to the amplifier; second, calculate R_1 and R_2 by using Equations (5.7-8) and (5.7-7).

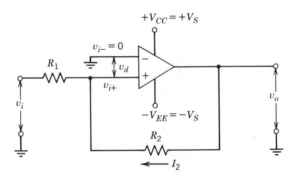

FIGURE 5.7-4 IC op-amp as a noninverting Schmitt trigger circuit.

5.8 IC 710 HIGH-SPEED DIFFERENTIAL COMPARATOR AS SCHMITT TRIGGER CIRCUIT

Complete Circuit of 710 Differential Comparator

The circuit of the 710 high-speed differential comparator is shown in Figure 5.8-1. The input stage is a difference amplifier $(Q_1 - Q_2)$ with a constant current source (Q_3). Two inputs are available—one for reference voltage V_R and one for input signal v_1. If v_i is applied as V_1 and V_R as V_2, output v_o will be out of phase with v_i, while if v_i is applied as V_2 and V_R as V_1, output v_o will be in phase with v_i. The circuit involving Q_3 serves to provide a large incremental (ac) resistance to minimize the common-mode voltage gain while the reasonable dc emitter currents (I_{E1} and I_{E2}) are still allowed. Using typical transistor parameters, the incremental resistance seen looking into the collector of Q_3 is found to be of the order of $250 \, k\Omega$. Since this incremental resistance is so high, the common-mode rejection ratio (CMRR = full differential voltage gain/common-mode voltage gain) of the input stage is also high. In any event, CMRR $\simeq 10^4$ ($= 80 \, dB$ min).

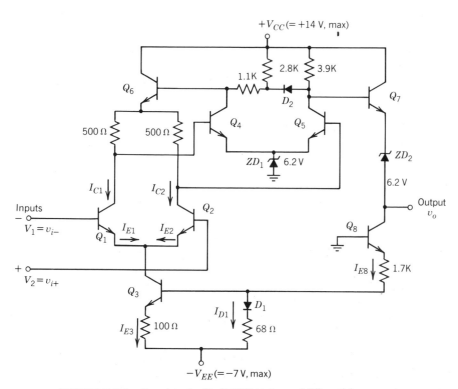

FIGURE 5.8-1   Complete circuit of IC 710 high-speed differential comparator.

The amplified signal from Q_2 is applied to Q_5, which provides additional gain. Q_7 is an emitter-follower output stage. Zener diodes ZD_1 and ZD_2 provide the necessary dc voltages to allow the dc coupling needed and to arrange proper output voltage levels.

Q_4 and Q_6 provide positive feedback to increase the gain from input to collector of Q_2. The increase of this gain will minimize the range over which the input must swing to carry the output from one limiting voltage to the other. To see the situation of such feedback, let V_R be applied as V_2 and v_i as V_1. If v_i then increases, the collector voltage of Q_2 will also increase. The collector voltage of Q_1 will decrease, as will the base voltage of Q_4. Therefore the collector of Q_4 and the base of Q_6 will rise. Finally the emitter voltage of Q_6 will increase and the collector of Q_2 will rise farther. Hence, when an input signal drives the collector of Q_2 in either direction, the feedback serves to push the collector of Q_2 further in the same direction. Thus the feedback is positive, which increases the gain from input to collector of Q_2.

Inverting Schmitt Trigger Using 710 Comparator

The IC 710 op-amp is manufactured especially for the voltage comparator function of a Schmitt trigger. The functional diagram of the circuit is shown in Figure 5.8-2a. The reference voltage is connected to the noninverting terminal of the op-amp to establish the triggering level. Input signals above the reference level cause the output to switch, and signals below the level force the output to return to its original state. The hysteresis zone is a few millivolts. The circuit of Figure 5.8-2b, however, has a controlled hysteresis zone, as expressed in the following:

$$V_H = \text{UTL} - \text{LTL} = \frac{\Delta v_o R_1}{R_1 + R_2}. \tag{5.8-1}$$

The 710 comparator has a maximum output high voltage of 4 V and a maximum output low voltage of 0 V. Hence $v_o = 4\,\text{V} - 0\,\text{V} = 4\,\text{V}$.

Now consider the operation of the circuit shown in Figure 5.8-2a when the sawtooth wave of Figure 5.8-2c is applied to the inverting input, the reference voltage $V_R = 1\,\text{V}$ is applied to the noninverting input, and the maximum output is 4 V. For $0 < t < t_1$, the input rises from 0 to 1 V. Since the reference is constant at 1 V, the input is at a lower level than the reference. The op-amp provides an output of 4 V due to its very high gain. A small difference in voltage between the inverting and noninverting input terminals causes the op-amp to saturate and deliver the maximum output of 4 V. For $t_1 < t < t_2$, the inverting input is more positive than the reference voltage, and so the output is at 0 V. For $t_2 < t < t_5$, the input is again less than the reference, and so the output is at 4 V. The output waveform is shown in Figure 5.8-2d.

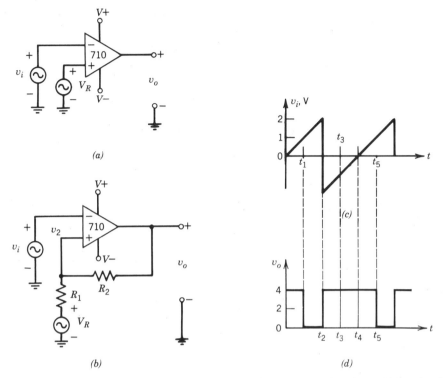

FIGURE 5.8-2 The op-amp comparator used as an inverting Schmitt trigger: (*a*) functional diagram; (*b*) practical Schmitt trigger circuit; (*c*) input and (*d*) output waveforms of the circuit of Figure 5.8-2*a*.

5.9   IC Op-AMPS AS VOLTAGE COMPARATORS

Introduction

A voltage comparator has two input voltages, and the change in its output level occurs at the moment the two inputs become equal. The op-amp is ideally suited for this application because of its differential input and single-end output. If a reference voltage V_R is applied to one input terminal and an unknown voltage v_i to the other, the value of output saturation voltage indicates whether v_i is greater or less than V_R. The simplest comparator with V_R at ground potential is known as a *zero-crossing detector* since the output voltage changes polarity every time the input voltage passes through about 0 V. It is assumed that the difference voltage $V_d = \pm 70\,\mu\text{V}$ required for saturation is so small that it can be neglected. The reference voltage may also be set to values other than zero, resulting in both positive and negative voltage-level detectors.

Zero-Crossing Detectors

In the zero-crossing detector of Figure 5.9-1a, $V_R = 0$ V is applied to the + input, and so $V_{i+} = V_R = 0$. V_i is applied to the − input, and so $V_{i-} = v_i$. In this case, the difference voltage is

$$v_d = V_{i+} - V_{i-} = 0 - v_i = -v_i. \tag{5.9-1}$$

Whenever v_i is slightly less (about $-70 \, \mu$V) than 0 V, v_d will be positive and v_o will go to $+V_{\text{sat}}$. On the other hand, if v_i is slightly greater than 0 V, v_d will be negative and v_o will go to $-V_{\text{sat}}$. If v_i is a randomly varying signal, the output v_o will switch from $-V_{\text{sat}}$ to V_{sat} whenever v_i crosses zero in the negative direction and will switch from $+V_{\text{sat}}$ to $-V_{\text{sat}}$ whenever v_i crosses zero in the positive direction. The voltage waveforms are shown in Figure 5.9-1b. This comparator is called an *inverting zero-crossing detector* since it inverts the polarity of the input

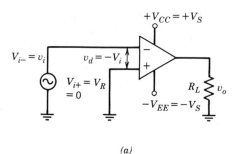

(a)

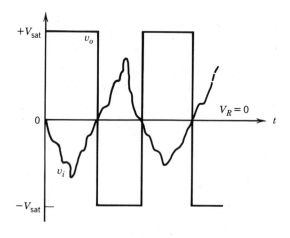

(b)

FIGURE 5.9-1   (*a*) An inverting zero-crossing detector and (b) its voltage waveforms.

signal. Hence it can be used to turn off a load device when v_i is over 0 V and turn it back on when v_i is under 0 V.

If the $V_R = 0$ V is applied to the − input and v_i to the + input, the comparator is still a zero-crossing detector, but the output saturation voltages do not reverse polarity. In this case, the difference voltage is

$$v_d = V_{i+} - V_{i-} = v_i - 0 = v_i. \tag{5.9-2}$$

Whenever v_i is slightly greater (about 70 μV) than 0 V, v_d becomes positive and v_o will go to $+V_{sat}$. On the other hand, if v_i is slightly less than 0 V, v_d will be negative and v_o will switch from $+V_{sat}$ to $-V_{sat}$, as shown in Figure 5.9-2b. This comparator is called a *noninverting zero-crossing detector* since the output will change saturation levels every time v_i crosses 0 V to the same polarity as v_i (noninverting). These output voltages can be used to turn various load devices on or off in response to a changing input signal. Hence the voltage comparator is often employed as a voltage-sensitive switch.

Nonzero-Crossing Detectors

Nonzero-crossing detectors can be constructed by applying a nonzero reference voltage to one of the input terminals. Figure 5.9-3a shows a noninverting voltage-level detector, in which $V_R = 3$ V is applied to the − input and the randomly varying signal v_i is applied to the + input. For v_o to switch to $+V_{sat}$, v_d should be positive ($v_d > 0$) and requires

$$v_d = V_{i+} - V_{i-} > 0; \quad \text{or} \quad V_i > V_R.$$

Thus

$$v_i - 3 \text{ V} > 0 \quad \text{or} \quad v_i > 3 \text{ V}.$$

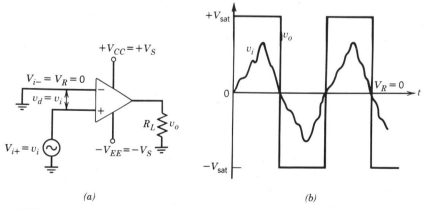

(a)   *(b)*

FIGURE 5.9-2   (a) A noninverting zero-crossing detector and (b) its voltage waveforms.

As long as v_i is greater than 3 V, v_d will be positive and v_o will be at $+V_{sat}$. If v_i is less than 3 V, however, v_d will be negative and v_o will be at $-V_{sat}$. Therefore Figure 5.9-3a represents a voltage-sensitive switch that can turn on a load device whenever v_i is over 3 V and turn it off whenever v_i is under 3 V.

If $+V_R$ is applied to the $+$ input, and v_i to the $-$ input, then v_o will switch to $-V_{sat}$ if

$$v_d = +V_R - v_i < 0 \quad \text{or} \quad v_i > +V_R,$$

and v_o will switch to $+V_{sat}$ if $v_i < +V_R$. Under these conditions the comparator operates as an inverting voltage-level detector.

If $-V_R$ is applied to one of the two inputs, then for a noninverting voltage-level detector, V_o will switch to $+V_{sat}$ if

$$v_d = v_i + V_R > 0 \quad \text{or} \quad v_i > -V_R,$$

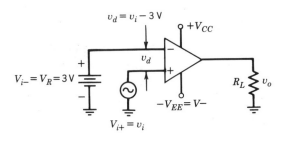

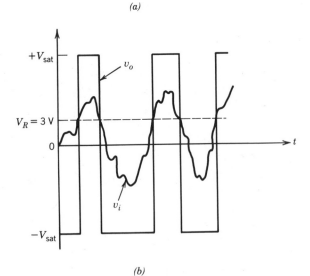

FIGURE 5.9-3 (a) $+3$-V voltage level detector (noninverting) and (b) its voltage waveforms.

and v_o will switch to $-V_{\text{sat}}$ if $v_i < -V_R$; for an inverting voltage-level detector, v_o will switch to $-V_{\text{sat}}$ if

$$v_d = -V_R - v_i < 0 \quad \text{or} \quad v_i > -V_R,$$

and v_o will switch to $+V_{\text{sat}}$ if $v_i < -V_R$.

Protecting the Output Load Device

Some devices connected to the output of the comparator will be damaged if the full output saturation voltage $\pm V_{\text{sat}}$ is applied to them. The value of $\pm V_{\text{sat}}$ exceeds the *pn* junction reverse-bias voltages of devices such as light-emitting diodes (LEDs), transistors, silicon controlled rectifiers (SCRs), digital logic, and so on. To protect an output device from a large reverse voltage, a diode can be inserted between the output terminal and the device so that one output polarity (either $+V_{\text{sat}}$ or $-V_{\text{sat}}$) will be dropped across the diode rather than across the device. If the load resistor R_L were replaced by an LED, as shown in Figure 5.9-4, the LED would light whenever $v_o = +V_{\text{sat}}$. Because the maximum reverse bias an LED can withstand is about 3 V, the diode protects the LED when $v_o = -V_{\text{sat}}$.

At the output of the comparator in Figure 5.9-5, two LEDs are placed in parallel but in opposite directions. In this case, no protective diode is required,

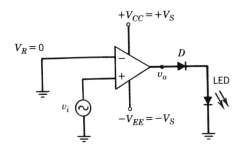

FIGURE 5.9-4 Example of the diode protecting the load LED at the comparator's output. LED lights when v_i is positive.

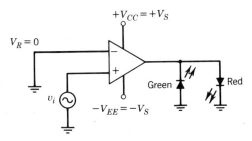

FIGURE 5.9-5 Two LEDs placed in parallel but in opposite directions for protecting each other at the output of the comparator.

since one LED will protect the other. When v_i is positive, v_o tries to go to $+V_{sat}$ and the red LED will light. When v_i is negative, v_o tries to go to $-V_{sat}$ and the green LED will light. The forward-bias voltage of the "on" LED is only about 1.5 V and less than the maximum permissible 3-V reverse bias for the "off" LED. Hence v_o only goes to 1.5 V as the LED conducts and draws the maximum output current of 25 mA.

REFERENCES

1. Tocci, R. J., *Fundamentals of Pulse and Digital Circuits*, 2nd ed., Charles E. Merrill Publishing, Columbus, Ohio, 1977, Chap. 11.

2. Bell, D. A., *Solid State Pulse Circuits*, 2nd ed., Reston Publishing Company, Reston, Virginia, 1981, Chap. 6.

3. Millman, J. and H. Taub, *Pulse, Digital, and Switching Waveforms*, McGraw-Hill, New York, 1965, Chap. 10.

4. Hnatek, E. R., *Applications of Linear Integrated Circuits*, John Wiley & Sons, New York, 1975, Chap. 5.

5. *µA Linear IC Data Book*, Fairchild Camera and Instrument Corporation, 313 Fairchild Drive, Mountain View, California, 1982, pp. 4-55–4-63 (µA741), pp. 5-3–5-9 (µA710).

6. Schoenwetter, H. K., A Sensitive Analog Comparator, *IEEE Transactions on Instrumentation and Measurement*, Vol. IM-31, No. 4, December 1982, pp. 266–269.

QUESTIONS

5-1. What is a voltage comparator?

5-2. Briefly describe regenerative action in the circuit of Figure 5.2-1.

5-3. What is a Schmitt trigger? Sketch its transistor circuit and briefly explain its operation.

5-4. Define the terms *upper trigger level*, *lower trigger level*, and *hysteresis zone*.

5-5. Sketch the hysteresis zone of a transistor Schmitt trigger. List the principal characteristics and uses of this basic circuit.

5-6. Plot the input and output waveforms to illustrate the waveshaping of the Schmitt trigger circuit.

5-7. Sketch the IC 7413 Schmitt trigger and its input and output waveforms. Briefly explain its operation.

5-8. Sketch the circuit of an op-amp used as an inverting Schmitt trigger circuit. Briefly explain how it functions.

5-9. Sketch the circuit of an op-amp used as a noninverting Schmitt trigger circuit. Briefly explain how it functions.

5-10. Briefly explain the positive feedback in the µA710 op-amp comparator.

5-11. Sketch the circuit of a voltage comparator used as an inverting zero-crossing detector. Also sketch the input and output waveforms. Briefly explain.

5-12. Sketch the circuit of a voltage comparator used as a noninverting +2-V voltage level detector. Also sketch the input and output waveforms. Briefly explain.

PROBLEMS

5-1. In the Schmitt trigger of Figure 5.3-1a, $V_{CE(sat)} = 0.2$ V, $V_{BE(sat)} = 0.8$ V, $h_{FE1} = h_{FE2} = 100$, and the circuit parameters are given as indicated. Determine: (a) If switching occurs, (b) the UTL, (c) the LTL, and (d) the output levels.

Answer. (a) Switching occurs [calculate using Equations (5.4-20) and (5.4-21)]; (b) UTL = 2.76 V; (c) LTL = 1.54 V; (d) 10 V and 2.16 V.

5-2. Design a Schmitt trigger circuit with UTL = 4 V. Use the silicon *npn* transistors with $h_{FE(min)} = 100$ and $I_C = 2$ mA. The available supply is $V_{CC} = 10$ V.

Answer. $R_E = 1.5$ kΩ; $R_{L2} = 3$ kΩ; $R_2 = 22$ kΩ; $R_{L1} + R_1 = 30$ kΩ. Select $R_{L1} = R_{L2} = 3$ kΩ. (Refer to Figure 5.5-1.)

5-3. Design a Schmitt trigger circuit with UTL = 4 V and LTL = 3 V. Use the silicon transistors with $h_{FE(min)} = 100$ and $I_C = 2$ mA. The available supply is $V_{CC} = 10$ V. (With the exception of the LTL, the circuit is exactly as specified for Problem 5-2.)

Answer. $R_{L1} = 2.2$ kΩ; $R_1 = 27$ kΩ.

5-4. The Schmitt trigger circuit (Figure 5.5-1) designed in Example 5.5-2 is to be triggered at a maximum frequency of 500 kHz. Determine the largest speed-up capacitor (C_1) that may be used.

Answer. 27 pF.

5-5. Referring to example 5.7-1 design an inverting Schmitt trigger circuit with UTL = 0 V and LTL = −2 V. The available supply is ±10 V rather than ±15 V.

Answer. $R_2 = 39$ kΩ; $R_1 = 150$ kΩ.

5-6. Referring to Example 5.7-1 and Figure 5.7-4 design a noninverting Schmitt trigger circuit with trigger levels of ±2 V. The available supply is ±12 V, and the op-amp is 741.

Answer. $R_2 = 220$ kΩ; $R_1 = 39$ kΩ.

5-7. Calculate UTL and LTL and the hysteresis zone for $\pm V_{sat} = 14$ V, $R_1 = 50$ kΩ, and $R_2 = 5$ kΩ in Figure 5.7-2.

Answer. UTL = 1.273 V; LTL = −1.273 V; $V_H = 2.546$ V.

5-8. In the Schmitt trigger circuit of Figure 5.7-2, the hysteresis zone $V_H = 100$ mV, $R_2 = 200\,\Omega$, and $\pm V_{sat} = \pm 14$ V. Find R_1.

Answer. 56 kΩ.

5-9. The 741 op-amp has a typical open-loop gain $A_v = 200{,}000$. (a) What value of difference voltage v_d will make output $+ V_o = + V_{sat} = 14$ V? (b) What value of v_d will make output $- V_o = - V_{sat} = -14$ V?

Answer. $\pm 70\ \mu$V.

5-10. Design a voltage comparator that will go to $+ V_{sat}$ whenever v_i exceeds -3 V.

5-11. Design a voltage comparator with an output that switches from $- V_{sat}$ to $+ V_{sat}$ whenever v_i drops below -5 V. The available supply is ± 12 V.

Hint. An inverting negative voltage detector with $V_R = -5$ V is required. A voltage divider $R_1 - R_2$ is connected across $- V_{EE} = -12$ V with negative polarity on the tap of potentiometer R_1. Let $R_2 = 10\,\text{k}\Omega$; $R_1 = $?

6

ASTABLE AND MONOSTABLE MULTIVIBRATORS AND BLOCKING OSCILLATORS

6.1 TRANSISTOR ASTABLE MULTIVIBRATOR CIRCUITS

Operation of Basic Astable Multivibrator

A *transistor astable multivibrator* circuit is essentially a nonsinusoidal two-stage oscillator in which one stage is on while the other is cut off until a point is reached at which the conditions of the stages are reversed. The output at the collector of each transistor is a square wave. In the collector-coupled astable multivibrator of Figure 6.1-1, each transistor has a bias resistor (R_{B1} or R_{B2}) and each is capacitor coupled to the collector of the other transistor. An increase in the collector current of Q_1 causes a decrease in the collector voltage, which, when coupled through C_2 to the base of Q_2, causes a decrease in the collector current of Q_2. The resultant rising voltage at the collector of Q_2, when coupled through C_1 to the base of Q_1, drives Q_1 further into conduction. This regenerative process occurs rapidly, driving Q_1 into heavy saturation (on) and Q_2 into cutoff. Q_2 is maintained in a cutoff condition by C_2 (which was previously charged to V_{CC} through R_{C1} with negative potential on C_2's right side) until C_2 discharges through R_{B2} toward V_{CC}. When the junction of C_2 and R_{B2} reaches a slight positive voltage, however, Q_2 begins to start into conduction and the regenerative process reverses. Q_2 then reaches a saturation condition, Q_1 is cut off by the reverse bias applied to its base through C_1, and the junction of C_1 and R_{B1} starts charging toward V_{CC}. The two transistors are therefore alternately cut off. The result is a square-wave oscillation, the frequency of which is determined by the time constants of the components. The multivibrator described previously oscillates regularly between the two states [(Q_1 on, Q_2 off)

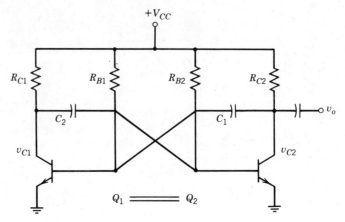

FIGURE 6.1-1 Collector-coupled astable multivibrator.

and (Q_2 on, Q_1 off)], and so is called an *astable multivibrator*. For Q_1 on and Q_2 off, C_1 is charged to $V_{CC} - V_{BE(sat)}$, positive on the right side. For Q_2 on and Q_1 off, C_2 is charged to $V_{CC} - V_{BE(sat)}$, positive on the left side.

Waveforms

The waveforms of the astable multivibrator are shown in Figure 6.1-2. It is seen that before time t_1, transistor Q_1 is on and its collector voltage is $V_{CE(sat)}$; also Q_2 is off and its collector voltage is V_{CC}; thus capacitor C_1 is charged to $V_{CC} - V_{BE(sat)}$. At t_1, v_{B2} rises above ground, causing Q_2 to switch on. Collector current I_{C2} now causes Q_2 collector voltage to fall to $V_{CE(sat)}$. Since C_1 will not discharge instantaneously, the base voltage of Q_1 becomes

$$V_{B1} = V_{C2} - \text{(charge voltage on } C_1)$$

$$= V_{CE(sat)} - [V_{CC} - V_{BE(sat)}] \simeq -V_{CC}.$$

Thus Q_1 is biased off with its emitter grounded and its base at about $-V_{CC}$. Consequently, as Q_1 switches off at t_1, C_2 charges through R_{C1}, causing the collector voltage of Q_1 to exponentially rise to V_{CC}.

During the interval between t_1 and t_2, Q_2 remains biased on at $v_{B2} = V_{BE(sat)}$ and v_{B1} rises from $-V_{CC}$ toward $+V_{CC}$, since C_1 discharges via resistance R_{B1}. When v_{B1} rises above ground, Q_1 begins to switch on. The falling value of collector voltage v_{C1} is coupled to Q_2 base via capacitor C_2, thus causing Q_2 to switch off. As Q_2 turns off, its collector voltage v_{C2} rises and C_1 is recharged via R_{C2} and Q_1 base. This raises a large current into Q_1 base, making it switch on very rapidly. Therefore v_{C1} falls very fast at switch-on. The switch-over process is reversed when C_2 discharges sufficiently to allow Q_2 base to rise above ground.

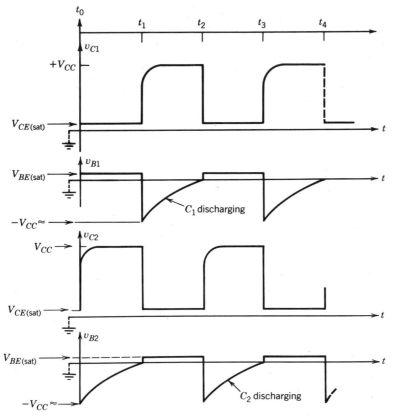

FIGURE 6.1-2 Waveforms of the collector-coupled astable multivibrator.

Pulse Width and Output Frequency

The output pulse is taken from either transistor. The pulse width T_W is equal to the time during which the transistor is off. This is the time taken by the capacitor C_1 or C_2 to discharge approximately from V_{CC} to 0 V. It may be derived from the capacitor-voltage exponential equation:

$$v_c = V - (V - V_C)e^{-t/RC}, \qquad (6.1\text{-}1)$$

where v_C = capacitor voltage at instant t,
  V = charging voltage,
  V_C = initial charged voltage across the capacitor,
  e = exponential constant = 2.71828,
  t = time from beginning of charge,
  R = charging resistance,
  C = capacitance being charged = $C_1 = C_2$.

$$e^{-t/RC} = \frac{V - v_c}{V - V_C};$$

$$e^{t/RC} = \frac{V - V_C}{V - v_c};$$

$$t = RC \ln\left(\frac{V - V_C}{V - v_c}\right). \qquad (6.1\text{-}2)$$

In Equation (6.1-2), $t = T_W$ = pulse width, $R = R_{B1} = R_{B2}$, V = supply voltage $= V_{CC}$, and V_C = initial capacitor voltage $= -V_{CC}$. (This is taken as negative since capacitor C_1 or C_2 would ultimately charge with reversed polarity to about $+V_{CC}$ if transistor switch-over did not occur.) The final capacitor charge voltage at switch-over is $v_c = 0$ V. Then Equation (6.1-2) becomes

$$T_W = RC \ln\left[\frac{V_{CC} - (-V_{CC})}{V_{CC} - 0}\right]$$

$$= RC \ln - \frac{2V_{CC}}{V_{CC}}$$

$$= RC \ln 2 \simeq 0.69 RC,$$

or

$$\text{Pulse width} = T_W \simeq 0.69\ RC. \qquad (6.1\text{-}3)$$

For $R = 180\,\text{k}\Omega$ and $C = 0.01\ \mu\text{F}$,

$$T_W = 0.69 \times 180\ \text{k}\Omega \times 0.01\ \mu\text{F} = 1.242\ \text{ms},$$

and the output frequency is

$$f = \frac{1}{T} = \frac{1}{2T_W} = \frac{1}{2 \times 1.242\ \text{ms}} = 402.6\ \text{Hz}.$$

Example 6.1-1.   Design a 2-kHz square-wave generator using the silicon transistors with $h_{FE(\text{min})} = 100$ and $I_C = 2$ mA. The available supply voltage is 10 V, and the load current is to be $I_C/100$ or 20 μA.
   Solution.

$$R_C = R_{C1} = R_{C2} \simeq \frac{V_{CC}}{I_C} = \frac{10\ \text{V}}{2\ \text{mA}} = 5\ \text{k}\Omega.$$

Use 4.7 kΩ standard value (Appendix 1).

$$I_{B(min)} = \frac{I_C}{h_{FE(min)}} = \frac{2\text{ mA}}{100} = 20\ \mu\text{A};$$

$$R = R_{B1} = R_{B2} = \frac{V_{CC} - V_{BE(sat)}}{I_B} = \frac{10\text{ V} - 0.8\text{ V}}{20\ \mu\text{A}} = 460\text{ k}\Omega.$$

Use 470 kΩ standard value.

$$\text{Pulse width} = T_W = \frac{1}{2f} = \frac{1}{2 \times 2\text{ kHz}} = 0.25\text{ ms.}$$

$$C = C_1 = C_2 = \frac{T_W}{0.69\text{ R}} = \frac{0.25\text{ ms}}{0.69 \times 470\text{ k}\Omega} = 0.0007709\ \mu\text{F.}$$

Use 750 pF standard value (Appendix 1).

Astable Multivibrator as Voltage-to-Frequency Converter

The period $T\ (=2T_W = 1/f)$ can be changed by connecting R_{B1} and R_{B2} to an auxiliary voltage V (the collector supply remains V_{CC}), as shown in Figure 6.1-3a. If V is varied, then T changes in accord with the equation

$$T = \frac{1}{f} = 2RC \ln\left(1 + \frac{V_{CC}}{V}\right), \tag{6.1-4}$$

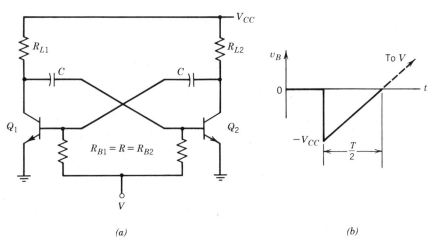

(a)   *(b)*

FIGURE 6.1-3 *(a)* Astable multivibrator with R_{B1} and R_{B2} connected to an auxiliary voltage V; *(b)* approximate waveform of v_B in the circuit of *(a)* if $V_{BE1} = V_{BE2} = 0$.

provided that V is large compared with the junction voltages. Thus this circuit becomes a voltage-to-frequency converter. In order to derive Equation (6.1-4), the forward-biased junction voltages are assumed to be zero. The approximate waveform of v_B is shown in Figure 6.1-3b. Applying Equation (2.2-1) to this waveform, we have

$$v_B = V - (V_{CC} + V)e^{-t/RC}.$$

At $t = T/2$, $v_B = 0$.

$$e^{T/2RC} = \frac{V + V_{CC}}{V} = 1 + \frac{V_{CC}}{V}.$$

Thus we obtain Equation (6.1-4).

A Complementary-Pair Multivibrator Circuit

The multivibrator in Figure 6.1-4 uses complementary-pair transistors so that both transistors are on only during the pulse interval itself. This conserves energy from the power supply. When the power supply is initially applied to this circuit, the capacitor C_1 is charged via R_1. Once C_1 is charged to a point where the base voltage of Q_1 reaches about -700 mV, this transistor is turned on. In this case, some of the Q_1 collector current passes to the base of Q_2 and turns it on, thus bringing the right side of C_1 to the negative polarity of V_{CC}. The charge on C_1 is discharged via Q_1, turning it off once sufficient charge has left C_1 to reduce its voltage below the turn-on voltage of Q_1. This, in turn, switches

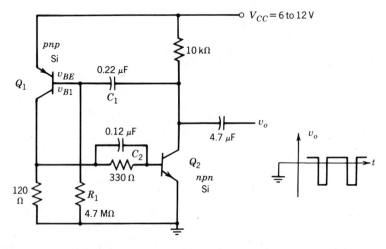

FIGURE 6.1-4 A complementary-pair astable multivibrator designed for minimal consumption of power.

off Q_2, and the process is repeated. The pulse repetition rate is determined by R_1 and C_1 as well as by V_{CC}.

6.2 EMITTER-COUPLED ASTABLE MULTIVIBRATOR AS HORIZONTAL OSCILLATOR IN A TELEVISION RECEIVER

Occasionally the frequency of an astable multivibrator has to be synchronized to some external frequency. For example, the emitter-coupled astable multivibrator in Figure 6.2-1 is employed as a horizontal oscillator that should be controlled by the sync pulse for a television receiver. As the supply voltage is

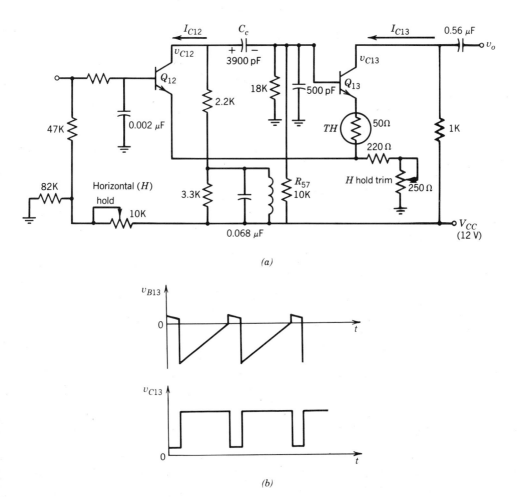

FIGURE 6.2-1 (a) Emitter-coupled astable multivibrator used as a horizontal oscillator in a television receiver; (b) the Q_{13} base and collector voltage waveforms in approximate plots.

switched on, the collector current of Q_{12} rises with the horizontal hold potentiometer properly adjusted. The reduced value of v_{C12} coupled to the base of Q_{13}
causes its emitter current to fall, thus raising the v_{BE} of Q_{12}. This regeneration
further raises I_{C12} and reduces I_{C13} until Q_{12} is in saturation and Q_{13} is at
cutoff. The condition of Q_{13} is maintained by the coupling capacitor C_c (which
was previously charged with its negative potential on the right side) until C_c is
discharged. When the Q_{13} base is slightly positive, Q_{13} starts conducing, causing
I_{C12} to reduce. The regenerative process is reversed until Q_{12} is cut off and Q_{13}
is saturated.

The approximate waveforms of Q_{13} are shown in Figure 6.2-1b. When Q_{12}
is cut off and Q_{13} is saturated, capacitor C_c is rapidly charged through Q_{13}
with the negative polarity of C_c on the right side. It is charged until Q_{13} is cut
off and Q_{12} is saturated. C_c then slowly discharges through R_{57}. Since Q_{13} is
off, the reduced emitter voltage (V_E) forces Q_{12} to saturate. Note that Q_{13} is
saturated for a short time and cut off for a long time.

The frequency of the emitter-coupled multivibrator can be varied by controlling its input base voltage. The method applies when the oscillator is controlled
by an *automatic frequency control* (AFC) circuit for horizontal synchronization.
A decrease of forward voltage at the input base raises the oscillator frequency.
A lower value of $+V_{BE12}$ reduces the collector current I_{C12} when Q_{12} conducts.
The result is a smaller drop in collector voltage V_{C12} and less negative drive at
the base of Q_{13}. Then less time is needed for C_c to discharge down to cutin
(V_γ) for conduction in Q_{13}. With a shorter cutoff time for Q_{15}, the multivibrator
frequency is increased.

6.3   ASTABLE MULTIVIBRATOR (CLOCK) BUILT WITH TTL ICs

The astable multivibrator can be built with the IC SN7400 NAND gates, as
shown in Figure 6.3-1a, or with IC SN7402 NOR gates, as shown in Figure
6.3-1b. The resistors R and capacitors C are externally connected. The frequency

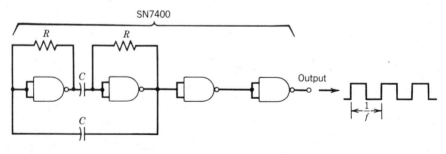

(a)

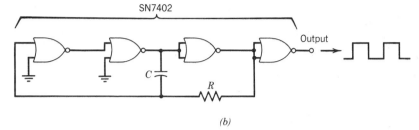

(b)

FIGURE 6.3-1 Clock circuit built with TTL IC units: (a) NAND-gate clock; (b) NOR-gate clock. The SN7400 consists of quadruple 2-input positive NAND gates, and the SN7402 consists of quadruple 2-input positive NOR gates.

of the NAND-gate clock is

$$f \simeq \frac{0.7}{RC},$$ (6.3-1)

and the frequency of the NOR-gate clock is

$$f \simeq \frac{0.455}{RC}.$$ (6.3-2)

6.4 TRANSISTOR MONOSTABLE (ONE-SHOT) MULTIVIBRATOR CIRCUITS

Collector-Coupled Monostable Multivibrator

The one-shot or monostable multivibrator is used to generate a rectangular waveform (pulse) having a width determined essentially by the values of a few circuit components. It has only one permanently stable state and one quasi-stable (unstable) state. One transistor is normally on and the other transistor is normally off. The condition can be reversed by application of a triggering pulse, which turns on the normally off transistor and switches off the normally on transistor. The circuit under the reversed condition is in the quasi-stable state, where it remains for a predetermined duration (T_W) before returning to its original stable state.

The collector-coupled monostable multivibrator is shown in Figure 6.4-1. In the normal dc conditions of the circuit, base current I_{B2} is provided from V_{CC} to Q_2 through resistance R. Therefore Q_2 is normally on (in saturation). With the Q_2 collector near ground potential, V_{B1} is likely to be negative. Thus Q_1 is normally off with Q_2 normally on. Under this condition, the capacitor voltage $V_C = V_{CC} - V_{B2} = V_{CC} - 0.8$, positive on the left side.

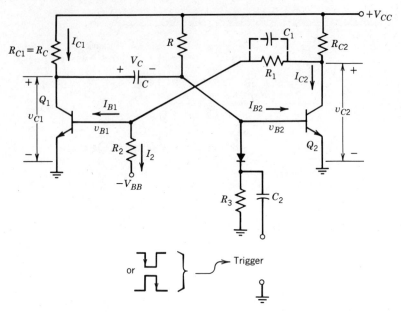

FIGURE 6.4-1   A collector-coupled monostable multivibrator. A diode can be connected from Q_2 emitter to ground in forward direction to protect the base-emitter voltage of Q_2 against excessive bias.

The waveforms of the one-shot multivibrator are shown in Figure 6.4-2. A negative trigger applied to the base of Q_2 at time $t = 0$ causes the transistor to turn off. The voltage (v_{C2}) at the collector of Q_2 increases, forcing Q_1 to turn on. As Q_1 turns on and saturates, its collector drops from V_{CC} to $V_{CE(sat)}$. Since the voltage across a capacitor (C) cannot change instantaneously, the voltage (v_{B2}) at the base of Q_2 experiences a drop of the same magnitude. Thus at $t = 0+$, the negative value of v_{B2} is given by

$$v_{B2} = V_{CE(sat)} - V_C = V_{CE(sat)} - (V_{CC} - 0.8)$$
$$= 0.8 - [V_{CC} - V_{CE(sat)}]$$

or

$$v_{B2} = 0.8 - I_{C1}R_{C1} \quad \text{(for Si)}, \tag{6.4-1}$$

and Q_2 is kept off with Q_1 on. Before Q_1 is triggered on, $v_{B1} = -V_{BE}, v_{C1} = V_{CC}$, $v_{B2} = 0.8$ V, and $V_{C2} = V_{CE(sat)} = 0.2$ V. When Q_1 is triggered on with Q_2 off, $v_{B1} = 0.8$ V, $v_{C1} = V_{CE(sat)} = 0.2$ V, $v_{B2} = v_{C1} - V_C \simeq -V_C = -(V_{CC} - 0.8)$, and $v_{C2} \simeq V_{CC}$.

While Q_2 is kept off, v_{B1}, v_{C1}, and v_{C2} remain constant, but v_{B2} is varied since C discharges through R. Voltage V_C across capacitor C is initially positive on its

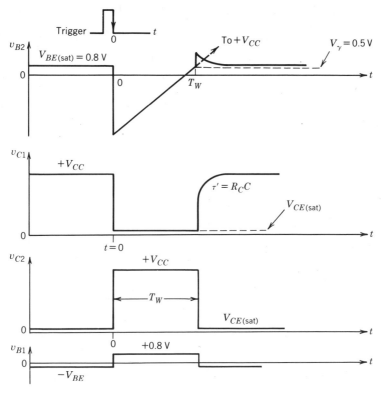

FIGURE 6.4-2   Waveforms of the one-shot circuit in Figure 6.4-1.

left side and negative on its right side. The current flowing into the right side of C will tend to discharge C and then recharge it with reversed polarity. Hence v_{B2} begins to rise toward ground level. When C is discharged to $v_C \simeq 0$ V, the base-emitter junction of Q_2 starts to be forward biased again. At this point I_{C2} again starts to flow and v_{C2} begins to fall. v_{C2} falls, causing v_{B1} to fall, and so v_{C1} rises, causing v_{B2} to rise. Therefore, as $v_C \simeq 0$ V, Q_1 rapidly switches off and Q_2 quickly comes on again. At this time, speed-up capacitor C_1 transmits all the Q_2 collector voltage change to the Q_1 base and then discharges, resulting in a negative spike at Q_1 base. As Q_1 switches off and Q_2 turns on again, C is rapidly recharged to V_C via R_{C1} and Q_2 base. The circuit has now returned to its original stable state. This stable condition is maintained until another pulse triggers the circuit. Since the width T_W of the output pulse is equal to the time during which Q_2 is off, it is derived in the same manner as for the astable circuit in Figure 6.1-1. Thus the pulse width is

$$T_W = 0.69\,RC. \qquad (6.1\text{-}3)$$

The value of R is chosen to be less than $h_{FE(\min)}R_{C2}$ in order to make Q_2 saturate in its stable state. The values of R_1 and R_2 are chosen so that Q_1 is off in the stable state and is saturated in the quasi-stable state. These requirements are met by applying the rule of thumb that $I_2 \simeq I_{C2}/10$. Since $I_{C1} \simeq I_{C2}$, I_{B1} becomes $I_{C2}/h_{FE(min)}$. Then I_2 is $[h_{FE(min)}/10]I_{B1}$. Making $I_2 \simeq I_{C2}/10$ will result in R_1 and R_2 each being 5 to 10 times R_{C2}. In order to avoid switching delays, the value of $R_C (= R_{C1} = R_{C2})$ should be much less than R.

Consider the effect of reverse saturation current I_{CBO} on the pulse width T_W. During the interval when Q_2 is off, a nominally constant current I_{CBO} flows out of the base of Q_2. Assume now that capacitor C were disconnected from the junction of resistor R with the base of Q_2. Then the voltage at the base of Q_2 with C disconnected would be, not V_{CC}, but $V_{CC} + I_{CBO}R$. It therefore appears that capacitor C, in effect, charges through R from a source $V_{CC} + I_{CBO}R$. Hence, since I_{CBO} increases with temperature, the time T_W will decrease. The initial voltage V_{C1} in the stable state is $V_{CC} - I_{CBO}R_C$, where I_{CBO} is now the collector current of Q_1 when it is off. When the multivibrator is triggered, $v_{C1} = V_{CC} - (I_{C1} + I_{CBO})R_C$; if $I_{CBO} = 0$, $v_{C1} = V_{CC} - I_{C1}R_C$. Thus the drop in v_{C1} when the circuit is triggered is smaller by $I_{CBO}R_C$ than it would otherwise be. This is a second way in which I_{CBO} affects T_W. However, since $R_C \ll R$, this second effect is negligible in comparison with the first. To take account of the effect of I_{CBO}, let us assume that $V_{CE(sat)} = V_{BE(sat)} = V_\gamma = 0$. During the interval T_W, Q_2 is off and the change in v_{B2} can be calculated from the simplified equivalent circuit shown in Figure 6.4-3. At $t = 0$, v_{B2} is zero and plunges to $-V_{CC}$. Using the superposition method, we find

$$v_{B2} \simeq (V_{CC} + I_{CBO}R)(1 - e^{-t/\tau}) - V_{cc}e^{-t/\tau},$$

where $\tau = (R + R_o)C \simeq RC$. At $t = T_W$, v_{B2} is again zero. Then

$$0 = (V_{CC} + RI_{CBO}) - (2V_{CC} + RI_{CBO})e^{-T_{W/s}},$$

from which we obtain

$$T_W \simeq \tau \ln 2 - \tau \ln \frac{1 + I_{CBO}R/V_{CC}}{1 + I_{CBO}R/2V_{CC}} = \tau \ln 2 - \tau \ln \frac{1 + \phi}{1 + \phi/2}, \quad (6.4\text{-}2)$$

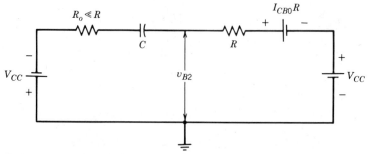

FIGURE 6.4-3 Simplified equivalent circuit used for calculating the pulse width T_W affected by I_{CBO}.

where $\phi = I_{CBO}R/V_{CC}$. From Equation (6.4-2) we see that the delay time T_W decreases as the temperature increases.

One method of temperature-compensating a monostable multivibrator is to connect R, not to $+V_{CC}$, but to a voltage source V whose value decreases as the temperature increases. The voltage V may be obtained from a voltage divider r_1-r_2 across V_{CC} in which the value of r_1 increases with the temperature as it is connected to the $+V_{CC}$ terminal. In this case, V is provided from the tap between r_1 and r_2.

Example 6.4-1. Design a collector-coupled one-shot multivibrator using silicon transistors with $I_C = 2$ mA and $h_{FE(min)} = 50$. The available supply is ± 10 V. The output pulse width is to be 200 μs.

Solution. The circuit is as shown in Figure 6.4-1. For Q_2 saturated,

$$R_{C2} = \frac{V_{CC} - V_{CE(sat)}}{I_C} = \frac{10 \text{ V} - 0.2 \text{ V}}{2 \text{ mA}} = 4.9 \text{ k}\Omega.$$

Use 4.7 kΩ standard value.

$$I_{B2(min)} = \frac{I_C}{h_{FE(min)}} = \frac{2 \text{ mA}}{50} = 40 \text{ }\mu\text{A}.$$

$$R = \frac{V_{CC} - V_{BE(sat)}}{I_{B2}} = \frac{10 \text{ V} - 0.8 \text{ V}}{40 \text{ }\mu\text{A}} = 230 \text{ k}\Omega.$$

Use 220 kΩ standard value ($< h_{FE(min)}R_{C2} = 235$ kΩ). For Q_1 saturated (on), $R_{C1} = R_{C2} = 4.7$ kΩ. $I_{B1} = 40$ μA. Let $I_2 \simeq I_C/10 = 2$ mA/10 = 200 μA.

$$V_{R2} = V_{B1} - V_{BB} = 0.8 \text{ V} - (-10 \text{ V}) = 10.8 \text{ V}.$$

$$R_2 = \frac{V_{R2}}{I_2} = \frac{10.8 \text{ V}}{200 \text{ }\mu\text{A}} = 54 \text{ k}\Omega.$$

Use 56 kΩ standard value.

$$I_{B1} + I_2 \simeq 40 \text{ }\mu\text{A} + 200 \text{ }\mu\text{A} = 240 \text{ }\mu\text{A}.$$

$$R_{C2} + R_1 = \frac{V_{CC} - V_{B1}}{I_{B1} + I_2} = \frac{10 \text{ V} - 0.8 \text{ V}}{240 \text{ }\mu\text{A}} = 38.3 \text{ k}\Omega.$$

$R_1 = 38.3$ k$\Omega - 4.7$ k$\Omega = 33.6$ kΩ; use 33 kΩ standard value. When Q_2 is on, $V_{B1} = V_{C2} - V_{R1} = 0.2$ V $- V_{R1}$.

$$V_{R1} = \frac{R_1}{R_1 + R_2}(V_{C2} - V_{BB}) = \frac{33 \text{ K}}{33 \text{ K} + 56 \text{ K}}[0.2 \text{ V} - (-10 \text{ V})] = 3.78 \text{ V}.$$

$$V_{B1} = 0.2 - 3.78 = -3.58 \text{ V} < V_{BE(off)} = 0 \text{ V}.$$

Hence Q_1 is indeed biased off while Q_2 is on. Also -3.58 V is less than the typical limit of -5 V for a reverse-biased base-emitter junction.

For $T_W = 200 \ \mu s$ and $R = 220 \ k\Omega$,

$$C = \frac{T_W}{0.69 \ R} = \frac{200 \ \mu s}{0.69 \times 220 \ k\Omega} = 1300 \text{ pF}.$$

Emitter-Coupled Monostable Multivibrator

The emitter-coupled monostable multivibrator is shown in Figure 6.4-4. In the stable state Q_1 is cut off and Q_2 is in saturation. The bias voltage V_{B1} is obtained from the divider R_1-R_2 across the supply voltage V_{CC}. The feedback is provided through a common emitter resistor R_E. The signal at the collector of Q_2 is not directly involved in the regenerative loop. Therefore this collector makes an ideal point from which to obtain an output voltage waveform. The base of Q_1 is a good point at which to inject a triggering signal, since this electrode is coupled to no other in the circuit. Thus the trigger source cannot load the circuit. When Q_2 goes OFF, the emitter resistance will serve to stabilize the collector current I_{C1}. Hence the time T_W of the quasi-stable state may be controlled through I_{C1}. The current I_{C1} can be adjusted through the bias voltage V_{B1}, and it turns out that T_W varies rather linearly with V_{B1}. The circuit is designed so that the maximum level of V_{B1} is less than the normal level of V_{B2}. When Q_1 is triggered on, the voltage drop at Q_1 collector and the charge on C cause V_{B2} to be pushed below V_{B1}. Q_2 remains off until C discharge allows V_{B2} to rise above

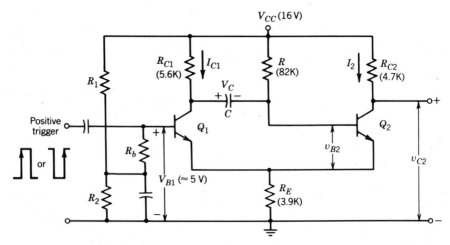

FIGURE 6.4-4 An emitter-coupled monostable multivibrator.

V_{B1} again. The time for this to occur depends upon the actual voltage level of V_{B1}. This time is also the output pulse width. The emitter-coupled monostable circuit makes an excellent gate waveform generator, whose width T_W is easily and linearly controllable by means of an electrical signal voltage.

Monostable Multivibrator as Micropower Switch

A micropower switch controlled by a monostable multivibrator is useful when it is desired to turn the implanted circuit on for a fixed period of time, following which it will turn itself off without the need for a trigger pulse. The circuit for this switch is shown in Figure 6.4-5. It can be divided into three basic sections: the switching circuit (Q_3-Q_4), the multivibrator control circuit (Q_1-Q_2), and the trigger circuit (Q_5-Q_7). At least one transistor must be on when the switch is off. In this case, that transistor is Q_1, and the overall current of the circuit when the switch is off is kept low by having a large collector resistor R_1. When $R_1 = 6.8$ MΩ, an off current of only 0.27 μA is obtained at a power supply voltage of 1.35 V. Light triggering may be achieved by using a short pulse of light, such as one that could be produced by an electronic photoflash light. It is not necessary to have a turn-off triggering pulse. While the switch is off, Q_1 is on, and thus the off current is given by

$$I_{\text{OFF}} = \sum_{i=1}^{4} I_{CE0i} + \frac{V_{CC}}{R_1} + \frac{V_{CC} - V_{BE}}{R_T}. \tag{6.4-3}$$

It is possible to obtain off currents as low as 0.27 μA with this circuit.

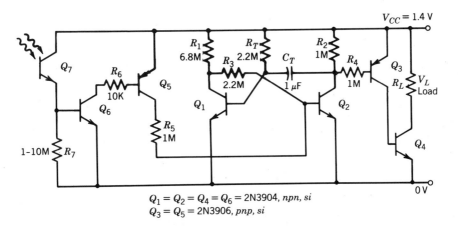

$Q_1 = Q_2 = Q_4 = Q_6 = $ 2N3904, *npn, si*
$Q_3 = Q_5 = $ 2N3906, *pnp, si*

FIGURE 6.4-5 Circuit diagram for a monostable micropower switch. (From Santic, A., S. Vamvakas, and M. R. Neuman, Micropower Electronic Switches for Implanted Instrumentation, *IEEE Transactions on Biomedical Engineering*, Vol. BME-29, No. 8, August 1982, pp. 583–589. Copyright © 1982 IEEE.)

It is desired to minimize the on current I_{ON} to the monostable circuit so as to maintain maximum overall efficiency of the switch. We do not want I_{CE0} to result in a significant voltage drop across the collector resistance R_C, and so we select 10% of power supply voltage as a maximum value for the voltage drop across R_C, or $I_{CE0}R_C \leqslant 0.1\,V_{CC}$. Thus

$$R_1 = R_C \leqslant \frac{0.1\,V_{CC}}{I_{CE0}}. \tag{6.4-4}$$

In the monostable switch circuit, Q_1 must have a maximum cutoff current (I_{CE0}) of 20 nA; hence $R_1 \simeq 0.135\,\text{V}/20\,\text{nA} = 6.8\,\text{M}\Omega$. Resistor R_3 can be found from the following equation:

$$R_1 + R_3 \leqslant \frac{V_{CC} - V_{BE}}{I_{B2}}, \tag{6.4-5}$$

where I_{B2} is the base current of Q_2. This current, in turn, can be determined from the β of Q_2 and its collector current.

Transistor Q_5 amplifies trigger pulses from the photodetector. This transistor operates as an unbiased switch and should have high gain so that the photodetector output will drive the transistor Q_2 from cutoff to near saturation conditions. The input pulses to the amplifier Q_5 must be greater than $V_{BE(\text{sat})}$ to accomplish this. There is a wide margin separating trigger signal pulses from other random noise pulses. The phototransistor Q_7 used to drive amplifier Q_6 can be replaced by two silicon photovoltaic cells connected in series that directly drive Q_5. The series connection ensures a high enough voltage to turn on Q_5. It was possible to trigger the switch at a distance of 250 cm with a small electronic photoflash light having a 0.5 ms, 2 W-S flash duration. A 0.6-W incandescent lamp manually switched on and off could also trigger the phototransistor circuit at a distance of 250 cm.

6.5   MONOSTABLE MULTIVIBRATOR INTEGRATED CIRCUIT

Figure 6.5-1 shows a popular TTL monostable multivibrator—the 74121. Operating from a 5-V supply, the unit provides two complementary outputs (Q and $\bar{Q}$), and has three input terminals. Two of the inputs (A_1 and A_2) typically require -1.4 V to effect triggering. The other input (B) triggers the circuit when a typical level of $+1.55$ V is applied. The output pulse width is $T_W \simeq 0.7\,R_T C_T$, where R_T is the timing resistor and C_T is the timing capacitor. The Q output goes from its normally low state to the high output state. Triggering can occur when the A inputs are both grounded (or go low) and the B input goes positive (or is lifted high).

Assume that the B input is unconnected (lifted high) and that a 100-kHz, 5-V clock is applied to the A inputs that are connected together. Then the resulting

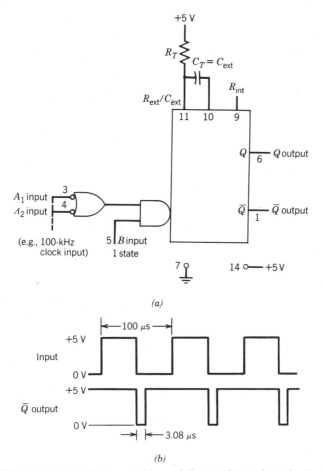

FIGURE 6.5-1   (a) 74121 monostable multivibrator; (b) waveforms for illustration.

$\bar{Q}$ output changes from its normally high voltage level each time the input clock goes negative (low), as shown in Figure 6.5-1b. If $R_T = 2.2$ kΩ and $C_T = 0.002$ μF, the output pulse width is $T_W = 0.7\,R_T C_T = 3.08$ μs. Pulse widths of up to 28 sec are possible. The output pulse width is around 30 to 35 ns when no external components are employed and R_{int} (terminal 9) is connected directly to V_{CC}.

6.6   IC OP-AMP ASTABLE MULTIVIBRATOR

Symmetrical Square-Wave Generator

Integrating the output of the Schmitt comparator of Figure 5.7-3 by means of an RC low-pass network and applying the capacitor voltage to the inverting terminal in place of the external signal result in an astable multivibrator or square-wave generator, as shown in Figure 6.6-1a. The operating frequency can be changed by varying feedback resistor R_f. Frequency stability is held very high by the zener diodes used and capacitor C. The magnitude and symmetry of the waveform depend on the matching of the two zener diodes.

Assume that capacitor C has been initially charged by voltage V_C and is now exponentially charged toward V_o $(= V_Z + V_D)$. Then the voltage across C is given by

$$v_C(t) = V_C + (-V_C + V_o)(1 - e^{-t/RC}). \tag{6.6-1}$$

Consider the circuit of Figure 6.6-1a. When the positive output voltage holds itself across the series resistors R_1 and R_2, it will charge capacitor C through R_f.

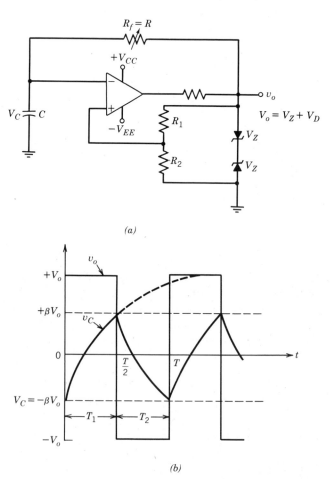

(a)

(b)

FIGURE 6.6-1   (*a*) Op-amp astable multivibrator and (*b*) its waveforms.

When the charging causes the voltage across C to be higher than the voltage across R_2 (or $v_C > V_{R2}$), the output voltage v_o is immediately converted from positive to negative peak so that capacitor C discharges. When v_C is less than V_{R2}, v_o will be immediately converted to positive. Therefore oscillations will be sustained in such a manner.

In the multivibrator circuit, the feedback factor is $\beta = R_2/(R_1 + R_2)$. From the waveforms shown in Figure 6.6-1b, we see that the initially charged voltage is

$$V_C = -\beta V_o = -\beta(V_Z + V_D),$$

where V_o is the square-wave peak voltage. Thus, referring to Equation (6.6-1), the voltage across capacitor C can also be expressed as

$$v_C(t) = -\beta V_o + (\beta V_o + V_o)(1 - e^{-t/RC}). \tag{6.6-2}$$

The waveforms in Figure 6.6-1b indicate that charging capacitor C up to $+\beta V_o$ requires the time $T/2$; as the capacitor voltage reaches $+\beta V_o$, the output state inverts immediately. Thus

$$v_C\left(\frac{T}{2}\right) = +\beta V_o = -\beta V_o + (\beta V_o + V_o)(1 - e^{-T/2RC}), \tag{6.6-3}$$

or

$$v_C\left(\frac{T}{2}\right) = V_o - V_o(1 + \beta)e^{-T/2RC} = \beta V_o.$$

Solving for T:

$$(1 + \beta)e^{-T/2RC} = 1 - \beta; \frac{1 + \beta}{1 - \beta} = e^{T/2RC};$$

hence

$$T = 2R_f C \ln \frac{1 + \beta}{1 - \beta}. \tag{6.6-4}$$

If $\beta = 0.475$ is chosen, then $\ln(1.475/0.525) = \ln 2.81 = 1$; hence the period is

$$T = 2R_f C, \tag{6.6-5}$$

and the oscillation frequency is

$$f = \frac{1}{2R_f C}. \tag{6.6-6}$$

The frequency range of 10 Hz to 10 kHz is usually chosen for proper operation of the symmetrical square-wave generator.

If an asymmetrical square wave is desired, the feedback resistor is replaced by the two diode-resistor combinations, as in the circuit of Figure 6.6-2. During the interval when the output is positive, diode D_1 conducts but D_2 is off. Therefore the circuit reduces to that in Figure 6.6-1a except that output V_o is reduced by the diode drop. Because the period is independent of V_o, T_1 is given by $T/2$ in Equation (6.6-4). During the interval when the output is negative, D_1 is off and D_2 conducts. Therefore the discharge-time constant is now $R_f'C$, and, consequently, T_2 is given by $T/2$ in Equation (6.6-4) with R_f replaced by R_f'. If $R_f' = 2R_f$, then $T_2 = 2T_1$.

Example 6.6-1. Using a 741 IC op-amp, design a symmetrical square-wave generator to produce an output frequency of 400 Hz and an output amplitude of ± 14 V (see Figure 6.6-3). Assume the feedback factor $\beta = 50\%$.

Solution. $V_o = \pm V_{sat} = \pm 14$ V.
$V_{sat} \simeq V_{CC} - 1; V_{CC} \simeq V_{sat} + 1 = +15$ V.
$-V_{sat} \simeq -V_{EE} + 1; -V_{EE} \simeq -V_{sat} - 1 = -15$ V.

$$V_{R2} = \pm \beta V_{sat} = \frac{\pm V_{sat}}{2}$$

$$= \pm \frac{14 \text{ V}}{2} = \pm 7 \text{ V}.$$

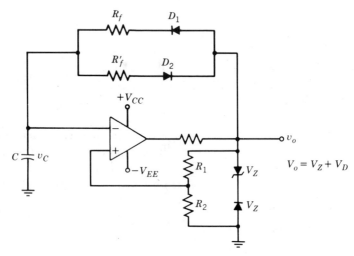

FIGURE 6.6-2 The astable op-amp multivibrator producing asymmetrical square waves.

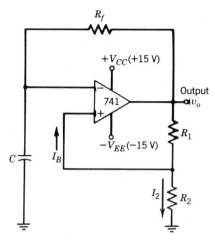

FIGURE 6.6-3   Example 6.6-1.

Make $I_2 \gg I_B$; $I_{B(\text{max})} = 500$ nA (see Section 5.7).
Let $I_2 = 100 \times I_{B(\text{max})} = 100 \times 500$ nA $= 50$ μA.

$$R_2 = \frac{V_{R2}}{I_2} = \frac{7 \text{ V}}{50 \ \mu\text{A}} = 140 \text{ k}\Omega.$$

Use 150 kΩ standard value.

$$R_1 = R_2 = 150 \text{ k}\Omega.$$

Let $I_{R_{f(\text{min})}} = 100 \times I_{B(\text{max})} = 50$ μA.

$$V_{R_{f(\text{min})}} = V_{\text{sat}} - \beta V_{\text{sat}} = 14 \text{ V} - 7 \text{ V} = 7 \text{ V}.$$

$$R_f = \frac{V_{R_{f(\text{min})}}}{I_{R_{f(\text{min})}}} = \frac{7 \text{ V}}{50 \ \mu\text{A}} = 140 \text{ k}\Omega.$$

Use 150 kΩ standard value.

$$T = \frac{1}{400 \text{ Hz}} = 2.5 \text{ ms.}$$

Substituting this into Equation (6.6-4), we get

$$2.5 \text{ ms} = 2R_f C \ln \frac{1 + 0.5}{1 - 0.5} = 2 \times 150 \text{ k}\Omega \times C \times 1.1.$$

Hence

$$C = \frac{2.5 \text{ ms}}{2 \times 150 \text{ k}\Omega \times 1.1} = 0.007576 \ \mu\text{F}.$$

Use 7500 pF standard value.

6.7 IC OP-AMP MONOSTABLE MULTIVIBRATORS

The op-amp astable multivibrator of Figure 6.6-1 is modified in Figure 6.7-1 to operate as a monostable multivibrator by adding a diode D_1 clamp across capacitor C and by introducing a negative trigger pulse through diode D_2 to the noninverting terminal. The circuit is in its stable state with the output at $v_o = +V_o$ and with the capacitor clamped at diode D_1 on voltage $V_D \simeq 0.7$ V. The feedback voltage is

$$\beta v_o = \frac{R_2 v_o}{R_1 + R_2}, \tag{6.7-1}$$

and $\beta V_o > V_D$. If the magnitude of the trigger amplitude is greater than $\beta V_o - V_D$, it will cause the comparator to switch to an output $v_o = -V_o$. As indicated in Figure 6.6-1b, capacitor C will now charge exponentially with a time constant $\tau = RC$ through R toward $-V_o$ because D_1 becomes reverse biased. When the capacitor voltage v_C becomes more negative than $-\beta v_o$, the comparator output swings back to $+V_o$. The capacitor now begins charging toward $+V_o$ through R until v_C reaches V_D and C becomes clamped again at $v_C = V_D$. Following Equation (2.2-1), the capacitor voltage for $t > 0$ is expressed as

$$v_C = -V_o + (V_D + V_o)e^{-t/\tau}. \tag{6.7-2}$$

When $v_C = -\beta V_o$, $t = T_W$, then Equation (6.7-2) becomes

$$T_W = RC \ln \frac{1 + V_D/V_o}{1 - \beta}. \tag{6.7-3}$$

If $V_o \gg V_D$ and $R_1 = R_2$ so that $\beta = \frac{1}{2}$, then $T_W = 0.69$ RC.

The duration T_p of the trigger pulse v_t should be much less than the delay time or width T_W of the generated pulse. The diode D_2 serves to avoid malfunctioning if any positive noise spikes are present in the triggering line. The capacitor voltage v_C does not reach its quiescent value $v_C = V_D$ until time $T'_W > T_W$. Consequently there is a recovery time $T'_W - T_W$ during which the circuit may not be triggered again.

Another monostable circuit using the IC op-amp is shown in Figure 6.7-2. Since the noninverting terminal has a positive input, the output is saturated

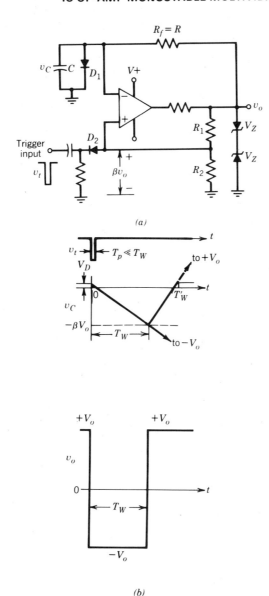

FIGURE 6.7-1 IC op-amp monostable multivibrator or pulse generator: (a) circuit diagram; (b) waveforms (including triggering pulse v_t, capacitor waveform v_C, and output pulse v_o).

near the V_{CC} level. The capacitor C is normally charged positive on its right side and negative on its left side. When a large enough positive-going spike is coupled to the inverting terminal through C_1, the inverting terminal voltage is lifted above the level of the noninverting terminal. Then the output quickly switches to approximately $-V_{sat}$ or $-(V_{EE} - 1)$. This pushes the noninverting

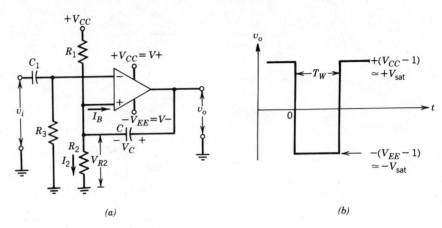

(a)   (b)

FIGURE 6.7-2 (*a*) Another op-amp monostable circuit and (*b*) its output waveform.

terminal down to $-(V_{sat} + V_C)$, thus ensuring that the output remains negative until C discharges. As soon as the output goes negative, C starts to discharge through R_1 and R_2. Ultimately C will charge positive on its left side and negative on its right side. When the voltage on the left side of C rises above the voltage level at the inverting terminal, the noninverting terminal again has a positive input. Now v_o rapidly returns to approximately $(V_{CC} - 1)$, and the circuit has returned to its original condition. Therefore a negative output pulse is generated when the circuit is triggered. The output pulse width is given by

$$T_W = RC \ln \left(\frac{V - V_C}{V - v_C}\right), \qquad (6.1\text{-}2)$$

where $R = R_1 \| R_2$,
V = charging voltage = V_{CC},
V_C = initial charge voltage across C,
v_C = capacitor voltage at instant t.

Example 6.7-1.   A one-shot multivibrator uses a 741 op-amp with a supply voltage of ± 12 V and a 1.4-V triggering spike. The output pulse width is to be 250 μs. Design a suitable circuit.

Solution.   Design the circuit referring to Figure 6.7-2. To use a 1.4-V input spike, let $V_{R2} = 1$ V. The 741 op-amp has a maximum input bias current $I_{B(max)} = 500$ nA (see Section 5.7). Let $I_2 = 100\, I_{B(max)} = 100 \times 500$ nA $= 50$ μA.

$$R_2 = \frac{V_{R2}}{I_2} = \frac{1\text{ V}}{50\ \mu\text{A}} = 20\text{ k}\Omega.$$

Use 18 kΩ standard value.

$$I_2 = \frac{1 \text{ V}}{18 \text{ k}\Omega} = 56 \ \mu\text{A}.$$

$$V_{R1} = V_{CC} - V_{R2} = 12 - 1 = 11 \text{ V}.$$

$$R_1 = \frac{V_{R1}}{I_2} = \frac{11 \text{ V}}{56 \ \mu\text{A}} = 196 \text{ k}\Omega.$$

Use 180 kΩ standard value.

$$R_3 = R_1 \| R_2 = 18\text{K} \| 180\text{K} \simeq 16.36 \text{ k}\Omega.$$

Use 15 kΩ standard value. When V_o is positive, the initial charged voltage across C is

$$V_C = V_{R2} - V_o \simeq V_{R2} - (V_{CC} - 1) \simeq 1 - 12 + 1 = -10 \text{ V}.$$

When V_o is negative, the final charge voltage across C at switch-over is

$$v_C \simeq -(V_{EE} - 1) = 12 - 1 = 11 \text{ V}.$$

Charging voltage $= V = V_{R2} - (-V_o) = 1 + 11 = 12 \text{ V}.$
Charging resistance $= R = R_1 \| R_2 = 18\text{K} \| 180\text{K} \simeq 16.36 \text{ k}\Omega.$

$$C = \frac{T_W}{R \ln\left(\dfrac{V - V_C}{V - v_C}\right)} = \frac{250 \ \mu\text{s}}{16.36 \text{ k}\Omega \ \ln\left[\dfrac{12 \text{ V} - (-10 \text{ V})}{12 \text{ V} - 11 \text{ V}}\right]} = \frac{250 \ \mu\text{s}}{16.36 \text{ k}\Omega \times 3.09}$$

$$= 0.004945 \ \mu\text{F}.$$

Use 5000 pF standard value.

6.8 555 IC TIMER

The 555 monolithic IC timer is a very versatile unit. It can be used as a monostable multivibrator, an astable multivibrator, a ramp generator, a sequential timer, and so on. Although the 555 timer is considered an analog circuit, it can also function as an astable multivibrator to generate clock pulses, or as a monostable (one-shot) multivibrator to provide accurate time delays. Moreover it is capable of supplying up to 200 mA of load current and can be operated over a wide voltage range, from 4.5 V to 16 V. A block diagram of this integrated circuit is shown in Figure 6.8-1. The 555 characteristics are listed in Table 6.8-1.

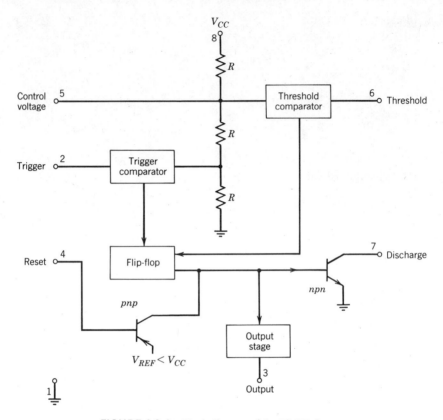

FIGURE 6.8-1 Block diagram of the IC 555 timer.

To use the 555 timer as an astable or a one-shot, we must understand the operation of the basic 555 circuit.

The trigger comparator compares a voltage of $V_{CC}/3$ to the voltage at the trigger input. When the trigger voltage drops to a value of $V_{CC}/3$ or less, the comparator produces an output that sets the flip-flop to the low output state. Then the flip-flop cuts the discharge transistor off and drives the output positive through the output inverting buffer.

The threshold comparator continually compares a voltage of $2V_{CC}/3$ to the voltage at the threshold input. When the threshold voltage equals or exceeds $2V_{CC}/3$, the comparator produces a voltage that sets the flip-flop output to the high state. The flip-flop then drives the base of the discharge transistor so that the transistor can saturate when the collector circuit is completed. The output pin is now low. A low voltage on the reset input can also drive the flip-flop to this state.

A 555 timer that has been wired to act as a timed switch is shown in Figure 6.8-2. The voltages as a function of time at trigger pin 2, threshold pin 6, and

TABLE 6.8-1
Major characteristics of 555 Timer[a]

Parameter	Conditions	Minimum	Typical	Maximum
Supply voltage, V_{CC}, V		4.5		16
Supply current, I_S, mA	$V_{CC} = 5.0$ V, $R_L = \infty$		3.0	6.0
	$V_{CC} = 15$ V, $R_L = \infty$		10	15
	Low state (I_S is typically 1.0 mA less when V_{out} is high)			
Timing error:	$R_A, R_B = 1$ k to 100 k;			
Initial accuracy, %	R_A for one shot;		1.0	
Drift with	R_A and R_B for astable.		50	
temperature, ppm/°C				
Drift with V_{CC}, %/V	$C = 0.1$ μF (tested at $V_{CC} = 5$ V and $V_{CC} = 15$ V)		0.1	
Threshold voltage, $X V_{CC}$			$\frac{2}{3}$	
Trigger voltage, V	$V_{CC} = 15$ V		5.0	
	$V_{CC} = 5.0$ V		1.67	
Trigger current, μA			0.5	
Reset voltage, V		0.4	0.7	1.0
Reset current, mA			0.1	
Threshold current, μA	At $V_{CC} = 15$ V,		0.1	0.25
	$(R_A + R_B)_{max} = 20$ MΩ			
Control voltage level, V	$V_{CC} = 15$ V	9.0	10	11
	$V_{CC} = 5.0$ V	2.6	3.33	4.0
Output voltage drop	$V_{CC} = 15$ V, $I_{sink} = 10$ mA		0.1	0.25
(low), V	$I_{sink} = 50$ mA		0.4	0.75
	$I_{sink} = 100$ mA		2.0	2.5
	$I_{sink} = 200$ mA		2.5	
	$I_{sink} = 5.0$ mA		0.25	0.35
Output voltage drop	$I_{source} = 200$ mA, $V_{CC} = 15$ V		12.5	
(high), V	$I_{source} = 100$ mA, $V_{CC} = 15$ V	12.75	13.3	
	$V_{CC} = 5.0$ V	2.75	3.3	
Rise time of output, ns			100	
Fall time of output, ns			100	

[a]The 556 dual timer is a pair of 555s. $T_A = 25°C$, $V_{CC} = +5$ V to $+15$ V, unless otherwise specified.

output pin 3 are shown in Figure 6.8-3. When switch S is closed (with pin 2 grounded), as in (a), the input voltage at pin 2 goes from high ($\simeq V_{CC}$) to low (0 V); in (c) the output voltage at pin 3 goes from low to high; and in (b) the internal switch of pin 7 opens to let capacitor C charge through resistor R_A. When the threshold input at pin 6 reaches $2V_{CC}/3$, the flip-flop output is set to high state, so that the output at pin 3 goes low, as in (c), and the internal switch of pin 7 closes to let capacitor C discharge, as in (b). The period of high

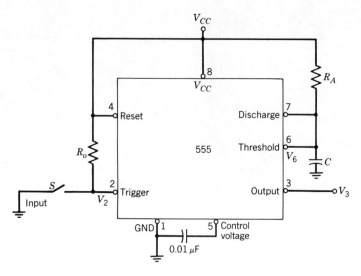

FIGURE 6.8-2   The 555 timed switch.

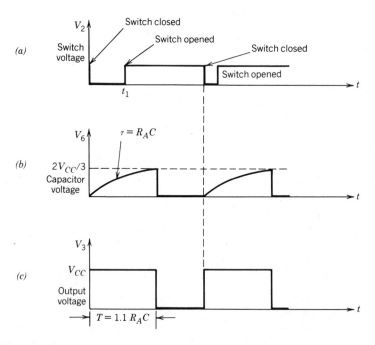

FIGURE 6.8-3   The voltages (V_2, V_6, and V_3) as a function of time at trigger pin 2, threshold pin 6, and output pin 3 for the 555 timed switch of Figure 6.8-2.

output voltage at pin 3 is the pulse width

$$T = 1.1 R_A C, \qquad (6.8\text{-}1)$$

which will be derived in the next section.

6.9   555 TIMER AS MONOSTABLE MULTIVIBRATOR

Operation Principle

To use the 555 timer as a one-shot, the configuration of Figure 6.9-1 can be employed. It is triggered electronically by applying a suitable external voltage to pin 2, whereas the timed switch in Figure 6.8-2 is activated by using an input switch S. In Figure 6.9-1, the 0.01-μF capacitor is connected to the control voltage output to filter out the noise supply. With the trigger input held at a voltage greater than $V_{CC}/3$ and the threshold input at V_{CC}, the output is at the low level of 0 V. The discharge transistor is saturated, and the capacitor C is at the 0-V level. When the trigger input is dropped below $V_{CC}/3$, the flip-flop changes states. shutting off the discharge transistor and driving the output positive to approximately V_{CC}. As the capacitor charges toward V_{CC}, the threshold input reaches the voltage $2V_{CC}/3$ and the flip-flop changes back to the original state, dropping the output to 0 V and again discharging the capacitor. The period T is the pulse width, which is the time taken for the capacitor to charge from 0 V to a voltage of $2V_{CC}/3$ with a final (target) voltage of V_{CC}. This

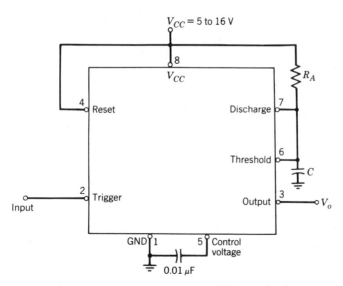

FIGURE 6.9-1   555 one-shot multivibrator circuit.

time can be found by writing the general equation for capacitor voltage, plugging in the correct initial and final values, and solving for T:

$$v_C = v_i + (v_f - v_i)(1 - e^{-t/\tau}), \tag{6.9-1}$$

where v_i is the initial voltage, v_f is the final voltage, and τ is the time constant $R_A C$. For this case $v_i = 0$, $v_f = V_{CC}$, and

$$v_C = \frac{2V_{CC}}{3} = V_{CC}(1 - e^{-T/R_A C}). \tag{6.9-2}$$

Solving for T we find

$$T = R_A C \ln 3 = 1.1 R_A C. \tag{6.8-1}$$

The trigger pulse must be shorter than T for Equation (6.8-1) to be valid. If the trigger input remains low after the period is over, the output remains high. When the input trigger pulse is longer than T, an $R'C'$ differentiator should be used to ensure that the pulse reaching the trigger terminal is short enough. Resistor R' connects from V_{CC} to pin 2, and capacitor C' is inserted between the input line and pin 2. The trigger input is then held positive until the leading edge of the trigger pulse arrives. This transition is coupled through the capacitor to initiate the period. Pin 2 then charges toward V_{CC} with a time constant largely determined by the $R'C'$ product of the trigger circuit. This value must be smaller than that of the one-shot period T. The range of T extends from 10 μs to several hours, and T changes by varying the elements R_A and C. For example,

$$T = 1.1 R_A C = 1.1(100 \text{ k}\Omega)(0.001 \text{ } \mu\text{F}) = 110 \text{ } \mu\text{s};$$

$$T = 1.1(10 \text{ M}\Omega)(100 \text{ } \mu\text{F}) = 1100 \text{ s}.$$

The output of the 555 monostable multivibrator has only one (mono-) stable value, 0 V. A trigger signal applied to pin 2 will produce a fixed-width pulse. A significant application of a one-shot multivibrator is waveshaping. As shown in Figure 6.9-2, a wide variety of input signals can be fed into the 555 one-shot, but the output will always have the same pulse height and width. Thus the one-shot often serves as an interface between an input transducer whose output may vary and subsequent circuits that require a standard input pulse for their operation.

Design Referring to the 555 Timer Characteristics

Design of the one-shot circuit in Figure 6.9-1 involves the choice of R_A and C. As indicated in Equations (6.9-2) and (6.8-1), the time required for C to charge

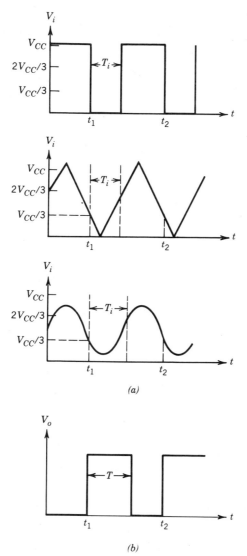

(a)

(b)

FIGURE 6.9-2 Input signals ($T_i < T$) and output waveform of the one-shot multivibrator in Figure 6.9-1. For proper triggering, the input signal must go from high (more than $2V_{CC}/3$) to low (less than $V_{CC}/3$) and then back to more than $2V_{CC}/3$ in a time less than the output pulse width: (a) input voltages, $T_i < T$; (b) output voltage.

through $2V_{CC}/3$ determines the output pulse width:

$$T_W = T = 1.1 \, R_A C. \qquad (6.8\text{-}1)$$

C must normally be chosen as small as possible to ensure that the discharge transistor has no difficulty in discharging it rapidly. However C must not be so small that it is affected by stray capacitance. If C is to be as small as possible, then the charging current must also be as small as possible. The minimum charging current occurs when the capacitor voltage is at its maximum level—that is, when $v_C = 2V_{CC}/3$. At this instant

$$V_{R_A} = V_{CC} - \tfrac{2}{3}V_{CC} = \tfrac{1}{3}V_{CC}$$

and the capacitor charging current is

$$I_{C(min)} = \frac{\tfrac{1}{3}V_{CC}}{R_A}$$

or

$$R_A = \frac{V_{CC}}{3I_{C(min)}}. \tag{6.9-3}$$

The value of the $I_{C(min)}$ chosen should be much greater than the threshold current I_{thres} that flows into terminal 6.

Example 6.9-1. A 555 one-shot circuit with $V_{CC} = 16$ V is to have a 2-ms output pulse width. Design a suitable circuit.

Solution. From Table 6.8-1, $I_{thres} = 0.25\ \mu$A max.

$$I_{C(min)} \gg I_{thres}.$$

Let $I_{C(min)} = 1000\ I_{thres} = 1000 \times 0.25\ \mu$A $= 250\ \mu$A. From Equation (6.9-3),

$$R_A = \frac{V_{CC}}{3I_{C(min)}} = \frac{16\text{ V}}{3 \times 250\ \mu\text{A}} = 21.3\text{ k}\Omega.$$

Use 22 kΩ standard value. From Equation (6.8-1),

$$C = \frac{T}{1.1\ R_A} \frac{2\text{ ms}}{1.1 \times 15\text{ k}\Omega} = 0.0826\ \mu\text{F}.$$

Use 0.082 μF standard value.

6.10   555 TIMER AS ASTABLE MULTIVIBRATOR

Figure 6.10-1 shows a 555 timer wired to make an astable multivibrator that incorporates a capacitor (C) and two resistors $(R_A$ and $R_B)$. Notice that the capacitor voltage (v_C), pin 6, is connected directly to the input trigger, pin 2.

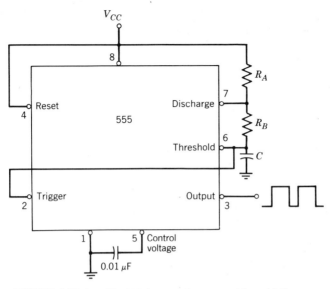

FIGURE 6.10-1   The 555 timer used as an astable multivibrator.

The voltage v_C can swing between $V_{CC}/3$ and $2V_{CC}/3$. It rises toward $2V_{CC}/3$ with a final voltage of V_{CC} and a time constant of $(R_A + R_B)C$. When it reaches $2V_{CC}/3$, the threshold comparator causes the flip-flop to change state (see Figure 6.8-1). The discharge transistor saturates, and the capacitor voltage heads toward ground with a time constant of $R_B C$. When the voltage drops to $V_{CC}/3$, the trigger comparator changes the state of the flip-flop and shuts off the discharge transistor. The capacitor again charges toward V_{CC}. The output and capacitor voltage waveforms of the 555 are shown in Figure 6.10-2.

The duration of the positive portion of the waveform is t_1 and is calculated from Equation (6.9-1), with $v_i = V_{CC}/3$, $v_f = V_{CC}$, and $\tau = (R_A + R_B)C$. This gives

$$v_C = \frac{2V_{CC}}{3} = \frac{V_{CC}}{3} + \left(V_{CC} - \frac{V_{CC}}{3} \right)(1 - e^{-t_1/\tau}).$$

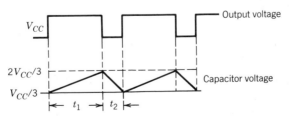

FIGURE 6.10-2   Typical waveforms for a 555 astable multivibrator.

Solving for t_1 we get

$$t_1 = (R_A + R_B)C \ln 2 = 0.7(R_A + R_B)C. \qquad (6.10\text{-}1)$$

Similarly the portion of t_2 can be found to be

$$t_2 = 0.7 R_B C. \qquad (6.10\text{-}2)$$

The duty cycle of the astable is the ratio of the duration of the positive portion of the period to the total period, or

$$\text{Duty cycle} = \frac{t_1}{t_1 + t_2} = \frac{R_A + R_B}{R_A + 2R_B}. \qquad (6.10\text{-}3)$$

If $R_A \gg R_B$, the duty cycle will be a maximum value of 100%. If $R_B \gg R_A$, the duty cycle will approach a minimum value of 50%.

The ratio of the time the output transistor is on (low voltage) to the total period is called the *ON duty cycle*:

$$\text{ON duty cycle} = \frac{t_2}{t_1 + t_2} = \frac{R_B}{R_A + 2R_B}. \qquad (6.10\text{-}4)$$

If the resistor R_B in Figure 6.10-1 is shunted by a diode with its anode connected to pin 7 and cathode to pin 6, then the capacitor can be charged and discharged through the separate resistors; in this case, R_A and R_B can be varied separately to give nearly a full range of output duty cycle from 0 to 100%.

Example 6.10-1. (a) Design a 555 astable multivibrator to generate an output pulse with PRF (pulse-repetition frequency) = 4 kHz and a duty cycle of 60%. Use $V_{CC} = 15$ V. (b) Analyze the circuit designed in part (a) to determine the actual PRF and duty cycle.

Solution.

(a) $t_1 + t_2 = \dfrac{1}{\text{PRF}} = \dfrac{1}{4\text{ kHz}} = 250\ \mu\text{s}.$

$t_1 = (\text{duty cycle})(t_1 + t_2) = \dfrac{60}{100} \times 250\ \mu\text{s} = 150\ \mu\text{s}.$

$t_2 = 250\ \mu\text{s} - 150\ \mu\text{s} = 100\ \mu\text{s}.$

$I_{c(\min)} \gg I_{\text{thres}} = 0.25\ \mu\text{A}$ (from Table 6.8-1). Let $I_{C(\min)} = 1$ mA.
From Equation (6.9-3)

$$R_A + R_B = \frac{V_{CC}}{3 I_{C(\min)}} = \frac{15\text{ V}}{3 \times 1\text{ mA}} = 5\text{ k}\Omega.$$

From (6.10-1)

$$C = \frac{t_1}{0.7(R_A + R_B)} = \frac{150\ \mu s}{0.7 \times 5\ k\Omega} = 0.0428\ \mu F.$$

Use 0.04 μF standard value. From Equation (6.10-2),

$$R_B = \frac{t_2}{0.7C} = \frac{100\ \mu s}{0.7 \times 0.04\ \mu F} = 3.571\ k\Omega.$$

Use 3.3 kΩ standard value.

$$R_A = (R_A + R_B) - R_B = 5\ k\Omega - 3.3\ k\Omega = 1.7\ k\Omega.$$

Use 1.8 kΩ standard value.

(b)   From Equation (6.10-1),

$$t_1 = 0.7(R_A + R_B)C = 0.7 \times (1.7\ k\Omega + 3.3\ k\Omega) \times 0.04\ \mu F = 140\ \mu s.$$

From Equation (6.10-2),

$$t_2 = 0.7\ R_B C = 0.7 \times 3.3\ k\Omega \times 0.04\ \mu F = 92.4\ \mu s.$$
$$t_1 + t_2 = 140\ \mu s + 92.4\ \mu s = 232.4\ \mu s.$$

$$PRF = \frac{1}{t_1 + t_2} = \frac{1}{232.4\ \mu s} = 4.303\ kHz.$$

$$Duty\ cycle = \frac{t_1}{t_1 + t_2} \times 100\% = \frac{140\ \mu s}{232.4\ \mu s} \times 100\% = 60.2\%.$$

6.11   ASTABLE BLOCKING OSCILLATORS

Basic Blocking Oscillator

An astable blocking oscillator conducts for a short period of time and is blocked out (cut off) for a much longer period. A basic blocking oscillator is shown in Figure 6.11-1. The base coil (winding 1-2) and collector coil (winding 3-4) must be connected for regenerative feedback. This feedback through winding 1-2 and capacitor C_1 causes current through the transistor to rise rapidly until saturation is reached. The transistor is then cut off until C_1 discharges through resistor R_1. The output waveform is a pulse, the width of which is primarily determined by winding 1-2. The time between pulses (the blocking time) is determined by the time constant $R_1 C_1$.

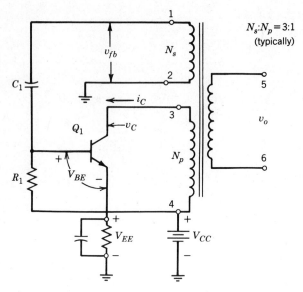

FIGURE 6.11-1   Basic blocking oscillator.

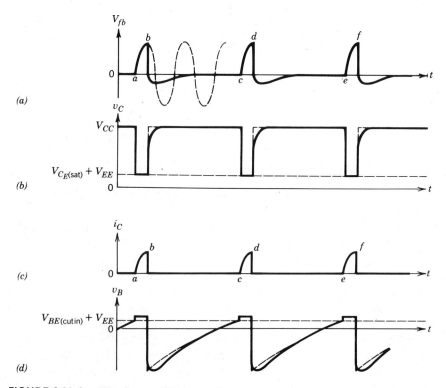

FIGURE 6.11-2   Waveforms of the base-coil voltage (v_{fb}), collector voltage and base voltage in the basic blocking oscillator.

The stray winding capacitance determines a resonant frequency. When the oscillator is cut off, $v_C = V_{CC}$. When the base (C_1) discharges to the level of $V_{BE(cutin)} + V_{EE}$, current i_C starts to flow, inducing a voltage on the base that aids current flow. The circuit shuts off at the peak value of the cycle. A pulse-repetition rate of 400 means that there are 400 on and off conditions per second.

The waveforms are shown in Figure 6.11-2. The transistor is on during intervals ab and cd and off during bc and de. From a to b, energy is stored up in the magnetic field. When cut off at b, this energy must be dissipated and produces a high *back voltage*. This release of stored energy in the transformer causes an oscillation within the base coil that dies out rapidly. The negative loop on the base-coil voltage waveform adds to the waveform of v_B and causes the rounding effect. In the waveform of v_{fb}, the area under positive loop equals that under negative loop. The natural resonant frequency determines the time intervals between a and b and between c and d.

Hartley-Type Blocking Oscillator

The simple blocking oscillator shown in Figure 6.11-3 is a modified Hartley circuit. When the power switch is on, the transistor current i_C rises from zero. The regenerative feedback through L_B and C_1 causes i_C to rise rapidly until saturation is reached. The transistor is then cut off since its base is reverse biased by C_1, which has charged during the earlier feedback process. The time between pulses is determined by the time constant $R_1 C_1$.

This simple oscillator can be used as a single-channel biotelemeter with a resistance-type transducer. If a thermistor is used as R_1, then resistance R_1 decreases when the ambient temperature increases. Thus this oscillator is easily operated for temperature telemetry. Featuring a single transistor, the circuit draws little power from a small mercury or silver-oxide battery, and the telemeter lifetime approaches battery shell life. A disadvantage of the blocking oscillator telemeter is that its transmitting range is just a few feet. On the receiving end, a standard broadcast AM receiver may be used. A blocking rate or number of clicks can be counted and translated into animal temperature by prior calibration.

Blocking Oscillator with Inverter Output

A blocking oscillator with an inverter output is shown in Figure 6.11-4. When power is initially applied to this circuit, capacitor C_1 is charged through the 1.2-MΩ resistor. Once it is charged to a voltage that forward biases the base-emitter junction of Q_1, this transistor is turned on. In this case, a collector current flows through the transformer primary and an emitter current flows through resistor R_1 and the base of Q_2. The collector current in Q_1 causes a current to be induced in the secondary of the transformer that discharges C_1 and switches off Q_1. The process then repeats itself to generate a pulse train.

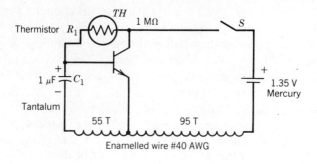

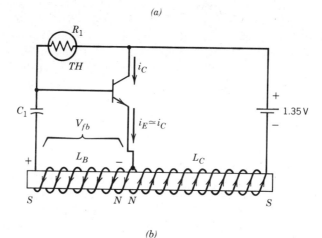

(b)

FIGURE 6.11-3 Hartley-type blocking oscillator: (*a*) a practical circuit; (*b*) the circuit with magnetic-flux linkages around the tapped coil at the moment when the power switch is closed, showing the regenerative (positive) feedback for oscillation start. Use Lenz's law and right-hand rule to determine the polarity of the feedback voltage V_{fb}.

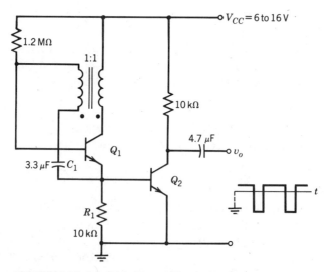

FIGURE 6.11-4 A blocking oscillator with an inverter output.

The pulse width is primarily determined by the transformer characteristics, and the pulse-repetition rate depends on R_1, C_1, V_{CC}, and the base turn-on voltage of Q_1.

6.12   MONOSTABLE BLOCKING OSCILLATOR (WITH EMITTER TIMING)

A triggered blocking oscillator (with emitter-timing) and the equivalent circuit are shown in Figure 6.12-1. To simplify the circuit, let us assume that a triggering signal is momentarily applied to the collector to lower its voltage. By transformer action and with the indicated winding polarities, the base will rise in potential.

Applying Kirchhoff's voltage law to the outside loop (see Figure 6.12-1b), including both the collector and base meshes gives

$$V = \frac{V_{CC}}{n + 1},\qquad(6.12\text{-}1)$$

where V is the voltage drop across the collector winding during the pulse. Since the voltage drop across R is

$$V_{EN} = nV = (i_C + i_B)R,$$

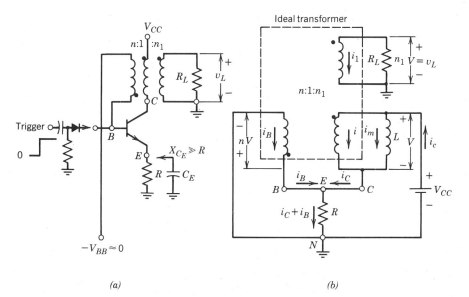

FIGURE 6.12-1   (a) A monostable (triggered) blocking oscillator with emitter timing; (b) the equivalent circuit from which to calculate the current and voltage waveforms.

the emitter current that is constant is given by

$$-i_E = i_C + i_B = \frac{nV}{R} = \frac{n}{n+1}\frac{V_{CC}}{R}. \tag{6.12-2}$$

Since the sum of the ampere turns in the ideal transformer is zero,

$$i - ni_B + n_1 i_1 = 0. \tag{6.12-3}$$

The current in the load circuit is

$$i_1 = -\frac{n_1 V}{R_L}. \tag{6.12-4}$$

Since V is a constant, the magnetizing current is given by

$$i_m = \frac{Vt}{L}. \tag{6.12-5}$$

From Kirchhoff's current law at the collector mode, we have

$$i = i_C - i_m = i_C - \frac{Vt}{L}. \tag{6.12-6}$$

Substituting from Equations (6.12-4) and (6.12-6) into Equation (6.12-3), we have

$$i_C - \frac{Vt}{L} - ni_B - \frac{n_1^2 V}{R_L} = 0. \tag{6.12-7}$$

Solving Equations (6.12-2) and (6.12-7) and using Equation (6.12-1), we obtain

$$i_B = \frac{V_{CC}}{(n+1)^2}\left(\frac{n}{R} - \frac{n_1^2}{R_L} - \frac{t}{L}\right) \tag{6.12-8}$$

and

$$i_C = \frac{V_{CC}}{(n+1)^2}\left(\frac{n^2}{R} + \frac{n_1^2}{R_L} + \frac{t}{L}\right). \tag{6.12-9}$$

Notice that the collector-current waveform is trapezoidal with a positive slope; the base current is also trapezoidal, but it has a negative slope; and the emitter current is constant during the pulse. These current waveforms and the voltage waveforms are pictured in Figure 6.12-2. If the damping is inadequate, the backswing may oscillate, as indicated by the dashed curve in Figure 6.12-2e, and regeneration will start again at the point marked X.

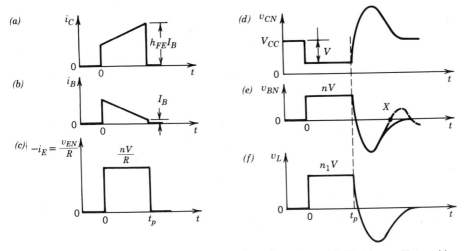

FIGURE 6.12-2   The current and voltage waveforms in a monostable blocking oscillator with emitter timing (Figure 6.12-1).

At $t = 0+$, $i_C < h_{FE}i_B$, the operating point on the collector characteristics of Figure 6.12-3 is at point P, and the transistor is in saturation. As time passes, i_C increases and the operating point moves up the saturation line in Figure 6.12-3. While i_C grows with time, i_B is decreasing and ultimately point P' is reached at $t = t_p$, where $i_B = I_B$ and

$$i_C = h_{FE}I_B. \tag{6.12-10}$$

At this point P' the transistor comes out of saturation and enters its active region. Because the loop gain exceeds unity in the active region, the transistor is quickly driven to cutoff by regenerative action, and the pulse ends. Since the regeneration that terminates the pulse starts when the transistor comes out of

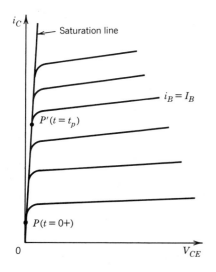

FIGURE 6.12-3   Collector characteristics. The path of the collector current is along the saturation line from P to P'. The pulse ends at P'. where the transistor comes out of saturation.

saturation, the pulse width t_p is determined by the condition given by Equation (6.12-10).

Applying Equation (6.12-10) to Equations (6.12-9) and (6.12-8) with $t = t_p$ we find

$$h_{FE}i_B = \frac{h_{FE}V_{CC}}{(n+1)^2}\left(\frac{n}{R} - \frac{n_1^2}{R_L} - \frac{t_p}{L}\right) = \frac{V_{CC}}{(n+1)^2}\left(\frac{n^2}{R} + \frac{n_1^2}{R_L} + \frac{t_p}{L}\right);$$

$$(h_{FE} + 1)\frac{t_p}{L} = \frac{h_{FE}n}{R} - \frac{n^2}{R} - (h_{FE} + 1)\frac{n_1^2}{R_L}.$$

Hence

$$t_p = \frac{nL}{R}\frac{h_{FE} - n}{h_{FE} + 1} - \frac{n_1^2 L}{R_L}. \tag{6.12-11}$$

Usually $\frac{1}{5} \leqslant n \leqslant 1$ and $h_{FE} \gg n$. Then Equation (6.12-11) becomes

$$t_p \simeq \frac{nL}{R} - \frac{n_1^2 L}{R_L}. \tag{6.12-12}$$

Therefore the pulse width t_p is independent of h_{FE} and depends only upon passive elements n, L, R, and so on. We may conclude that the blocking oscillator of Figure 6.12-1 is a simple circuit that yields a pulse of very stable duration.

A positive trigger pulse (wider than the blocking oscillator pulse) may be applied through a diode to the base, as shown in Figure 6.12-1a. A small capacitor may be used across the emitter resistor R ($\ll X_{C_E}$) to improve the rise time of the oscillator pulse.

REFERENCES

1. Bell, D. A., *Solid State Pulse Circuits*, 2nd ed., Reston Publishing Company, Reston, Virginia, 1981, Chap. 8.

2. Comer, D. J., *Electronic Design with Integrated Circuits*, Addison-Wesley, Reading, Massachusetts, 1981, Chap. 3.

3. Millman, J. and H. Taub, *Pulse, Digital, and Switching Waveforms*, McGraw-Hill, New York, 1965, Chaps. 11 and 16.

4. Pettit, J. M. and M. M. McWhorter, *Electronic Switching, Timing, and Pulse Circuits*, 2nd ed., McGraw-Hill, New York, 1970, Chap. 7.

5. *μA Linear IC Data Book*, (555 Single Timer and 556 Dual Timer), Fairchild Camera and Instrument Corporation, 313 Fairchild Drive, Mountain View, California, 1982, pp. 9-3–9-14.

6. Santic, A., S. Vamvakas, and M. R. Neuman, Micropower Electronic Switches for Implanted Instrumentation, *IEEE Transactions on Biomedical Engineering*, Vol. BME-29, No. 8, August 1982, pp. 583–589.

QUESTIONS

6-1. Draw the circuit of a collector-coupled astable multivibrator. Sketch the waveforms and explain the operation of the circuit.

6-2. Explain why the collector-coupled astable multivibrator can be used as a voltage-to-frequency converter.

6-3. Draw the circuit of the complementary-pair astable multivibrator. Explain the operation of the circuit.

6-4. Sketch the circuit of the emitter-coupled astable multivibrator used as a horizontal oscillator in a television receiver. Explain the operation of the circuit.

6-5. Explain why the frequency of the emitter-coupled astable multivibrator can be varied by controlling its input base voltage.

6-6. Sketch the circuit and waveforms of a collector-coupled monostable multivibrator. Explain the operation of the circuit.

6-7. Sketch the circuit of an emitter-coupled monostable multivibrator. Explain the operation of the circuit. Show how the circuit may be modified to provide pulse-width control.

6-8. Draw a circuit to show an additional triggering transistor Q_3 connected across Q_1 in the collector-coupled monostable multivibrator. Explain the function of each component in the additional stage (Q_3).

6-9. Sketch the circuit of an IC op-amp used as an astable multivibrator that generates symmetrical square waves. Explain the operation of the circuit.

6-10. Sketch the block diagram of a 555 IC timer. Briefly explain the function of each component.

6-11. (a) Sketch the circuit of a 555 astable multivibrator. (b) Show the waveforms of the capacitor and output voltages (v_C and v_o). (c) Explain how the circuit functions.

6-12. Sketch the circuit of a basic blocking oscillator. Briefly explain its operation.

6-13. Sketch the circuit of an astable blocking oscillator with an inverter output. Explain the operation of the circuit.

PROBLEMS

6-1. Derive an expression for the output pulse width of a collector-coupled astable multivibrator.

6-2. Verify Equation (6.1-4).

6-3. Design an astable multivibrator to generate a 4-kHz output square wave.

The available supply is 12 V, and the load current is to be 40 μA. Make I_C = 100 × (load current). Use transistors with $h_{FE(min)}$ = 100.

Answer.   3.3 kΩ; 270 kΩ; 680 pF.

6-4.   Derive an expression for the output pulse width of a collector-coupled monostable multivibrator.

6-5.   Design a collector-coupled monostable multivibrator using silicon transistors with I_C = 3 mA and $h_{FE(min)}$ = 60. The available supply is ±9 V. The output pulse width is to be 350 μs.

Answer. R_{C1} = R_{C2} = 2.7 kΩ; R_1 = 22 kΩ; R_2 = 33 kΩ; R = 150 k; C = 3300 pF.

6-6.   A 74121 TTL monostable multivibrator is used to produce an output pulse width of 3 μs. Choose suitable external components (let R_T = 3.3 kΩ), and show how they must be connected to the circuit.

6-7.   Design an astable multivibrator using a 741 op-amp to generate a symmetrical square-wave output with an amplitude of approximately ±8 V and a frequency of approximately 200 Hz.

Answer. R_1 = R_2 = R_f = 82 kΩ; C = 0.027 μF.

6-8.   Using a 741 operational amplifier, design a monostable multivibrator to give an output pulse width of 500 μs. The circuit is to be triggered by a 1.2-V input spike, and the available supply is ±10 V.

Answer. R_1 = 180 kΩ; R_2 = 22 kΩ; R_3 = 18 kΩ; C = 9100 pF.

6-9.   Design a 555 one-shot circuit to give a 1.5-ms output pulse width. Use V_{CC} = 15 V.

Answer.   22 kΩ; 0.06 μF.

6-10.   Verify Equations (6.10-1) and (6.10-2).

6-11.   Derive the expression for the duty cycle of the astable multivibrator when resistor R_B in Figure 6.10-1 is shunted by a diode with its anode connected to pin 7 and cathode to pin 6.

6-12.   (a) Using a supply voltage of 14 V, design a 555 astable multivibrator to produce an output pulse with its PRF = 8 kHz and duty cycle = 70%. (b) Analyze the circuit designed in part (a) to determine the actual pulse width and duty cycle.

6-13.   A common-collector triggered blocking oscillator is shown in Figure p6-13. The junction saturation voltages are neglected. The magnetizing inductance of the emitter winding is L. Find expressions for the pulse width and pulse amplitude.

Answer.

$$t_p = nL\left[\left(\frac{1}{1 + h_{FE}}\right)\frac{n-1}{R} + \frac{n-1}{nR} - \frac{n_1^2}{nR_L}\right].$$

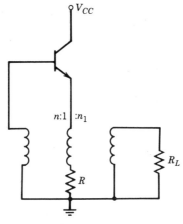

FIGURE P6-13

Pulse amplitude is $\dfrac{n_1}{n}(nV) = n_1 V.$

6-14. If the capacitor between the ground and the junction of R_1, R_2, and R_b is removed from ground and connected to the Q_1 emitter in Figure 6.4-4, then the Q_1 stage becomes an emitter follower with the boostrapping effect. In this case the effective resistance R_{eff} due to the biasing arrangement becomes extremely large if the Q_1 voltage gain A_v approaches unity. (a) Draw this modified emitter follower circuit; (b) verify that the effective resistance is given by

$$R_{\text{eff}} = \frac{V_{\text{in}}}{I_{R_b}} = \frac{R_b}{1 - A_v}.$$

Hint. $I_{R_b} = V_{R_b}/R_b;\ V_{R_b} = V_{\text{in}} - A_v V_{\text{in}}.$

6-15. The IC CMOS 4011 contains quadruple 2-input NAND gates. This IC will operate as an astable multivibrator (clock) if it is externally connected as follows: (1) each of the NAND gates is connected as an inverter; (2) the four inverters are connected in cascade; (3) one end of a resistor R is connected between the first and the second inverter, and one end of a capacitor C is connected between the second and the third inverter; both the other ends of R and C are connected to the input of the first inverter. (a) Sketch the circuit of the clock; (b) explain how it oscillates; (c) list the factors determining the frequency.

Reference: *CMOS Data Book*, Fairchild, 1980, page 7-19.

7
BISTABLE MULTIVIBRATORS OR FLIP-FLOPS

7.1 BASIC BISTABLE MULTIVIBRATOR CIRCUITS

The NOT-Gate Latch

The *bistable multivibrator* is known by several other names, including *flip-flop*, *binary*, and *latch*. The circuit of Figure 7.1-1, containing a pair of coupled inverters (NOT gates), is the basic structure of a most important and basic logic circuit, which is referred to as a *static latch*. The logic levels at the two accessible terminals Q and $\bar{Q}$ are complementary to each other. In the following discussion, we use positive logic in which H (the higher voltage level) represents logic 1 and L (the lower voltage level) represents logic 0.

Without any external intervention, the latch may persist indefinitely in one of two stable states. Assume $Q = L$; then the output of the bottom inverter is indeed $\bar{Q} = H$. If the latch is initially established in the $Q = L$ and $\bar{Q} = H$ state, it will remain in that state indefinitely. The second stable state of the latch is the one in which $Q = H$ and $\bar{Q} = L$. The latch can be used to *store*—that is, register or remember a logic bit. The two states of a latch are termed the *set state* and the *reset state*. The set state is the one in which $Q = H$ ($\bar{Q} = L$). The reset state is the one in which $Q = L$ ($\bar{Q} = H$). The reset state is often called the *clear state*. To put a latch in the set state, we need only connect the Q terminal temporarily to an external point that is at high voltage level. This temporary connection drives Q high and $\bar{Q}$ low, and when the connection is removed, the latch will remain in the state with $Q = H$ and $\bar{Q} = L$. Correspondingly, a temporary connection of the Q terminal to an external point at the low voltage level will reset (or clear) the latch. The latch will remain in the reset state indefinitely if not further disturbed.

234

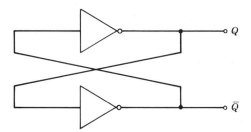

FIGURE 7.1-1   A static latch containing a pair of coupled inverters (NOT circuits).

Flip-flops form the basis of most sequential logic circuits that depend not only on the external inputs but also on the state of the system before the application of the input signals. Such sequential circuits involve feedback. There is no feedback in a combinational logic circuit whose outputs depend only on external inputs with no dependence on the state of the system.

Fixed-Bias Flip-Flop Circuit

The fixed-bias flip-flop shown in Figure 7.1-2 is a static latch containing a pair of inverters Q_1 and Q_2. The circuit has two stable states. Either Q_1 is on (saturated) and Q_2 is off or Q_2 is on and Q_1 is off. Each transistor is held in its original state by the condition of the other. The circuit is symmetrical. Each transistor is biased from the collector of the other. When either transistor is on, the other device is off.

Consider the circuit condition when Q_1 is on and Q_2 is off, as shown in Figure 7.1-2b. With Q_1 on in saturation, its collector voltage is $V_{CE(sat)}$ and Q_2 base is reverse biased through R'_1 and R'_2, causing Q_2 to be off. With Q_1 triggered off, Q_2 switches on (because its base is forward biased through R_{C1}, R'_1, and R'_2), while Q_1 remains off (because its base is reverse biased from the Q_2 collector). Each of the two stable states (Q_1 on, Q_2 off; and Q_1 off, Q_2 on) can be maintained indefinitely. The output swing at each collector is

$$V_W = V_{C1} - V_{C2} \simeq V_{CC}.$$

The speed-up capacitors C_1 and C_2 are also called *memory capacitors.* Consider what occurs when the on transistor (say Q_1) is triggered off for a brief instant. With both transistors off, each base voltage becomes $V_B \simeq V_{CC} -$ (charged voltage on the capacitor at the transistor base). When Q_1 is on and Q_2 is off, C_2 has a smaller charge than C_1, and thus V_{B2} is greater than V_{B1}; therefore Q_2 switches on before Q_1, and, in doing so, Q_1 becomes biased off. Once switchover occurs, C_2 becomes charged to a greater voltage than C_1. Consequently the charges on C_1 and C_2 enable them to remember which transistor was on and which was off.

In designing a flip-flop, I_C may be specified at a level much larger than the

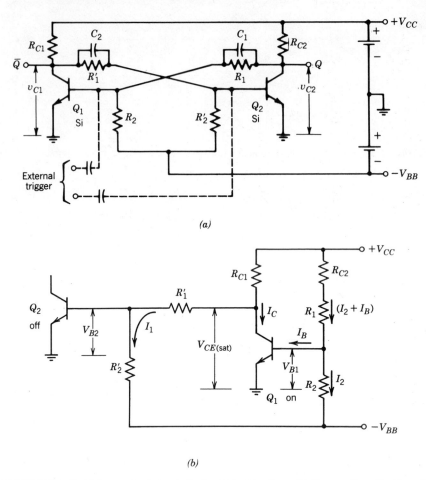

(a)

(b)

FIGURE 7.1-2 (a) Symmetrical circuit of a fixed-bias bistable multivibrator: $Q_1 = Q_2$, $R_1 = R'_1$, $R_2 = R'_2$, $R_{C1} = R_{C2} = R_C$, and $C_1 = C_2$; (b) Circuit condition when Q_1 is on and Q_2 is off.

output current. R_1 and R_2 should be selected small enough to provide a stable bias level but large enough so that they do not overload R_C. These requirements are met by applying the rule of thumb that bias curent $I_2 \simeq I_C/10$. During the turn-off time of the transistors, the voltages on the memory capacitors should not change significantly. If these capacitors are allowed to discharge by 10% of the difference between maximum and minimum capacitor voltages, then from the discussion of Equation (2.2-3),

$$C_1 = C_2 = C = \frac{t_{\text{off}}}{0.1 \, R} \simeq \frac{t_{\text{off}}}{0.1(R_1 \| R_2)}. \tag{7.1-1}$$

The time for the capacitors to discharge from maximum voltage to minimum voltage or vice versa is the recovery time t_{re} used to determine the maximum triggering frequency f_{max}. From Equation (2.2-3),

$$t_{re} = 2.3\,RC, \tag{7.1-2}$$

where $R = R_1 \| R_2$ and $C = C_1 = C_2$. The maximum triggering frequency is

$$f_{max} = \frac{1}{t_{re}} = \frac{1}{2.3(R_1 \| R_2)C}. \tag{7.1-3}$$

Example 7.1-1. Design a fixed-bias flip-flop to operate from a supply of ± 6 V. Use silicon *npn* transistors (e.g., 2N3904) with $I_C = 1$ mA, $h_{FE(min)} = 70$, and $t_{off} = 250$ ns (see Tables 4.1-3 and 4.4-1). Refer to Figure 7.1-2.

Solution. $V_{CE(sat)} = 0.2$ V.

$$R_C = R_{C1} = R_{C2} \simeq \frac{V_{CC} - V_{CE(sat)}}{I_C} = \frac{6\text{ V} - 0.2\text{ V}}{1\text{ mA}} = 5.8 \text{ k}\Omega.$$

Use 5.6 kΩ standard value.

$$I_{B(min)} = \frac{I_C}{h_{FE(min)}} = \frac{1\text{ mA}}{70} = 14.3 \ \mu\text{A}.$$

With Q_1 on, the voltage across R_2 is

$$V_{R2} = V_{BE1} - (-V_{BB}) = 0.8 - (-6) = 6.8 \text{ V}.$$

$$I_2 \simeq \frac{1}{10} I_C \simeq \frac{1\text{ mA}}{10} = 100 \ \mu\text{A}.$$

$$R_2 = \frac{V_{R2}}{I_2} = \frac{6.8\text{ V}}{100 \ \mu\text{A}} = 68 \text{ k}\Omega \text{ (standard value)}.$$

$$R_{C2} + R_1 = \frac{V_{CC} - V_{BE}}{I_2 + I_B} = \frac{6\text{ V} - 0.8\text{ V}}{100 \ \mu\text{A} + 14.3 \ \mu\text{A}} = 45.5 \text{ k}\Omega.$$

$R_1 = (R_{C2} + R_1) - R_{C2} = 45.5 \text{ k}\Omega - 5.6 \text{ k}\Omega = 39.9 \text{ k}\Omega$; use 39 kΩ standard value.

$$C_1 = C_2 = C \simeq \frac{t_{off}}{0.1(R_1 \| R_2)} = \frac{250\text{ ns}}{0.1(39\text{ k}\Omega \| 68\text{ k}\Omega)} = \frac{250\text{ ns}}{0.1(24.79\text{ k}\Omega)}$$
$$= 100.85 \text{ pF}.$$

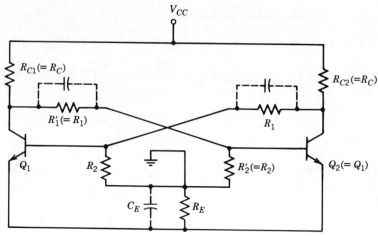

FIGURE 7.1-3 Symmetrical circuit of a self-bias flip-flop.

Use 100 pF standard value.

$$t_{re} = 2.3\,RC = 2.3 \times 24.79\,k\Omega \times 100\,pF = 5701.7\,ns = 5.7017\,\mu s.$$

$$\text{Maximum triggering frequency} = f_{max} = \frac{1}{t_{re}} \simeq 175.39\,kHz.$$

Self-Bias Flip-Flop

In the self-bias flip-flop of Figure 7.1-3, R_E provides the self-bias. C_E keeps the self-bias almost constant during the transition time. R_E also limits the collector current of the on transistor to any desired level, so that the transistor may be saturated or unsaturated.

When designing an unsaturated flip-flop circuit, make the voltage drop across R_E several times V_{BE} so that reasonably stable bias conditions can be maintained. In order to avoid device saturation, $h_{FE(max)}$ should be used in design calculation, and a minimum V_{CE} of about 3 V should be designed into the circuit. $(V_{CC} - V_{CE})$ should be divided equally between V_{RC} and V_{RE}, R_1 and R_2 should be carefully chosen, and the next smaller standard value should be selected when R_C is calculated. With bias current $I_2 \simeq I_C/10$ (see Figure 7.1-2b), R_1 and R_2 are small enough to provide a stable bias voltage, but not so small that they will overload R_C.

7.2 FLIP-FLOP COLLECTOR TRIGGERING CIRCUITS

Asymmetrical Collector Trigger Circuit

Flip-flop triggering circuits are normally designed to turn off the on transistor with asymmetrical or symmetrical triggering. Asymmetrical triggering is also referred to as *set-reset triggering,* since one of the two trigger inputs is used to set the circuit in one particular state and the other input is used to reset to the opposite state. Symmetrical triggering employs only one trigger input to make the two sides of the flip-flop turn on alternately. The flip-flop triggering circuit may use either collector triggering or base triggering; the latter requires less trigger energy but a more accurately controlled trigger amplitude.

An asymmetrical collector triggering circuit is shown in Figure 7.2-1. Its operation is described by using positive logic. Output Q $(\bar{Q})$ is taken from the collector of Q_2 (Q_1). When a negative-going step input is applied at the set terminal S, Q_2 will be off and Q_1 on, so that $Q = 1$ and $\bar{Q} = 0$. Likewise, when a negative-going step input is applied at the reset terminal R, Q_2 will be on and Q_1 off, so that $Q = 0$ and $\bar{Q} = 1$.

Assume Q_1 is on and Q_2 off. The voltage at Q_2 collector is approximately V_{CC}. The negative-going input applied at terminal R, coupled through C_3, forward biases D_1 with the result of voltage drop V_{D1} and pulls D_1 cathode down by ΔV. The anode of D_1 is pulled down by $\Delta V - V_{D1}$. Hence Q_2 collector voltage is changed from approximately V_{CC} to $[V_{CC} - (\Delta V - V_{D1})]$. The voltage change at Q_2 collector appears at Q_1 base. The initial Q_1 base voltage $V_{BE(sat)}$ falls by $\Delta V - V_{D1}$, causing the base of Q_1 to be pushed below its emitter voltage (see the waveforms in Figure 7.2-1). Thus Q_1 switches off and Q_2 turns on. When the negative-going step is applied at terminal R, C_3 immediately begins to charge through R_{C2}. Q_2 collector and Q_1 base voltages rise from their minimum

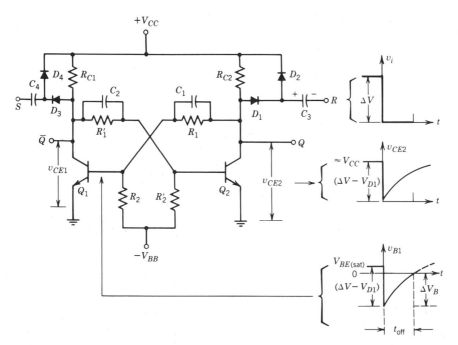

FIGURE 7.2-1 Asymmetrical collector triggering to turn off the ON transistor. The negative-going step is applied at terminal R when Q_1 is ON and Q_2 is OFF.

levels. It is best to choose C_3 as small as possible. The smallest suitable capacitor is one that will allow the base voltage to rise to the emitter voltage during the Q_1 turn-off time.

When the trigger input becomes positive (returning to its normal dc level), D_1 is reverse biased and the flip-flop state is unaffected. D_2 is then forward biased by the capacitor charge, and C_3 is quickly discharged through D_2. The triggering circuit is now ready to receive another negative-going input applied at terminal S.

Symmetrical Collector Trigger Circuit

A symmetrical collector trigger circuit is shown in Figure 7.2-2. With Q_1 on and Q_2 off, collector voltage v_{C1} is about zero and v_{C2} is approximately V_{CC}. When the negative-going trigger input V_i is applied, diode D_2 does not become forward biased, since the amplitude of V_i does not exceed the voltage drop across R_{C1}. However the negative-going input forward biases D_1, and so Q_1 is turned off. With Q_1 off, collector voltage v_{C1} rises to approximately V_{CC}, and v_{C2} drops to about zero because Q_2 is on. The next negative-going input causes the circuit

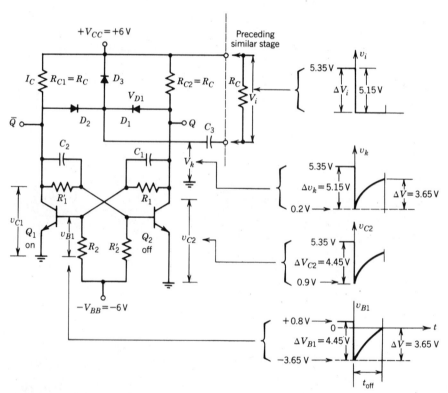

FIGURE 7.2-2 Symmetrical collector triggering. The flip-flop is triggered by the collector output of a preceding similar stage.

to return to its original state, since D_2 is forward biased, causing Q_2 base to be pushed below its emitter level. Diode D_3 is used to avoid the recovery problem; its low forward impedance ensures fast recovery, and its high reverse impedance avoids shunting the negative-going trigger input during the triggering period. Equation (7.1-1) can be used to calculate C_1 and C_2. Equation (6.1-2) can be used to calculate C_3.

Example 7.2-1. The fixed-bias flip-flop designed in Example 7.1-1 is to be triggered by the collector output of a preceding similar stage. Design a suitable symmetrical collector trigger circuit, referring to the waveforms of the triggering voltage as it appears at various points in the circuit shown in Figure 7.2-2.

Solution. $V_{C(on)} = V_{CE(sat)} = 0.2$ V.

$$V_{C(off)} = V_{CC} - V_{RC2} \text{ (for } Q_2 \text{ off).}$$

Voltage across $(R_{C2} + R_1) = V_{CC} - V_{BE(sat)} = 6$ V $- 0.8$ V $= 5.2$ V.

$$V_{RC2} = [V_{CC} - V_{BE(sat)}] \frac{R_{C2}}{R_1 + R_{C2}} = 5.2 \text{ V} \times \frac{5.6 \text{ K}}{39 \text{ K} + 5.6 \text{ K}} = 0.65 \text{ V.}$$

$V_{C(off)} = 6$ V $- 0.65$ V $= 5.35$ V.
The collector voltage of the flip-flop changes from 5.35 V to 0.2 V. This change is used as an input triggering voltage.

$$V_i = 5.35 \text{ V} - 0.2 \text{ V} = 5.15 \text{ V.}$$

At the junction of diode cathodes, $\Delta V_k = 5.15$ V.

$$\Delta V_{C2} = \Delta V_k - V_{D1} = 5.15 \text{ V} - 0.7 \text{ V} = 4.45 \text{ V.}$$

$$\Delta V_{B1} = \Delta V_{C2} = 4.45 \text{ V.}$$

During t_{off} (Q_1 off), $V_{B1} < V_{E1}$,

$$\Delta V = \Delta V_{B1} - V_{BE(sat)} = 4.45 \text{ V} - 0.8 \text{ V} = 3.65 \text{ V.}$$

Hence C_3 can charge 3.65 V during t_{off} under the following conditions:

Initial voltage $= V_{C3} = V_C \simeq 0$ V.
Final voltage $= V_{C3} = v_C = \Delta V = 3.65$ V.
Charging voltage $= V = \Delta V_i - V_{D1} = 5.15$ V $- 0.7$ V $= 4.45$ V.
Charging resistance $\simeq R_C = 5.6$ kΩ.
Turn-off time $= t_{off} = 250$ ns (typically).

From Equation (6.1-2)

$$C_3 = \frac{t_{\text{off}}}{R_C \ln(V - V_C)/(V - v_C)} = \frac{250 \text{ ns}}{5.6 \text{ k}\Omega \ln(4.45 \text{ V} - 0)/(4.45 \text{ V} - 3.65 \text{ V})} = 26 \text{ pF}.$$

Use 27 pF standard value.

7.3 *RS* FLIP-FLOPS

Basic Flip-Flop Building Blocks

For convenience in manipulating the flip-flop (FF), the simple inverter latch of Figure 7.1-1 is usually replaced by the NOR-gate or NAND-gate latch, as shown in Figures 7.3-1 and 7.3-2. The extra input terminals of the gates provide control terminals and allow additional avenues of access to the latch. The NOR-gate latch is realized by cross-coupling NOR gates in an *RS* flip-flop with active high inputs. The NAND-gate latch is realized by cross-coupling NAND gates in an $\bar{R}\bar{S}$ flip-flop with active low inputs. Either of these is called a *static latch*

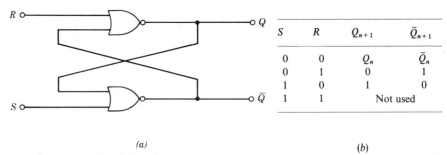

S	R	Q_{n+1}	$\bar{Q}_{n+1}$
0	0	Q_n	$\bar{Q}_n$
0	1	0	1
1	0	1	0
1	1	Not used	

(a)   (b)

FIGURE 7.3-1 An *RS* NOR-gate latch with active high inputs: (*a*) logic diagram; (*b*) truth (function) table.

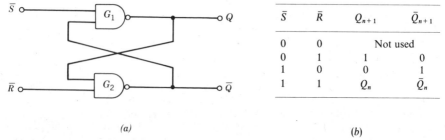

$\bar{S}$	$\bar{R}$	Q_{n+1}	$\bar{Q}_{n+1}$
0	0	Not used	
0	1	1	0
1	0	0	1
1	1	Q_n	$\bar{Q}_n$

(a)   (b)

FIGURE 7.3-2 An $\bar{R}\bar{S}$ NAND-gate latch with active low inputs: (*a*) logic diagram; (*b*) truth (function) table.

and is the basic building block for other types of FFs. The static latch is an example of a sequential circuit.

An *RS* Latch Using NOR Gates

The *RS* NOR-gate latch has two inputs (S, R) and two outputs $(Q, \bar{Q})$. Application of a pulse to input S (i.e., momentarily bringing S to 1 as $R = 0$) will cause the output Q to go to 1 until reset by bringing input R to 1. The second output $\bar{Q}$ is the inverse of Q. The FF outputs remain in a state indefinitely unless altered by an input pulse, however brief. The *RS* latch could have a circuit similar to the asymmetrically triggered flip-flop, as in Figure 7.2-1.

The truth table of Figure 7.3-1*b* summaries the behavior of the *RS* NOR-gate latch. When $R = S = 0$, the NOR gates are enabled and, so far as the other input of each gate is concerned, each gate is simply an inverter as demanded in a flip-flop. Hence with $R = S = 0$ the latch can be in either of its two possible states ($L = 0$ for $H = 1$). When $R = 1$ and $S = 0$, the upper gate is disabled (or inhibited) and $Q = 0$ while $\bar{Q} = 1$. This is the reset state; if R is now restored to $R = 0$, the latch will remain in the reset state. In a similar way, starting with $R = S = 0$, if S goes to $S = 1$ permanently or temporarily, the latch will go to the set state ($Q = 1$) or remain in the set state if it is already in that state. The case $S = R = 1$ is not used. There will be intervals when $S = R = 0$; intervals when $S = 0$, $R = 1$; and intervals when $S = 1$, $R = 0$. Let us number the intervals in order $1, 2, \ldots, n, n + 1, \ldots$ as they occur in time. In the truth table of Figure 7.3-1*b*, there are four cases listed: (1) if in the interval $n + 1$ we have $S = R = 0$, the state of the latch will be the same in that interval as it was in interval n; (2) in the interval $n + 1$ when $S = 0$ and $R = 1$, we shall find $Q_{n+1} = 0$ ($\bar{Q}_{n+1} = 1$) no matter what the logic levels of S and R were in the preceding time intervals; (3) In the interval $n + 1$ when $S = 1$ and $R = 0$ we shall find $Q_{n+1} = 1$ ($\bar{Q}_{n+1} = 0$) no matter what the previous history; and (4) if $S = R = 1$ simultaneously, neither Q and $\bar{Q}$ are used. This is the case for the following reasons: suppose that starting with $S = R = 1$, we now allow both S and R to become $S = R = 0$ simultaneously; then the resultant state of the latch will not be predictable; absolute simultaneousness is impossible, and if R (or S) should actually change first, we would pass first through the situation $S = 1$, $R = 0$ (or $R = 1$, $S = 0$). As a result the latch would go to the set (or reset) state and remain in that state when S and R become $S = R = 0$. We note incidentally that with $S = R = 1$, we would have $Q = \bar{Q} = 0$, and the implication in Figure 7.3-1*a* that the outputs are complementary would be incorrect.

When the *RS* latch is intended to be left alone to remember a bit, it will have $S = R = 0$. When we intend to store logic 0 (or 1) in the latch, we arrange that $S = 0$ (or 1) while $R = 1$ (or 0), at least temporarily.

When we say that a gate is *enabled*, we mean that it is enabled to perform the function for which it is intended. When the gate of a latch has an output independent of the input other than R or S, this gate is *disabled* or *inhibited*.

An $\bar{R}\bar{S}$ Latch Using NAND Gates

In the $\bar{R}\bar{S}$ NAND-gate latch of Figure 7.3-2a, the lower control terminal allows us to reset and the upper control terminal allows us to set. This situation is the opposite of that in the RS NOR-gate latch. The truth table of the $\bar{R}\bar{S}$ latch given in Figure 7.3-2b is easily verified. When $\bar{S} = \bar{R} = 1$, both gates are enabled and either of the states ($L = 0$ or $H = 1$) is possible, the state being determined by the logic levels on $\bar{S}$ and $\bar{R}$ that prevailed earlier. Starting with $\bar{S} = \bar{R} = 1$, if we change $\bar{R}$ from 1 to 0, gate G_2 will become disabled (or inhibited) and $\bar{Q}$ will become (or remain) $\bar{Q} = 1$. Now both inputs of G_1 are $\bar{S} = \bar{Q} = 1$, and Q becomes (or remains) $Q = 0$. This output $Q = 0$ is an input to gate G_2 that is thereby disabled. If $\bar{R} = 1$ and $\bar{S}$ is changed from 1 to 0 even temporarily, the latch will be set ($Q = 1$). A typical example of the $\bar{R}\bar{S}$ latch is the TTL 74279, which is called a quad SR latch.

Gated RS Latch and D-Gated Latch

The R and S inputs and the $\bar{R}$ and $\bar{S}$ inputs are often called *data inputs* because the information presented at these terminals determines what is stored in the latch. The gated RS latch in Figure 7.3-3 will allow us to connect it to, or to isolate it from, the data source. The signal applied at the Enable or G input is known as the *gate* or the *strobe*, and the circuit is termed a *gated* or *strobed latch*, or sometimes a *dynamic latch*. When Enable $= 0$, gates G_1 and G_2 are disabled and the latch is isolated from the data; thus the circuit will not allow the S or R inputs to affect the output. When Enable $= 1$, S and R become active-high set and reset inputs, and thus the circuit behaves like a normal RS latch.

In some applications the S and R inputs are always complementary: $S = \bar{R}$; that is, $S = 0$ when $R = 1$ and $R = 0$ when $S = 1$. Since pin connections on an IC can be minimized for this situation by employing a D-gated latch, this circuit has become a popular device. A typical example of the D-gated latch is the TTL 74LS375, which is called a *quad transparent latch*. Figure 7.3-4 shows the schematic for the D-gated latch. When Enable $= 1$, the input data D are transferred to output Q. So long as Enable is active, the latch output Q will follow the data input D. This feature of the operation is characterized by describing the latch as *transparent*.

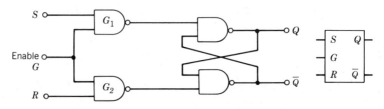

FIGURE 7.3-3 A gated RS latch.

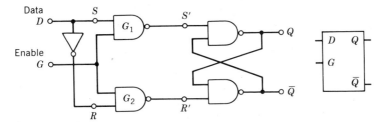

FIGURE 7.3-4   A *D*-gated latch.

Clocked *RS* Flip-Flop (FF)

The clocked *RS* flip-flop has three inputs: the set, the reset, and the clock (*CK*) or trigger (*T*). This FF allows us to set or reset it in synchronism with a pulse train. The clocked *RS* FF in Figure 7.3-5 is an edge-triggered *RS* FF. The gates G_3 and G_4 form a latch, whereas G_1 and G_2 become the control or steering gates that program the state of the FF after the clock pulse appears. The clocked *RS* FF can be operated for any of the conditions—no change, set, and reset. In order to understand the operation of this FF, let us begin with the assumption that it is in the reset state ($Q = 0$) and $S = R = CK = 0$. For this condition, $S' = R' = 1$; $Q = 0$ is an input of gate G_4, making $\bar{Q} = 1$. Since $\bar{Q} = 1$, both inputs to

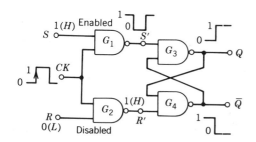

(a)

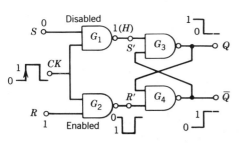

(b)

FIGURE 7.3-5   The clocked *RS* flip-flop making a transition (*a*) from the reset state to the set state, (*b*) from the set state to the reset state.

gate G_3 are 1, holding $Q = 0$; if a clock pulse is now momentarily applied to the CK input, then S' and R' remain 1 since G_1 and G_2 are disabled by $S = R = 0$. Hence there is no change in the state of the FF—it remains reset. Now let us make $S = 1$, leave $R = 0$, and apply a clock pulse. Since $S = 1$, S' goes to 0 when $CK = 1$, causing Q to be 1; both inputs to gate G_4 are now 1 ($R' = 1$ since $R = 0$), forcing $\bar{Q}$ to be 0. This $\bar{Q} = 0$ is an input of G_3, insuring that Q will remain $Q = 1$ after the clock pulse goes back to 0. The FF is now in the set state, as illustrated in Figure 7.3-5a. Next, let us make $S = 0$, $R = 1$, and apply a clock pulse. Since $R = 1$, $CK = 1$ produces $R' = 0$, causing $\bar{Q}$ to be 1; since $\bar{Q} = 1$, both inputs to G_3 are now 1, forcing Q to be 0. This $Q = 0$ is an input of G_4, insuring that $\bar{Q}$ will remain 1 after the clock pulse goes back to 0; the FF is now in the reset state, as illustrated in Figure 7.3-5b. As with the basic RS latch, Q and $\bar{Q}$ are not used when both S and R are 1 at the same time. The behavior of the clocked RS FF is summarized by the truth table of Figure 7.3-6.

Edge-Triggered *SR* Flip-Flop with Clear and Preset Asynchronous Inputs

Most commercially available flip-flops have two additional inputs: Clear (CLR), or direct reset, and Preset (PR), or direct set. Their functions are listed in Table 7.3-1.

When both Preset and Clear are *active* (i.e., when they are at the low voltage level for a positive-logic flip-flop or both are at logic 0), $Q = \bar{Q} = 1$, which is

CK	S	R	Q_{n+1}	$\bar{Q}_{n+1}$	Comments
1	0	0	Q_n	$\bar{Q}_n$	No change
1	0	1	0	1	Reset condition
1	1	0	1	0	Set condition
1	1	1	?	?	Indeterminate (not used)

FIGURE 7.3-6   Truth table for the clocked *RS FF.*

TABLE 7.3-1
Functions of Clear and Preset Inputs in Most Flip-Flops

Preset	Clear	Q	$\bar{Q}$	Remarks
0	1	1	0	As Preset $= 0$ (L), the flip-flop sets regardless of
1	0	0	1	what the other inputs (e.g., S, R, CK) are.
0	0	1^a	1^a	Similarly, as Clear $= 0$, the flip-flop resets regard-
1	1	No change		less of what the other inputs are. Essentially the Preset or Clear input "overrides" the other inputs.

[a]Not allowed.

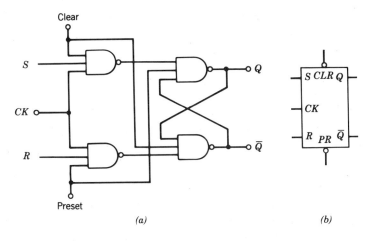

		Input			Output	
PR	*CLR*	*CK*	*S*	*R*	Q_{n+1}	$\bar{Q}_{n+1}$
0	1	x	x	x	1	0
1	0	x	x	x	0	1
0	0	x	x	x	1*	1*
1	1	1	0	0	Q_n	$\bar{Q}_n$
1	1	1	0	1	0	1
1	1	1	1	0	1	0
1	1	1	1	1	1*	1*

*Not allowed; x = regardless of state.

(c)

FIGURE 7.3-7 Edge-triggered *SR* flip-flop with Clear and Preset asynchronous inputs: (*a*) logic diagram; (*b*) logic symbol; (*c*) truth table.

not allowed for any flip-flops. These inputs are asynchronous and independent of the clock. While they are present, all other operations are inhibited. An *SR* edge-triggered flip-flop with Clear and Preset asynchronous inputs is shown in Figure 7.3-7. Using this circuit as a building block, other types of flip-flops with Clear and Preset can be constructed.

7.4 *JK* AND *D* FLIP-FLOPS

JK Flip-Flop

The *JK* flip-flop shown in Figure 7.4-1 has two data inputs, *J* and *K*, a single clock input (*CK*), and two outputs, *Q* and $\bar{Q}$. The operation of a *JK* flip-flop is identical to that of an *SR* flip-flop, except that the former allows the input

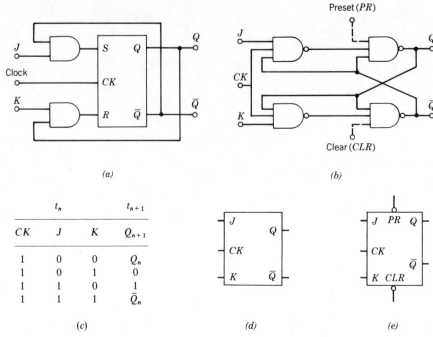

	t_n			t_{n+1}
CK	J	K		Q_{n+1}
1	0	0		Q_n
1	0	1		0
1	1	0		1
1	1	1		$\bar{Q}_n$

(c)   (d)   (e)

FIGURE 7.4-1   *JK* flip-flop: (*a*) obtained by modifying an *SR* flip-flop; (*b*) logic diagram; (*c*) truth table where Q_{n+1} (Q_n) means the state of *Q* output after (before) clocking; (*d*) symbolic representation without Clear and Preset; (*e*) symbolic representation with Clear and Preset.

$J = K = 1$. When $J = K = 1$, the state *Q* will change regardless of what the state *Q* was prior to clocking. Note that the operation of the *JK* flip-flop may be unstable. Because of the feedback connection Q $(\bar{Q})$ at the input to $K(J)$, the input will change during the clock pulse ($CK = 1$) if the output changes state. Hence, for the pulse duration t_p (while $CK = 1$), the output will oscillate back and forth between 0 and 1. At the end of the pulse ($CK = 0$), the value of *Q* is ambiguous. This unstable situation is called a *race-around condition*.

There are two categories of commercially available *JK* flip-flops: (1) edge-triggered *JK* flip-flops, such as the TTL 7470 JK positive edge-triggered *FF* and the 74LS73 dual *JK* negative edge-triggered *FF*; and (2) *JK* master-slave FFs, such as the TTL 7472, and so on.

JK Flip-Flop with J and K Inputs Tied Together

The *JK* flip-flop with *J* and *K* inputs tied together forms the *T* flip-flop, which is not offered as a separate circuit by manufacturers. The *T* flip-flop shown in Figure 7.4-2 has only one data input, *T*. When the input *T* is at a 0 level prior to a clock pulse, the *Q* output will not change with clocking. When the input *T* is at a 1 level, the *Q* output will be in the $\bar{Q}_n$ state after clocking (i.e., $Q_{n+1} = \bar{Q}_n$).

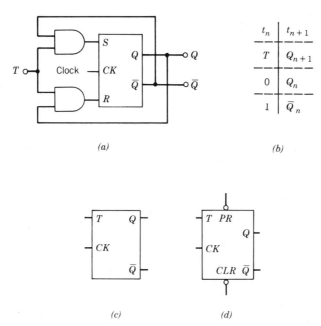

t_n	t_{n+1}
T	Q_{n+1}
0	Q_n
1	$\overline{Q}_n$

(a)   (b)

(c)   (d)

FIGURE 7.4-2 *T* flip-flop: (a) logic diagram; (b) truth table; (c) symbol 1 (without Clear and Preset); (d) symbol 2 (with Clear and Preset). If *T* input is high, positive transitions of *CK* input will cause the output to change state.

This is called *toggling*; thus the name *T flip-flop*. When the *T* input and the clock of a *T* flip-flop circuit are tied together, the flip-flop circuit generates two clock signals of the same frequency as the input clock signal.

Master-Slave *JK* Flip-Flops

The circuit of a master-slave flip-flop is basically two latches connected serially. The first latch is termed the *master*, and the second is termed the *slave*. A typical master-slave *JK* flip-flop and a master-slave *JK* flip-flop with Clear and Preset asynchronous inputs are shown in Figures 7.4-3a and 7.4-3b, respectively. Normal action in master-slave clocking consists of four steps, as shown in Figure 7.4-4. The main feature of this type of clocking is that the data inputs are never directly connected to the outputs at any time during clocking; this provides total isolation of outputs from data inputs.

The function (or truth) table operations for both the master-slave and the edge-triggered *JK* flip-flops are identical; the primary difference is the clocking operation. For an edge-triggered type, the data on the inputs is entered into the flip-flop and appears on the outputs on the same edge of the clock pulse. For a master-slave type, the data on the *J* and *K* inputs is entered on the leading edge of the clock but does not appear on the outputs until the trailing edge of the clock.

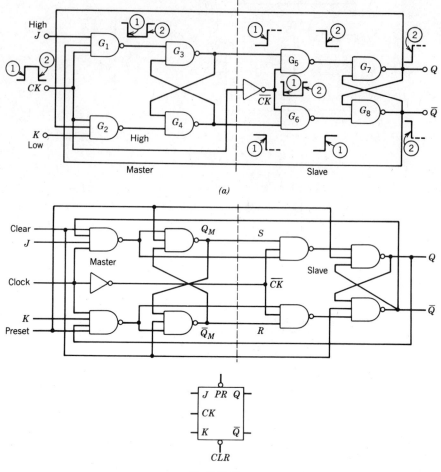

Truth Table

		Input			Output	
PR	CLR	CK	J	K	Q_{n+1}	$\bar{Q}_{n+1}$
0	1	×	×	×	1	0
1	0	×	×	×	0	1
0	0	×	×	×	1*	1*
1	1	⎍	0	0	Q_n (No change)	$\bar{Q}_n$
1	1	⎍	0	1	0 (Reset)	1
1	1	⎍	1	0	1 (Set)	0
1	1	⎍	1	1	$\bar{Q}_n$ (Toggle)	Q_n

*Not allowed; × = Regardless of state.

(b)

FIGURE 7.4-3 (a) A typical master-slave JK flip-flop, making the transition from the reset state to the set state; (b) master-slave JK flip-flop with Clear and Preset asynchronous inputs.

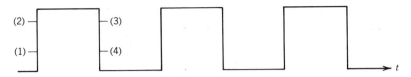

FIGURE 7.4-4   Master-slave clock. (1) Isolate slave from master. (2) Enable data inputs to master. (3) Disable data inputs. (4) Transfer data from master to slave.

Figure 7.4-3*a* exhibits the transitions for the flip-flop going from the reset state to the set state. Let us examine the operation of this circuit in detail. Assume that the flip-flop is reset. Let *J* be high and *K* be low, making a set condition on the inputs. The circled number associated with each transition indicates when that transition occurs with respect to the clock pulse. A circled 1 corresponds to the leading edge of the clock pulse and a circled 2 corresponds to the trailing edge. On the leading edge of the clock pulse, the circuit operates as follows:

1. The output of gate G_1 goes from high to low, since all its inputs are high.
2. The output of G_3 goes from low to high and the output of gate G_4 goes from high to low because of the low on the output of gate G_1.
3. The inverted clock ($\overline{CK}$) goes low, disabling both G_5 and G_6 gates. This insures that their outputs remain high.
4. The master section has been set, since the *J* input is high and the *K* input is low. The slave section (*SR*) has not changed state, and so the *Q* and $\bar{Q}$ outputs remain in the reset state.

On the trailing edge of the clock pulse, the circuit of Figure 7.4-3*a* operates as follows:

1. The master section remains in the set state.
2. The output of gate G_5 goes from high to low, since both its inputs are now high.
3. The output of gate G_7 goes from low to high and the output of gate G_8 goes from high to low because of the low on the output of gate G_5.
4. The flip-flop is now in the set state, since the *Q* output is high. It did not become set until the trailing edge of the clock pulse, although the master section was set on the leading edge.

In summary, during clock pulse, the output *Q* does not change but the output Q_M of the master section follows *JK* logic; at the end of the pulse, the value of Q_M is transferred to *Q*. Since Q_n (the output before clocking) is invariant for the duration t_p of the clock pulse, the unstable situation in the basic *JK* flip-flop of Figure 7.4-1*b* will be eliminated.

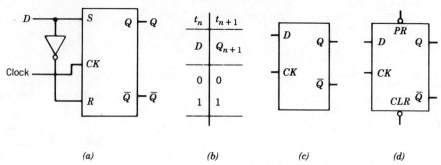

FIGURE 7.4-5 *D* flip-flop: (*a*) logic diagram; (*b*) truth table; (*c*) and (*d*) symbols.

D Flip-Flop

When the $S(J)$ input of an SR flip-flop (or JK flip-flop) is connected to the inverted R (or K) input of the same flip-flop, a D flip-flop is formed, as shown in Figure 7.4-5. This type of flip-flop has only one data input, D. The operation of the D flip-flop may be described as follows. When the D input is at a 0 level prior to a clock pulse, the Q output will be 0 after clocking. When the D input is at a 1 level, the Q output will be 1 after clocking. In other words, $Q_{n+1} = D_n$, where Q_{n+1} and D_n denote the value of the output Q and the input D at t_{n+1} (after clocking) and t_n (before clocking), respectively. For this type of flip-flop, the Q output is identical to the D input, except that there is one pulse timing delay; thus the name D flip-flop.

The clocked D flip-flop or simple D flip-flop is closely related to the JK flip-flop and has become quite useful since the development of the integrated circuit. This device eliminates the K gate external input connection by including an on-chip inverter from J input to K. This always forces K to equal $\bar{J}$. Therefore, if $J = 1$, a negative clock transition will result in $Q = 1$. If $J = 0$, the $Q = 0$ after the transition. The TTL 7475 contains four D flip-flops on a single chip.

7.5 BISTABLE MULTIVIBRATOR AS CONTROL OF MICROPOWER SWITCH

Bistable Switch Circuits

The bistable micropower switch with radio frequency (RF) and light triggering is shown in Figures 7.5-1 and 7.5-2. The central part of this circuit is a complementary silicon transistor pair bistable multivibrator (Q_1-Q_2). Both transistors are conducting when the switch is on and are cut off when the switch is in the off state. The switch itself consists of transistors Q_3 and Q_4. These transistors are complementary so that when the switch is in the off state, both transistors are

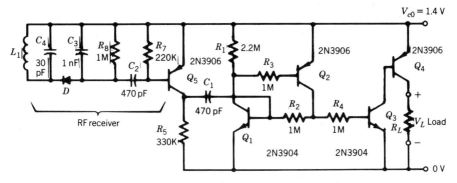

FIGURE 7.5-1   Circuit diagram for the radio-frequency-triggered bistable switch. R_L represents the load switched on and off by this circuit. (Figure 7.5-1 to 7.5-4 from Santic, A., S. Vamvakas, and M. R. Neuman, Micropower Electronic Switches for Implanted Instrumentation, *IEEE Transactions on Biomedical Engineering*, Vol. BME-29, No. 8, August 1982, pp. 583–589. Copyright © 1982 IEEE.)

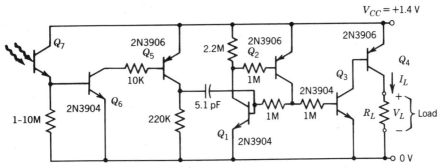

FIGURE 7.5-2   Circuit diagram for a light-triggered bistable switch. Q_7 is an *npn* phototransistor sensitive to visible or near-infrared light. When photovoltaic devices are substituted for Q_7, two or three should be connected in series in place of the phototransistor. (From Santic, A. S. Vamvakas, and M. R. Neuman, Micropower Electronic Switches for Implanted Instrumentation, *IEEE Transactions on Biomedical Engineering*, Vol. BME-29, No. 8, August 1982, pp. 583–589. Copyright © 1982 IEEE.)

cut off. Triggering transistor Q_5 is also arranged so that it does not conduct when the switch is in the off state. Thus current consumption will be very low when the circuit is in the off state. Since all transistors are cut off, the sensitivity of the circuit will be lower than if the transistors were biased in their active region.

The switch is turned on or off by altering the state of the bistable multivibrator. At the base of Q_1, a positive pulse turns the transistors to the on state and a negative pulse will turn them off again. At the base of Q_2, the opposite will be true. Triggering of both states with unipolar pulses at the same point is not possible with this circuit. In the case of this bistable switch, the base of Q_1 was chosen for triggering and the necessary positive or negative pulses were achieved by differentiating the envelope of the radio-frequency triggering signal. The two methods of doing this are illustrated in Figure 7.5-3.

Triggering Waveforms

A radio frequency or a light signal must be on continuously while the switch is on, as shown by signal A in Figure 7.5-3. The detected and low-pass filtered envelope of such a signal is waveform B. The derivative of B taken with an RC differentiator is signal C, which gives a positive pulse when the continuous-wave triggering signal begins and a negative pulse when it ends. Signal C can be fed directly to the base of Q_1 to trigger the bistable switch on and off.

The triangular wave trigger source D in Figure 7.5-3 can be used in the circumstances where it is not convenient to have the triggering signal continuously on while the bistable switch is on. The envelope E of the trigger signal consists of a sharp rise and slow fall for turning the bistable multivibrator on and a slow rise and sharp fall for turning it off. When this is differentiated, the biphasic pulses F are obtained. We see that a narrow high amplitude pulse is

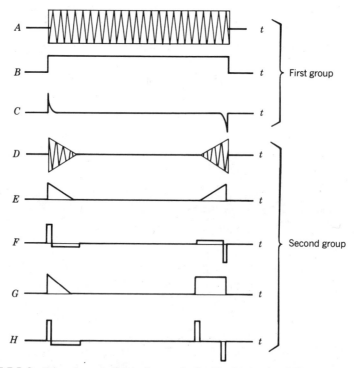

FIGURE 7.5-3 Triggering waveforms that can be used in the circuits of Figures 7.5-1 and 7.5-2. (From Santic, A., S. Vamvakas, and M. R. Neuman, Micropower Electronic Switches for Implanted Instrumentation, *IEEE Transactions on Biomedical Engineering*, Vol. BME-29, No. 8, August 1982, pp. 583–589. Copyright © 1982 IEEE.)

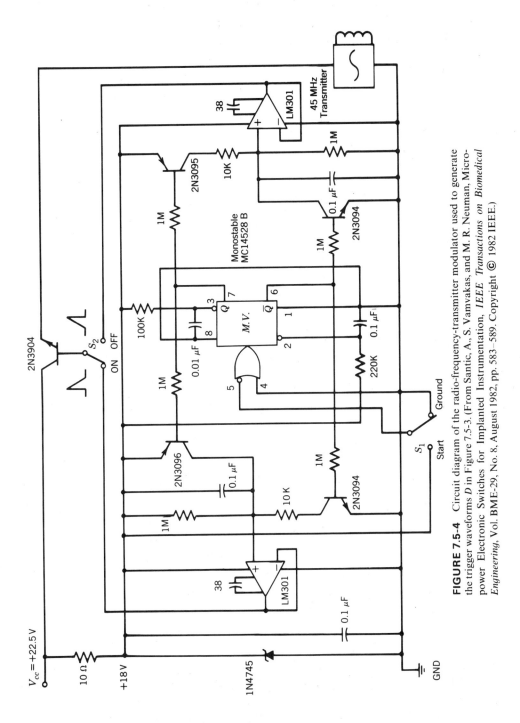

FIGURE 7.5-4 Circuit diagram of the radio-frequency-transmitter modulator used to generate the trigger waveforms *D* in Figure 7.5-3. (From Santic, A., S. Vamvakas, and M. R. Neuman, Micropower Electronic Switches for Implanted Instrumentation, *IEEE Transactions on Biomedical Engineering*, Vol. BME-29, No. 8, August 1982, pp. 583–589. Copyright © 1982 IEEE.)

generated for the on pulse, followed by a wide, low-amplitude, equal-area pulse, while the off trigger is just the opposite. In this way, the trigger signal need not be on continuously while the bistable circuit is in the on state. The signals G and H can also be used to control the bistable circuit, since all that is required to turn the circuit off is that the turn-off signal has a sharp falling edge.

The trigger waveforms D in Figure 7.5-3 are generated by the circuit of the radio-frequency transmitter modulator shown in Figure 7.5-4. This circuit consists of a 45-MHz free-running oscillator followed by a buffer and a class-B power amplifier that are modulated by a triangular signal produced by the ramp generator. The input power to the final amplifier stage is 0.8 W. The output tank circuit coil also serves as the magnetic signal source for the switch circuit. It consists of a 0.7-μH coil wound on a 2.5-cm diameter form. Due to the flywheel effect of the LC tank, the output modulated waveform has the shape shown in D in Figure 7.5-3.

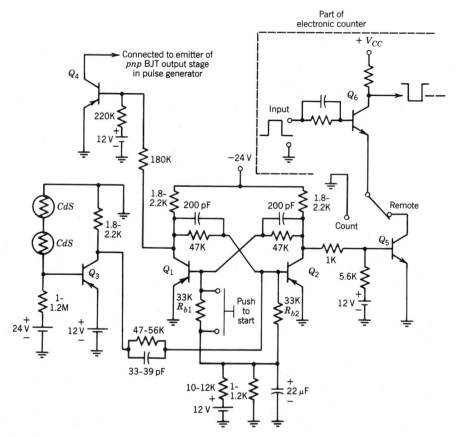

FIGURE 7.6-1 A bistable circuit used to control the start and stop of the pulse generator and electronic counter operated for a measurement in a preset time. Q_1, Q_2, and Q_3 are Ge low power transistors.

7.6   BISTABLE CONTROL CIRCUIT

A bistable control circuit is shown in Figure 7.6-1. It is used to control the start and stop of the pulse generator and electronic counter operated for a measurement in a preset time.

The central part of the circuit is the fixed bias bistable multivibrator Q_1-Q_2. The base of transistor Q_3 is connected with two series photoconductors (CdS) to ground. When the photoconductors are not illuminated, Q_1 is cut off due to its reverse-biased base-emitter junction. When the start button is pushed, resistor R_{b1} is shorted, forcing Q_1 to be cut off and Q_2 to be on. Due to the effect of this state, Q_4 is on, causing the pulse generator to start, and Q_5 is on, causing the electronic counter to count. The photoconductors are placed close to the last-digit light for the preset time. When the preset time for counting is reached, the photoconductors are illuminated, forcing Q_3 to saturate. Under this condition, Q_2 is off, forcing Q_5 to be off, and Q_1 is on, forcing Q_4 to be off. From the effect of this state, both the pulse generator and the counter stop working simultaneously.

REFERENCES

1. Bell, D. A., *Solid State Pulse Circuits*, 2nd ed., Reston Publishing Company, Reston, Virginia, 1981, Chap. 9.
2. Tocci, R. J., *Fundamentals of Pulse and Digital Circuits*, 2nd ed., Charles E. Merrill, Columbus, Ohio, 1977, Chap. 10.
3. Millman, J. and H. Taub, *Pulse, Digital, and Switching Waveforms*, McGraw-Hill, New York, 1965, Chap. 10.
4. Floyd, T. L., *Digital Fundamentals*, 2nd ed., Charles E. Merrill, Columbus, Ohio, 1982, Chap. 7.
5. Taub, H., *Digital Circuits and Microprocessors*, McGraw-Hill, New York, 1982, Chap. 4.
6. Santic, A., S. Vamvakas, and M. R. Neuman, Micropower Electronic Switches for Implanted Instrumentation, *IEEE Transactions on Biomedical Engineering*, Vol. BME-29, No. 8, August 1982, pp. 583–589.

QUESTIONS

7-1.   What is the difference between a static latch and a dynamic latch?

7-2.   Sketch the static latch containing a pair of coupled inverters.

7-3.   Sketch the fixed-bias flip-flop circuit using *npn* transistors. Explain its operation.

7-4.   Repeat Problem 7-3 for a circuit using *pnp* transistors.

7-5.   Repeat Problem 7-3 for a self-bias flip-flop circuit.

7-6.   Sketch the asymmetrical collector triggering circuit for a fixed-bias bistable multivibrator. Show the triggering waveforms and explain the circuit function.

7-7. Repeat Problem 7-6 for symmetrical collector triggering.

7-8. Sketch logic diagrams for the RS NOR-gate flip-flop and $\bar{R}\bar{S}$ NAND-gate flip-flop. Write a truth table for each circuit.

7-9. Sketch logic diagrams for the gated RS latch and D-gated latch. Explain how they work.

7-10. Sketch the logic diagram for the clocked RS flip-flop. Explain how it operates.

7-11. Sketch the logic diagram for the SR negative edge-triggered flip-flop with Clear and Preset asychronous inputs. Also write its truth table and explain its operation.

7-12. Sketch the logic diagram for the JK flip-flop. Also write its truth table and explain its operation.

7-13. Sketch the logic diagram for the T flip-flop. Also write the truth table and briefly explain its operation.

7-14. Draw a master-slave JK flip-flop system. Explain its operation and show that the race-around condition is eliminated.

7-15. Sketch the logic diagram for the D flip-flop and explain its operation.

7-16. Draw the circuit for the radio-frequency-triggered bistable switch and explain its operation.

7-17. Draw the circuit for the light-triggered bistable switch and explain its operation.

PROBLEMS

7-1. In the circuit of Figure 7.1-2a, $R_{C1} = R_{C2} = 1\,k\Omega$, $R_1 = R_1' = 10\,k\Omega$, $R_2 = R_2' = 47\,k\Omega$, $V_{CC} = +5\,V$, and $V_{BB} = -5\,V$. Verify that this circuit operates as a flip-flop, exhibiting two stable states. The transistors are silicon, each having an $h_{FE(min)} = 40$, $V_{CE(sat)} = 0.2\,V$, and negligible reverse saturation current.

7-2. Design a fixed-bias flip-flop to operate from a ± 5-V supply. Use silicon *npn* transistors (e.g., 2N3903) with $I_C = 2\,mA$, $h_{FE(min)} = 35$, and $t_{off} = 225$ ns. Refer to Figure 7.1-2.

Answer. $R_{C1} = R_{C2} = 2.2\,k\Omega$; $R_2 = 27\,k\Omega$; $R_1 = 12\,k\Omega$; $C_1 = C_2 = 270$ pF; $f_{max} = 193.83$ kHz.

7-3. The flip-flop designed for Problem 7-2 is to be triggered by the collector output of a previous similar stage. Design a suitable symmetrical collector triggering circuit.

Answer. $C_3 = 68$ pF.

7-4. (a) Verify that $R = S = 1$ in Figure 7.3-1 is not allowed. (b) Verify that $\bar{R} = \bar{S} = 0$ in Figure 7.3-2 is not allowed.

7-5. Sketch the Q output waveform if the inputs shown in Figure p7-5 are applied to the gated RS latch (Figure 7.3-3), which is initially reset. Explain the latch function referring to the waveforms.

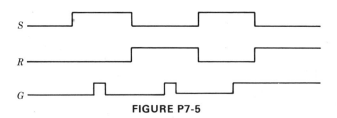

FIGURE P7-5

7-6. Sketch the Q output waveform if the inputs shown in Figure p7-6 are applied to the D-gated latch (Figure 7.3-4), which is initially reset. Explain the latch function referring to the waveforms.

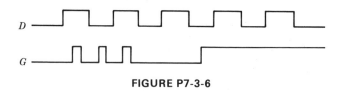

FIGURE P7-3-6

7-7. (a) Verify that an RS flip-flop becomes a T flip-flop if S is connected to $\bar{Q}$ and R to Q. (b) Verify that a D flip-flop becomes a T flip-flop if D is tied to $\bar{Q}$.

7-8. Sketch the output waveform from the T flip-flop of Figure p7-8. The wedge-shaped symbol on the CK input indicates that the output change will occur on the positive transition.

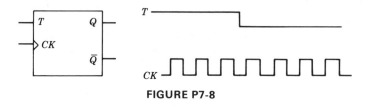

FIGURE P7-8

8

BASIC LOGIC GATES
AND LOGIC FAMILIES

8.1 INTRODUCTION

The basic circuits most commonly used in digital systems are the four basic logic gates: the OR, AND, NOT, and flip-flop. Flip-flops are the bistable multi-vibrators described in Chapter 7. For conventional digital systems, integrated circuits are used almost exclusively, with discrete digital circuits applied to high-power or other special designs. There are four broad classifications of integration: small-scale integration (SSI), medium-scale integration (MSI), large-scale integration (LSI), and very large-scale integration (VLSI). Logic elements that have in common a particular configuration and device are said to belong to a *logic family*. Some examples of presently useful families are transistor-transistor logic (TTL), emitter-coupled logic (ECL), complementary symmetry MOS (CMOS), and integrated injection logic (I^2L). All circuits of a given family should have compatible operating characteristics.

Logic gates are usually connected in complex combinations. The number of gates driven by a single gate is called the *fan-out* of that driving gate. We need to know how many gate inputs of driven gates we can connect to the output of a driving gate. Manufacturers generally provide this information by specifying a gate fanout. In TTL, provided each driven gate is driving gates of its same series, the fanout is 10 for the standard and high-power series and 20 for the low-power series. CMOS, because of its high input impedance, has the largest fan-out capability of any logic family. If I_L is the total output current that a gate can handle and I_i is the drive current for each input or unit load, then the fan-out is I_L/I_i. The *fan-in* of a gate is the number of inputs that can be connected to a gate.

A logic family or a series within a family is characterized by four parameters: (1) fan-out; (2) noise margins; (3) propagation delay; and (4) power dissipation.

The fan-out is often not an adequate parameter when we make interconnections between series. *Noise margin* is the maximum noise voltage that can be added to (or subtracted from) a digital signal before passing a threshold. *Propagation delay*, t_d, is the intrinsic delay existing in a logic gate or device, beginning with the application of an input and ending with the device's response. In other words, t_d is the time required for the gate to switch from its low output state to its high output state and vice versa. t_d results from unavoidable parasitic capacitance or inductance in an electronic circuit. NAND gate propagation delays in TTL are typically 10 ns. We can calculate the speed-power product from the propagation delay and power dissipation.

8.2 NOR GATES AND NAND GATES

NOR Gates

The OR gate and AND gate shown in Figures 3.7-1 and 3.7-2 were described in Section 3.7. Each of these circuits has a resistor R connected at the output terminal. If such a gate is to drive successive circuits, the input impedance of these following stages should be much greater than R in order to negligibly affect the operation of the gate. The output impedance of any stage that precedes the gate should be much less than R to minimize attenuation. Therefore one or two stages of amplification are commonly added to the gate in order to solve the loading problem resulting when three gates are arranged in cascade. If a single inverting amplifier is employed, the OR gate becomes a NOR gate as shown in Figure 8.2-1. If a noninverting amplifier is used, the OR gate characteristics remain unchanged. The NOR gate symbol contains an OR gate followed by a small circle that represents an inversion. The term NOR is a contraction of NOT-OR.

The inverting amplifier used as a logic gate is called an *inverter* (NOT circuit), as described in Section 4.2. An inverter has a single input and a single output that is the logical complement of the input. When the input is H, the output is L, or vice versa; that is, when the input is A, the output is $Y = \bar{A}$. The essential part of the logic symbol for an inverter is the small circle at the apex of the triangle form, as shown in Figure 4.3-1b. When the inversion is to be indicated on a

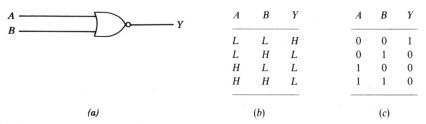

A	B	Y		A	B	Y
L	L	H		0	0	1
L	H	L		0	1	0
H	L	L		1	0	0
H	H	L		1	1	0

(a) (b) (c)

FIGURE 8.2-1 Positive logic NOR gate: (*a*) NOR gate symbol: (*b*) function table; (*c*) truth table for positive logic.

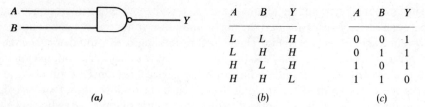

A	B	Y		A	B	Y
L	L	H		0	0	1
L	H	H		0	1	1
H	L	H		1	0	1
H	H	L		1	1	0
(a)				(b)		(c)

FIGURE 8.2-2   Positive logic NAND gate: (*a*) NAND gate symbol; (*b*) function table; (*c*) truth table for positive logic.

logic diagram that has other logic gates or symbols to which the circle can be affixed, the triangular form is omitted.

NAND Gate

The term NAND is a contraction of NOT-AND, and it implies an AND function with a complemented (inverted) output. Thus a NAND gate is an AND gate followed by an inversion, as shown in Figure 8.2-2. If negative logic is used, the NOR gate of Figure 8.2-1 becomes a NAND gate and the NAND gate of Figure 8.2-2 becomes a NOR gate.

All IC logic families supply inverters that can easily be employed with other gates to provide required logic functions. For instance, a NOR gate following two inverters can implement an AND function. Also a NAND gate following two inverters can implement an OR function. NOR and NAND gates are often employed as simple inverters: thus a given gate serves a variety of functions.

8.3   TRANSISTOR-TRANSISTOR LOGIC (TTL) CIRCUITS

The TTL family is one of the most popular logic families at the present time. This family has good fan-out figures and relatively high-speed switching. A significant improvement in TTL switching speed results from employing Schottky barrier diodes to clamp the base-collector junctions of all transistors to avoid heavy saturation. This diode-clamped arrangement is shown in Figure 8.3-1. A low forward voltage across the Schottky diode causes the diode to divert most of the excess base current around the base-collector junctions. The Schottky-clamped TTL gates possess propagation delay times of 2–3 ns.

The basic TTL gate is shown in Figure 8.3-2. When the low voltage level appears at one or more of the inputs, Q_1 will be saturated with a very small voltage appearing at its collector. In this case Q_2 and Q_3 are off, since at least $2V_{BE(on)}$ should appear at the base of Q_2 in order to turn Q_2 and Q_3 on. When Q_2 is off, the current through R_2 is diverted into the base of Q_4, which then drives the load as an emitter follower.

When all inputs are at the high voltage level, transistors Q_2 and Q_3 are turned on, clamping the collector of Q_1 to approximately $2V_{BE(on)}$. In this case the

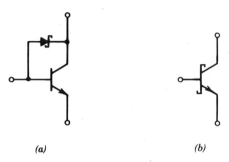

FIGURE 8.3-1   (*a*) Schottky-clamped transistor and (*b*) its symbol.

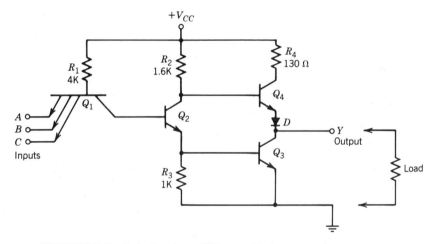

FIGURE 8.3-2   Basic three-input *TTL* gate with the totem-pole output stage.

base-collector junction of Q_1 appears as a forward-biased diode, whereas the base-emitter junctions are reverse-biased diodes. When Q_2 turns on, the base voltage of Q_4 drops, decreasing the current through the load. Since Q_3 is turning on to divert more current from the load, the load current tends to decrease even faster than it would if only Q_4 were present. Thus Q_2 and Q_3 are on with Q_4 off at the end of the transition. This circuit behaves as a positive logic NAND gate.

The arrangement of the output transistors (Q_2, Q_3, and Q_4) is termed a *totem pole*. The transistors function as follows: When Q_2 is off, R_3 biases Q_3 off and R_2 biases Q_4 on. Hence Q_4 provides active pull-up or low output impedance, resulting in the high state at the gate output. When Q_2 is on in saturation, base current supplied to Q_3 drives Q_3 into saturation. Therefore the output voltage is pulled down and Q_3 offers a low output impedance, resulting in the low state at the gate output. At this time, Q_4 is biased off by the voltage drop across R_2. As the emitter follower Q_4 turns on, its output impedance decreases.

Turning Q_4 off increases its output impedance and can lead to distortion of the load voltage, especially for capacitive loads. The totem-pole output stage overcomes this problem.

A standard method of improving the high-speed switching characteristics of TTL is to add clamping diodes to the input emitters of the gate to reduce transmission line effects by providing more symmetrical impedances. This improvement is shown in Figure 8.3-3 along with smaller resistors and an output Darlington connection. This gate has a typical propagation delay time of 6 ns.

TTL is available in five series; they are listed with their characteristics in Table 8.3-1. A tristate TTL logic gate has a control input as well as the usual input and output terminals.

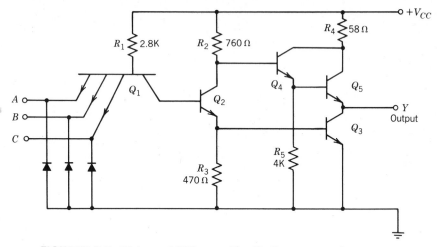

FIGURE 8.3-3 High-speed TTL gate with a Darlington connection at output.

TABLE 8.3-1
Typical Characteristics of TTL in Five Series (54/74 Family SSI)

Series	Power and Transistor Type	Propagation Delay Time t_d (ns)	Power Dissipation P_d (mW)	Noise Margin (V)
54LS/74LS	Low-power Schottky	9.5	2	0.4
54L/74L	Low-power standard	33	1	0.4
54S/74S	Standard-power Schottky	3	19	0.4
54/74	Standard-power standard	10	10	0.4
54H/74H	High-power standard	6	22	0.4

8.4 EMITTER-COUPLED LOGIC (ECL) FAMILY

The storage time of standard transistors is the chief limitation of the switching speed of logic gates. The *storage time* is the time required to reverse the forward bias on the collector-base junction of a saturated transistor. The good switching characteristics of the emitter-coupled logic (ECL) family result from the avoidance of saturation of any transistor within the gate.

Figure 8.4-1 shows a typical integrated-circuit ECL gate with two separate outputs, X and Y. For positive logic X is the OR output while Y is the NOR output. Frequently the positive supply voltage is taken as 0 V and V_{EE} as approximately -5 V. When the ECL family operates with a negative supply voltage of -5.2 V, the logic levels are approximately -0.75 V and -1.6 V. A fan-out as large as 25 is generally allowed. The noise margins are of the order of 0.3 V. The major merit of ECL is its high speed. The propagation delay time can be 2 ns or less. The power dissipation is about 25 mW per gate. The ECL is also called *current mode logic*. Although the output stages are emitter followers, they conduct reasonable currents for both logic level outputs and, therefore, minimize the asymmetrical output impedance problem. The ECL family has the disadvantage of requiring more input power than the TTL family.

In Figure 8.4-1, Q_5 has its base bias voltage provided by the divider composed of $R_5, D_1, D_2,$ and R_6. The diodes provide temperature compensation for changes in V_{BE5}. Q_5 operates as an emitter follower to provide a low impedance bias to the base of Q_4. With a constant bias voltage at the base of Q_4, the voltage drop across emitter resistor R_2 is also held constant so long as all inputs are low

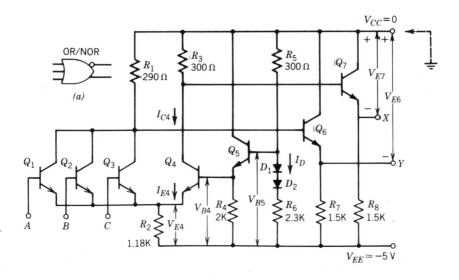

(b)

FIGURE 8.4-1   A three-input ECL OR/NOR gate: (a) symbol; (b) circuit.

enough to maintain Q_1, Q_2, and Q_3 in the off state. Under these conditions, the emitter and collector currents of Q_4 are kept constant and Q_4 is out of saturation. With Q_4 on, the output voltage V_{E7} is low while V_{E6} is high. When a positive voltage is applied to terminal A OR B OR C, the emitter voltage of Q_4 is pulled up above its base level. Thus Q_4 switches off as Q_1, Q_2, or Q_3 switches on. In this case the voltage at Q_6 base falls and that at Q_7 base rises, resulting in a low value of V_{E6} and a high value of V_{E7}. Consequently the gate functions as an OR gate when the output is derived from Q_7 emitter and as a NOR gate when the output is derived from Q_6 emitter.

8.5   MOSFET LOGIC

PMOS and NMOS

The MOS digital ICs use enhancement MOSFETs. The enhancement PMOSFET (or NOMOSFET) is normally off when its gate is at the same potentials as its substrate. When the gate is made negative (or positive) with respect to the substrate, a p type (or n type) channel is created from drain to source. NMOS circuits are very similar to PMOS circuits, with the important exception that all voltage polarities and current directions are reversed. NMOS is faster than PMOS due to the fact that charge carriers (electrons) in n-channel devices have greater mobility than those (holes) in p-channel FETs.

The major advantages of MOS logic over TTL and other bipolar logic families are (1) the improved packing density since on-chip resistors are not required, (2) the lower power dissipation resulting from high impedance active loads, and (3) the simplicity of the fabrication process. Many microprocessors are based on MOS logic to make their cost reasonable. Compared with TTL, MOS logic has a slower operating speed and a lower drive-current capacity.

CMOS Logic Family

CMOS switching was explained in Section 4.8. The CMOS logic family includes many SSI and MSI devices. CMOS offers an externally low power dissipation, possibly operates with a supply of 1 V to 18 V, and has excellent immunity.

An inverter like the CMOS switch shown in Figure 4.8-1 is the basic building block for a CMOS logic gate. The threshold voltage can be controlled during fabrication and is generally designed so that switching occurs at an input voltage of about $V_{DD}/2$. The output voltage levels of a CMOS logic gate are typically 4.99 V to 5.00 V for the 1 state ($V_{DD} = 5$ V) and less than 0.01 V for the 0 state, corresponding to the required input voltages of about 0.0 V to 1.5 V for a 0 and 3.5 V to 5.0 V for a 1, respectively. A CMOS NOR gate for positive logic is shown in Figure 8.5-1. If both inputs A and B are at the 0 logic level or ground, both PMOS devices are on while both NMOS devices are off. Thus output Y is very close to V_{DD}. If A moves to the 1 state while B is held at ground level, the

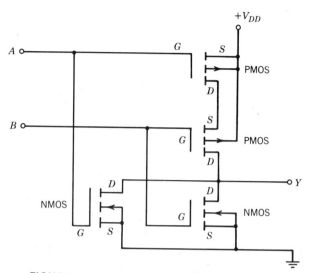

FIGURE 8.5-1   Positive logic NOR gate using CMOS.

upper PMOS device turns off while the corresponding NMOS device turns on; thus output Y is very close to ground. If B is at 1 level while A is at 0 level, the lower PMOS device shuts off while the corresponding NMOS device turns on again, resulting in an output 0. When both A and B are at 1 level, the output is at 0 level.

A CMOS NAND gate for positive logic is shown in Figure 8.5-2. When A, B, or both A and B are at the low logic level, at least one of the series NMOS

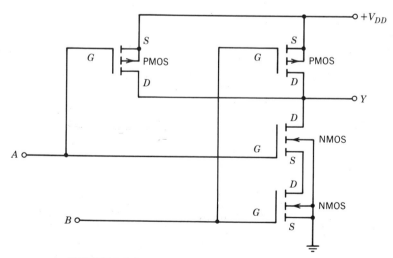

FIGURE 8.5-2   Positive logic NAND gate using CMOS.

devices is off and at least one of the parallel PMOS devices is on. The output Y is then high. If both A and B are raised to the high logic level, Y is equal to the low level. In this case, both PMOS devices are off while both NMOS devices are on.

Either CMOS or low-power TTL (74L) is suitable for applications where switching speed is not the chief consideration. CMOS has the following typical characteristics: propagation delay time = 25 ns; power dissipation per gate = 10 nW; noise margin (V) = $0.3\,V_{DD}$; fan-out > 50. A CMOS logic gate has excellent noise immunity and is at least five times smaller than the comparable TTL gate.

8.6   ANALYTIC EXPRESSIONS FOR MOSFET CHARACTERISTICS
Drain Characteristics and Transfer Curve

In an n-channel MOSFET, the charge carriers are negative, the conventional direction of current I_{DS} is from drain (D) to source (S), and so the voltage V_{DS} is positive, as is also the current I_{DS} (see Figure 8.6-1c). Figure 8.6-1b shows the transfer curve of an enhancement NMOSFET, which indicates $I_{DS} = 0\,\text{mA}$ until the gate-to-source voltage V_{GS} exceeds a threshold voltage V_T. There is no channel between source and drain at $V_{GS} = 0\,\text{V}$ in an enhancement NMOSFET, where both V_{DS} and V_T are positive, and a channel forms when $V_{GS} > V_T$. Figure 8.6-1a shows drain characteristics of the enhancement NMOSFET.

In an enhancement PMOSFET, the carriers are positive, and so V_{SD} and i_{SD} are positive. A channel forms when $V_{SG} > V_T$. In an NMOSFET $V_T (= V_{TN})$ is a particular value of V_{GS}, while in a PMOSFET $V_T (= V_{TP})$ is a particular value of V_{SG}.

Figure 8.6-2 shows the characteristics of a depletion NMOSFET. In such a device a channel exists when $v_{GS} = 0\,\text{V}$ and V_T is negative. A depletion NMOSFET operates in the depletion mode when v_{GS} is negative and in the enhancement mode when v_{GS} is positive. The depletion MOSFET has a behavior similar to the JFET characteristics.

Analytic Equations

Either the enhancement or depletion MOSFET may operate in the linear (nonsaturation) region or in the saturation region. In the linear region, there is a continuous channel between source and drain, and I_{DS} varies linearly with V_{DS} for fixed V_{GS}. At the source, the channel depth is proportional to $V_{GS} - V_T$. At the drain, the channel depth is proportional to $V_{GD} - V_T = V_{GS} - V_{DS} - V_T$. When $V_{GD} - V_T \leqslant 0$, or when

$$V_{DS} \geqslant V_{GS} - V_T, \tag{8.6-1}$$

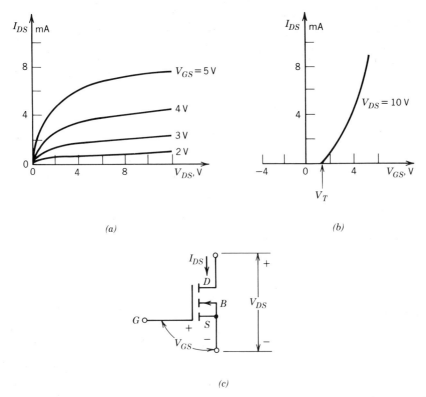

FIGURE 8.6-1 Enhancement NMOSFET: (*a*) drain characteristics; (*b*) transfer curve; (*c*) voltages and current defined.

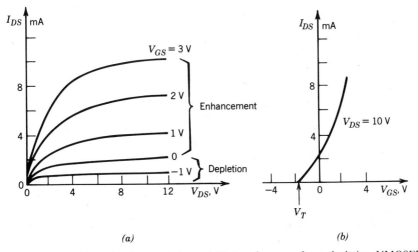

FIGURE 8.6-2 (*a*) Drain characteristics and (*b*) transfer curve for a depletion NMOSFET.

269

the channel is pinched off at the drain and the transistor is in saturation. Because of the channel pinch-off, the current I_{DS} remains nearly constant. In the linear region (below pinch-off), it is found that for an NMOSFET

$$I_{DS} = k[2(V_{GS} - V_T)V_{DS} - V_{DS}^2], \qquad V_{DS} \leqslant V_{GS} - V_T. \qquad (8.6\text{-}2)$$

In the saturation region (above pinch-off),

$$i_{DS} = k(V_{GS} - V_T)^2 \qquad V_{DS} \geqslant V_{GS} - V_T. \qquad (8.6\text{-}3)$$

The constant k is given by

$$k = \frac{\mu \varepsilon}{2t} \frac{W}{L}, \qquad (8.6\text{-}4)$$

where μ = mobility of carriers in channel (electrons in NMOSFETs)
   ε = dielectric constant of oxide insulating layer
   t = thickness of oxide under gate
  W = channel width
  L = channel length.

The dimensions of k are in amperes/volt^2, and typically k lies between 10^{-3} and 10^{-2}. An NMOSFET designed to act as a low resistance has a large W/L ratio and so has a large k, while an NMOSFET designed to act as a high resistance has a small W/L ratio and therefore a small k.

The equations for a PMOSFET in nonsaturation and saturation regions can be written in the same forms as Equations (8.6-2) and (8.6-3), with I_{SD}, V_{SD}, and V_{SG} instead of I_{DS}, V_{DS}, and V_{GS}, respectively.

Temperature Effects

Equations (8.6-2) and (8.6-3) for the current I_{DS} are affected by the temperature since k and V_T are temperature sensitive. The mobility μ decreases approximately inversely with the absolute temperature and therefore so does k (Equation 8.6-4). The temperature dependence of V_T is given by

$$\frac{dV_T}{dT} \simeq -2.5 \, \text{mV/°C}. \qquad (8.6\text{-}5)$$

When the temperature increases, I_{DS} reduces because of the decreased mobility and rises because of the lowering of the magnitude of V_T. It is found that the effect of μ may be five times greater than the effect of V_T. The overall effect of a temperature increase is generally a decrease of current i_{DS}.

8.7   ANALYSIS OF MOS INVERTER CHARACTERISTICS

A typical MOS inverter is shown in Figure 8.7-1. Here PMOS is indicated although NMOS can be employed. This inverter consists of a MOSFET switch driver (Q_D) driving a load MOSFET (Q_L). Both transistors are enhancement devices. In order to produce a reasonably sharp response of output to input it is necessary that the load channel be narrow and long relative to the driver; that is, the parameter λ should be large. Here λ is defined as

$$\lambda \equiv \frac{(W/L)_D}{(W/L)_L}, \tag{8.7-1}$$

where $(W/L)_D$ = width-to-length ratio of channel in driver Q_D and $(W/L)_L$ = width-to-length ratio for load Q_L.

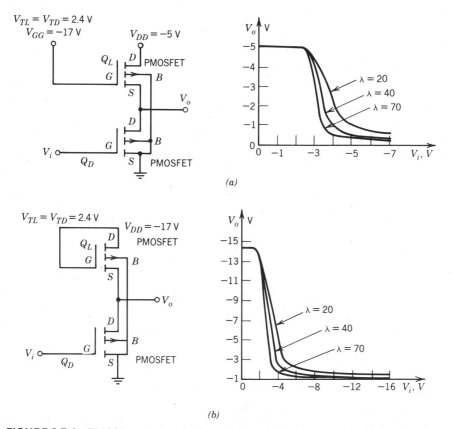

FIGURE 8.7-1   PMOS inverter transfer characteristics: (a) load MOSFET (Q_L) in linear region; (b) Q_L in saturation region.

We must expect the form of the input-output characteristic to depend principally on the parameter λ. The driver using the enhancement device is OFF when the gate voltage is at or near ground. When the driver is turned ON, it is biased to operate in the nonsaturation region (see Equation 8.6-2). Such is the case since the gate voltage will be at or near the supply voltage and the drain-to-source voltage will be at minimum magnitude.

Referring to Equations (8.6-2) and (8.6-3), we can write the equations to express the current I_{SD} for a PMOSFET in two operation regions. In the nonsaturation region (below pinch-off),

$$I_{SD} = k[2(V_{SG} - V_T)V_{SD} - V_{SD}^2], \qquad V_{SD} \leqslant V_{SG} - V_T. \qquad (8.7\text{-}2)$$

In the saturation region (above pinch-off),

$$I_{SD} = k(V_{SG} - V_T)^2, \qquad V_{SD} \geqslant V_{SG} - V_T. \qquad (8.7\text{-}3)$$

Since the transistors in Figure 8.7-1a are PMOSFETs with $V_T = 2.4$ V, we apply the criterion given in Equation (8.7-2) to determine whether we are in the nonsaturation or saturation region. To be in the nonsaturation region we require that $V_{SG} - V_{SD} \geqslant V_T$, or

$$V_D - V_G = V_{DG} \geqslant V_T. \qquad (8.7\text{-}4)$$

Since $V_{DG} = V_D - V_G = -5 - (-17) = 12$ V $> V_T = 2.4$ V,   the   load MOSFET is biased to operate in the nonsaturation region.

To determine whether the load transistor in Figure 8.7-1b is in the saturation or nonsaturation region, let us use the criterion given in Equation (8.7-3). To be in the saturation region we require that

$$V_D - V_G = V_{DG} \leqslant V_T. \qquad (8.7\text{-}4)$$

Since $V_{DG} = V_D - V_G = -17 - (-17) = 0$ V $< V_T = 2.4$ V,   the   load transistor is biased to operate in the saturation region. The load MOSFET may be either an enhancement device or a depletion device and may operate in either nonsaturation or saturation.

Note that as λ increases, the transition of the output between its high and low levels become sharper, as shown in Figure 8.7-1. The inverter using an enhancement MOSFET driver and a depletion MOSFET load will have the steepest transition region.

8.8   ANALYSIS OF CMOS INVERTER CHARACTERISTICS

About CMOS Inverter

A CMOS inverter is shown in Figure 8.8-1. A supply voltage V_{SS} ($= V_{DD}$) is applied from source to source. Since the source of NMOSFET is grounded,

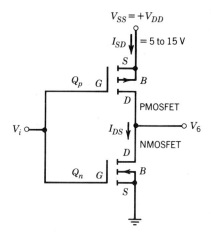

(a)

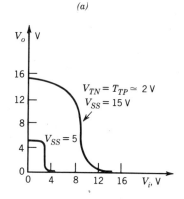

(b)

FIGURE 8.8-1 The CMOS inverter with its transfer characteristics.

V_{SS} must be a positive voltage $+V_{DD}$ and V_i swings between ground and V_{SS}. In the complementary-symmetry MOS inverter, we want the two transistors to be reasonably alike. It is usually arranged that the parameter k in Equations (8.6-2), (8.6-3), (8.7-2), and (8.7-3) is the same for the two transistors. The carrier mobility in the PMOSFET is smaller than the mobility in the NMOSFET by a factor of 2 or 3. Therefore to make the ks equal, the ratio W/L for the PMOSFET should be correspondingly larger by a factor of 2 or 3 than W/L for the NMOSFET (see Equation 8.6-4).

The CMOS inverter operates as follows: When input voltage V_i is less than threshold voltage V_{TN}, transistor Q_n is OFF and, as is usually the case, $V_{SG(p)} = V_{SS} - V_i > V_{TP}$, so that transistor Q_p is ON. With Q_n OFF no current can flow in Q_p, even though it is ON. With $I_{SD(p)} = 0$, we shall have $V_{SD(p)} = 0$. Thus $V_o = V_{SS}$. When V_i increases above V_{TN}, both Q_n and Q_p turn ON and output

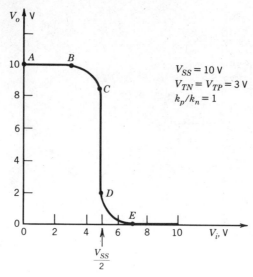

FIGURE 8.8-2   A plot illustrating the calculation of the CMOS-inverter transfer characteristic.

voltage V_o decreases. Ultimately, when V_i increases enough for Q_p to go OFF, $V_{SS} - V_i < V_{TP}$ and Q_n is ON; thus $V_o = 0$ V.

Calculation of Transfer Characteristic of CMOS Inverter

The transfer characteristic of a CMOS inverter is shown in Figure 8.8-2. It is plotted for a circuit in which $V_{SS} = 10$ V, $V_{TN} = V_{TP} = 3$ V, and $k_p/k_n = 1$. When $V_i < 3$ V, Q_n is OFF, Q_p is ON, and $V_o = 10$ V (see region AB). When $V_i > 7$ V, Q_p is OFF, Q_n is ON, and $V_o = 0$ V (see point E). In region BC ($V_i = 3.2$ to 5 V), Q_n is in saturation while Q_p is in the nonsaturation range, since

$$V_o > V_i - V_{TN} \quad \text{(see Equation 8.6-3)}$$

and

$$V_{SS} - V_o < V_{SS} - V_i - V_{TP} \text{ (see Equation 8.7-2)}.$$

Therefore the characteristic in region BC is calculated by equating Equations (8.6-3) and (8.7-2):

$$k_n(V_i - V_{TN})^2 = k_p[2(V_{SS} - V_i - V_{TP})(V_{SS} - V_o) - (V_{SS} - V_o)^2]. \quad (8.8\text{-}1)$$

In region DE, Q_p is in saturation while Q_n is in the nonsaturation range, since

$$V_{SS} - V_o > V_{SS} - V_i - V_{TP} \text{ (see Equation 8.7-3)}$$

and

$$V_o < V_i - V_{TN} \quad \text{(see Equation 8.6-2)}.$$

Thus the characteristic in region DE is calculated by equating Equations (8.6-2) and (8.7-3):

$$k_n[2(V_i - V_{TN})V_o - V_o^2] = k_p(V_{SS} - V_i - V_{TP})^2. \qquad (8.8\text{-}2)$$

In region CD both transistors Q_n and Q_p are in saturation. The characteristic between C and D is calculated by equating Equations (8.6-3) and (8.7-3):

$$k_n(V_i - V_{TN})^2 = k_p(V_{SS} - V_i - V_{TP})^2. \qquad (8.8\text{-}3)$$

With $k_p/k_n = 1$ and $V_{TN} = V_{TP}$, we have

$$V_i = V_{i(\text{sat})} = V_{SS}/2 = 5 \text{ V}. \qquad (8.8\text{-}4)$$

The simultaneous saturation of both transistors (Q_n and Q_p) defines the unique voltage $V_{i(\text{sat})}$, which is calculated from Equation (8.8-3) to be

$$V_{i(\text{sat})} = \frac{\sqrt{k_p/k_n}(V_{SS} - V_{TP}) + V_{TN}}{1 + \sqrt{k_p/k_n}}. \qquad (8.8\text{-}5)$$

The characteristic in region CD is a vertical line, indicating that the output voltage changes abruptly as V_i moves across the value $V_{SS}/2$. If k_p were not equal to k_n, then even if V_{TN} were equal to V_{TP}, complete symmetry would not prevail. In practical circuits, the transition from C to D would be sharp but not abrupt.

8.9   INTEGRATED-INJECTION LOGIC (I^2L)

A resistor-*npn*-transistor inverter has a base resistor and a collector resistor. If each of these resistors is replaced by a *pnp* transistor, then this inverter without resistors becomes a basic integrated injection logic (I^2L) circuit. Integrated circuit resistors can occupy many times the area of a transistor. At least twenty pure transistor inverters might be fabricated on the area normally occupied by one resistor-transistor inverter. Eliminating the resistors reduces the unwanted capacitance and thus improves the switching speed of each inverter. The typical characteristics of the I^2L unit are as follows: propagation delay time is 10 to 250 ns; power dissipation per gate is 6 nW to 70 μW; noise margin is 0.25 V; gate size is small.

Figure 8.9-1 shows two I^2L inverters. The first one has its output connected to the input of the other. Complementary bipolar transistors are used; the *pnp* injector transistor serves as a current source, and a multiple collector *npn*

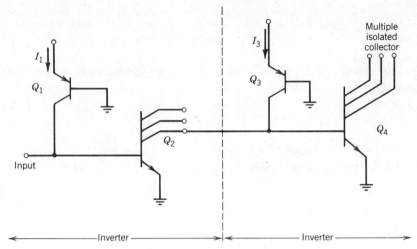

FIGURE 8.9-1   Two I^2L inverters.

transistor forms each inverter. The collector load for the first inverter is the input stage of the second inverter. With Q_2 off, I_3 flows into the base of Q_4 to drive it into saturation. With Q_2 on, I_3 is diverted through Q_2, and Q_4 is off. The term *integrated injection logic* is derived from the fact that the charge carriers constituting the currents are injected into the *pnp* transistor emitters.

The I^2L NOR gate, which is shown in Figure 8.9-2, consists of two inverters with their output terminals connected in parallel. With both inputs low, Q_2

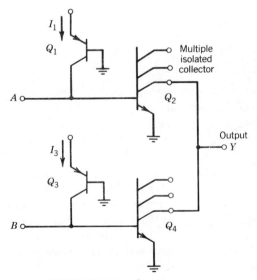

FIGURE 8.9-2 I^2L NOR gate.

and Q_4 are off and the output is high. If either A or B is driven high, one of the output transistors will be saturated, bringing the gate output down to low level.

The I^2L NAND gate is simply an inverter stage with several common input terminals, such as A, B, and C. If all the collectors of previous stage transistors connected to A, B, and C are open circuited, then all inputs are high, and so the inverter (or NAND) output is low. If any one of the transistors connected to A, B, or C is on, the input is low, and so the inverter (or NAND) output is high.

REFERENCES

1. Floyd, T. L., *Digital Fundamentals*, 2nd ed., Charles E. Merrill, Columbus, Ohio, 1982, Chap. 3 and Appendix A.

2. Strangio, C. E., *Digital Electronics: Fundamental Concepts and Applications*, Prentice-Hall, Englewood Cliffs, New Jersey, 1980. Chap. 1.

3. Taub, H., *Digital Circuits and Microprocessors*, McGraw-Hill, New York, 1981, Chap. 3.

4. Millman, J., *Microelectronics: Digital and Analog Circuits and Systems*, McGraw-Hill, New York, 1979, Chap. 4.

5. Schilling, D. L. and C. Belove, *Electronic Circuits, Discrete and Integrated*, McGraw-Hill, New York, 1979, Chap. 12.

6. Comer, D. J., *Electronic Design with Integrated Circuits*, Addison-Wesley, Reading, Massachusetts, 1981, Chap. 3.

QUESTIONS

8-1. Define the terms *fan-in* and *fan-out*.

8-2. What are the parameters by which a logic family is characterized?

8-3. Define *noise margin* and *propagation delay*.

8-4. How does a NOR gate differ from an OR gate?

8-5. Under what conditions is the output of a NOR gate high?

8-6. Construct a truth table for a three-input NOR gate.

8-7. Construct a truth table for a three-input NAND gate.

8-8. Sketch the circuit for a TTL NAND gate. Explain the operation of the circuit.

8-9. How does the totem pole function?

8-10. Sketch the complete circuit for an ECL OR/NOR gate. Explain the operation of the circuit.

8-11. Discuss the major advantages and disadvantages of ECL.

8-12. What are the advantages and disadvantages of MOS logic compared with TTL?

8-13. What are the advantages of CMOS logic?

8-14. Sketch the circuit of a CMOS NOR gate. Explain the operation of the circuit.

8-15. Sketch the circuit of a CMOS NAND gate. Explain the operation of the circuit.

8-16. Sketch the circuit of an I^2L inverter. Explain the operation of the circuit.

8-17. Sketch the circuit of an I^2L NAND gate. Explain the operation of the circuit.

8-18. Compare TTL, CMOS, and I^2L in terms of propagation delay time, power dissipation, and noise margin.

PROBLEMS

8-1. Determine the output waveform if the inputs shown in Figure p8-1 are applied to the three-input NOR gate.

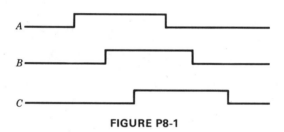

FIGURE P8-1

8-2. Determine the output waveform if the inputs shown in Figure p8-2 are applied to the three-input NAND gate.

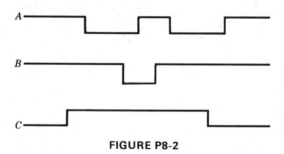

FIGURE P8-2

8-3. A logic gate draws 3.8 mA when its output is low and 2.2 mA when its output is high. Determine its average power dissipation if it operates on a 50% duty cycle with $V_{CC} = 5$ V.

Answer.   15 mW.

8-4. The ECL OR/NOR gate circuit in Figure 8.4-1 has a supply voltage $V_{EE} = -5$ V. Find the following currents and voltages when inputs A, B, and C are low: (a) I_D, V_{B5}, V_{B4}, V_{E4}, I_{E4}, I_{C4}. (b) V_{B7} and V_{E7} for emitter follower Q_7. Is the output low or high at Q_7 emitter? (c) V_{E6}. Is the output low or high at Q_6 emitter?

Answer. (a) $I_D \simeq 1.4$ mA; $V_{B5} \simeq 4.6$ V; $V_{B4} \simeq 3.9$ V; $V_{E4} \simeq 3.2$ V; $I_{E4} \simeq 2.7$ mA; $I_{C4} \simeq 2.7$ mA. (b) $V_{B7} \simeq -0.18$ V; $V_{E7} \simeq -1.5$ V = Low. (c) $V_{E6} \simeq -0.7$ V = High.

9

DIGITAL COUNTERS, REGISTERS, AND DATA CONVERTERS

9.1 COUNTING CHAIN OF n FLIP-FLOPS

The cascade of flip-flops (or binaries) shown in Figure 9.1-1 can be employed as a counting chain to count up to 16. Each flip-flop is a symmetrical collector trigger circuit, as shown in Figure 7.2-2. The counting chain consists of four flip-flops (FF1–FF4). Before the trigger pulse is applied, all the flip-flop stages must be reset to zero, at which *npn* transistors Q_2, Q_4, Q_6, and Q_8 are saturated (on) with a condition of 0000. Each time a negative-going input pulse is applied, the first flip-flop FF1 will change state. When Q_2 switches off (FF1 in 1 state), its collector voltage is a positive step, which does not affect the next flip-flop, FF2. When Q_2 switches on, its collector voltage is a negative step, which is coupled to FF2 to change the FF2 state. Similarly FF3 is triggered from FF2 and FF4 is triggered from FF3. When an *npn* transistor is on, its collector voltage is low and is designated by 0. On the other hand, when it is off, its collector voltage is high and is represented by 1. Thus the output $Y = 0$ when the left transistor of each flip-flop is on and $Y = 1$ when it is off.

Assume that all flip-flops are in state 0 and that the 16 successive input negative pulses are applied to the T input of FF1; then the waveforms shown in Figure 9.1-2 will appear at the output Y of the individual flip-flops. We can verify that the waveform chart of Figure 9.1-2 is correct by applying the following principles:

1. Flip-flop FF1 must make a transition at each externally applied pulse.

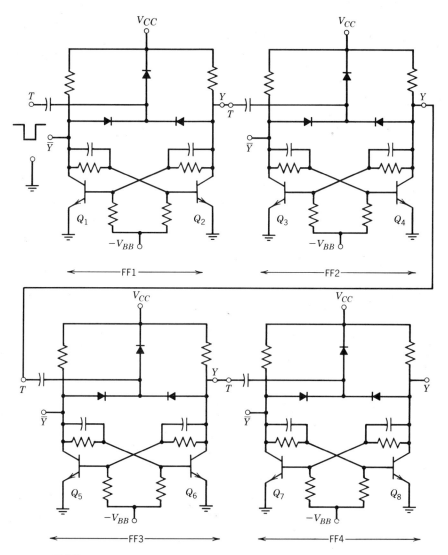

FIGURE 9.1-1   A cascade of four flip-flops used as a binary counting chain.

2. Each of the other flip-flops must make a transition when and only when the preceding flip-flop makes a transition from state 1 to state 0.

Thus the first external pulse applied to FF1 causes the flip-flop to make a transition from state 0 to state 1. As a result of this transition, a positive voltage step is applied at *T* input of FF2. Because of the arrangement of diodes in FF2 (see Figure 9.1-1), this positive step will not induce a transition in FF2. The overall result is that FF1 has changed its state to 1, and all other flip-flops remain

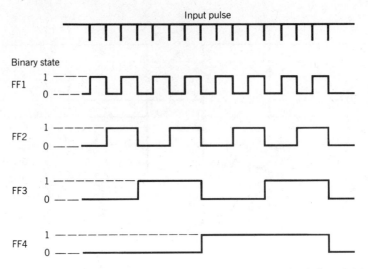

FIGURE 9.1-2 Waveform chart for the binary counting chain of Figure 9.1-1.

in state 0, as shown in Figure 9.1-2. The second externally applied pulse causes flip-flop FF1 to return from state 1 to state 0. Flip-flop FF2 now receives a negative step voltage to which the flip-flop is sensitive, and it responds by making a transition from state 0 to state 1. Then the flip-flop FF3 receives a positive step from FF2 and hence does not respond to the transition in FF2. The overall result of the application of two input pulses is that flip-flop FF2 is in state 1 while all other flip-flops are in state 0. The remainder of the waveform chart can be verified in a similar manner.

The states of the binaries (or flip-flops) listed in Table 9.1-1 may be verified directly by comparison with the waveform chart of Figure 9.1-2. For each flip-flop one output pulse appears for two input pulses. A chain of n flip-flops will count up to the number 2^n before it resets itself into its original state. Such a chain is called a *counter modulo* 2^n. The number 2^n is called a *dividing* or *scaling factor*.

If we should differentiate each of the waveforms of Figure 9.1-2, a positive pulse would appear at each transition from 0 to 1 and a negative pulse at each transition from 1 to 0. If we now count only the negative pulses (the positive pulses may be eliminated by, say, using a diode), then it appears that each binary divides the number of negative pulses applied to it by 2. The four binaries together accomplish a division by a factor $2^4 = 16$. A single negative pulse will appear at the output for each 16 negative pulses applied at the input. A chain of n flip-flops used for the purpose of dividing or scaling down the number of pulses is known as a *scaler*. Thus a chain of four flip-flops constitutes a scale-of-16 scaler, and so on. A scale-of-64 scaler followed by a mechanical register will be able to respond to the pulses of Geiger-Muller tubes, which occur at a rate 64 times greater than the maximum rate at which the mechanical register will

TABLE 9.1-1
States of the Binaries (Flip-Flops)[a]

Number of Input Pulses	Binary State			
	FF4	FF3	FF2	FF1
0	0	0	0	0
1	0	0	0	1
2	0	0	1	0
3	0	0	1	1
4	0	1	0	0
5	0	1	0	1
6	0	1	1	0
7	0	1	1	1
8	1	0	0	0
9	1	0	0	1
10	1	0	1	0
11	1	0	1	1
12	1	1	0	0
13	1	1	0	1
14	1	1	1	0
15	1	1	1	1
16	0	0	0	0
17	0	0	0	1

[a]This table gives binary-to-decimal equivalents for the 17 integers.

respond. The net could in such a case will be 64 times the reading of the mechanical register plus the count left in the scale-of-64 counter.

9.2   INTEGRATED-CIRCUIT BINARY COUNTERS

The JK flip-flop can be used as a binary counter, since it reverses states upon application of clock pulses when both JK inputs are at 1. For this application, as Figure 9.2-1 shows, the clock pulse becomes the input and the JK terminals are connected to a 1 voltage supply ($+V_{CC}$). Before counting, a pulse is applied to all the reset (clear) inputs, so that all the outputs (Q_A, Q_B, etc.) are at 0. If a series of clock pulses is applied to T input (clock), its output will reverse each time the clock pulse goes through one complete cycle, which for many types of flip-flop occurs on the falling edge of the clock pulse. Two complete input cycles are required to produce one complete output cycle. There is no requirement that the input pulse be symmetric or periodic. A package of four flip-flops with the JK inputs internally connected to 1 is known as a *four-bit binary* or *ripple counter* (Figure 9.2-1) with the clock input renamed the *toggle* or *count input T.*

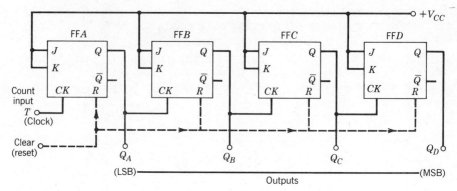

FIGURE 9.2-1 Basic four-bit binary or ripple (asynchronous) counter. Its waveform chart is similar to that shown in Figure 9.1-2.

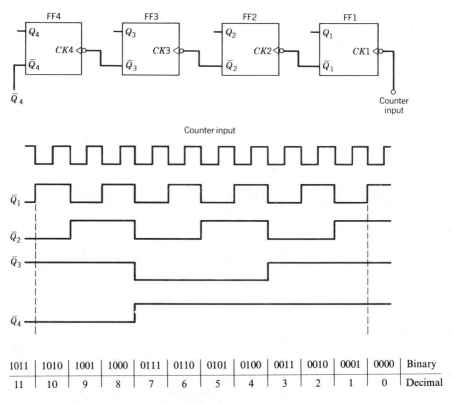

1011	1010	1001	1000	0111	0110	0101	0100	0011	0010	0001	0000	Binary
11	10	9	8	7	6	5	4	3	2	1	0	Decimal

FIGURE 9.2-2 Backward (or down) counter and waveform (timing) chart.

284

Decimal equivalents for several binary states are given in Table 9.1-1. A four-bit binary counter ranges from 0 to 15 decimal and has an output or carry pulse that can be applied to the toggle input of additional stages.

Some digital systems require a backward or down counter. A binary number is set into the counter, which then counts toward zero as input pulses occur. Figure 9.2-2 illustrates a simple method of constructing a backward counter. The waveform chart assumes that the original count was 1011 or decimal 11. After 11 negative transitions of the input signal, the counter includes a count of 0000. An example of an integrated circuit up-down counter is the 74193. It is a synchronous, four-bit, up-down counter with a clear input and borrow and carry outputs. Data can be loaded into the counter in parallel if a preset count is desired.

9.3 DECIMAL COUNTERS

Decade Counter with Initial Condition of 0110 (Decimal 6)

A *decimal* (or *decade*) counter is a scale-of-10 counter. It requires the application of a cascade of four flip-flops, in which a scale-of-16 must be modified to eliminate either the first six states or the last six states. When the first six states are to be eliminated, the counter should always have an initial condition of 0110

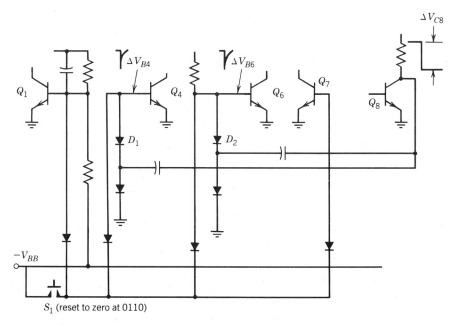

FIGURE 9.3-1   Resetting a decade counter to zero at 0110 (decimal 6) by pressing switch S_1 and resetting Q_4 and Q_6 to OFF by using the negative step voltage from Q_8 collector.

(decimal 6); to obtain this condition for a decade counter, the reset arrangement shown in Figure 9.3-1 can be used in combination with the circuit of Figure 9.1-1. When Q_8 switches on, a negative step voltage from its collector forward biases D_2 and D_1 and triggers Q_6 and Q_4 off. Figure 9.1-2 and Table 9.1-1 show that Q_8 switches on when the 16th input pulse is applied. Hence, at the end of the count of 16, the flip-flops are set to 0110. Before counting starts, the reset switch S_1 should be closed so that Q_1, Q_4, Q_6, and Q_7 are switched off; in this case, the decade counter is at the desired initial condition 0110. The collector waveforms and truth table for the decade counter are shown in Figure 9.3-2.

Binary-Coded Decimal (BCD or 8421) Counter

Figure 9.3-3 shows an example of a BCD decade counter. The circuit consists of a chain of four flip-flops and a two-diode OR gate. Its counting sequence

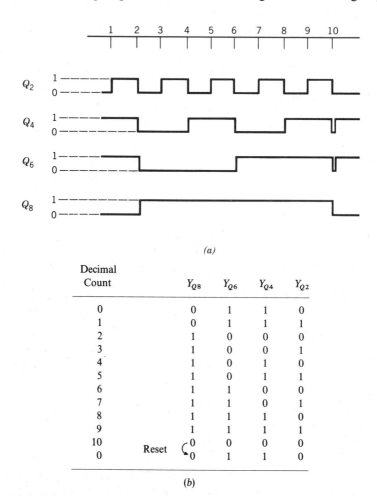

(a)

Decimal Count		Y_{Q8}	Y_{Q6}	Y_{Q4}	Y_{Q2}
0		0	1	1	0
1		0	1	1	1
2		1	0	0	0
3		1	0	0	1
4		1	0	1	0
5		1	0	1	1
6		1	1	0	0
7		1	1	0	1
8		1	1	1	0
9		1	1	1	1
10	Reset	0	0	0	0
0		0	1	1	0

(b)

FIGURE 9.3-2   (a) Waveforms of the decade counter collector and (b) its truth table.

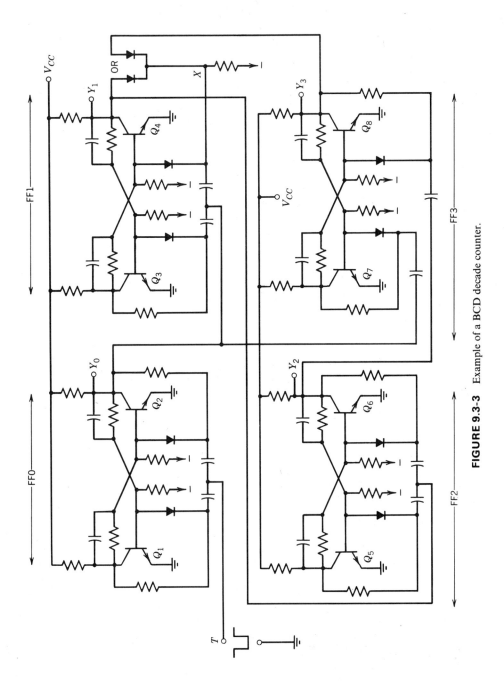

FIGURE 9.3-3 Example of a BCD decade counter.

corresponds to 8-4-2-1 code. FF3, FF2, FF1, and FF0 change state with the eighth, fourth, second, and first input trigger pulse, respectively. Decimal 8 = 1000 binary, 4 = 100, 2 = 10, and 1 = 1 (see Table 9.1-1). This counter requires the following counting steps: 0000, 0001, 0010, 0011, 0100, 0101, 0110, 0111, 1000, 1001, and then return to 0000. Each of these four-digit groups indicates the states of the four flip-flops. The 8.4-2-1 code consists of straight binary counting up to 1001 = 9; the remaining states, 10 to 15, are merely omitted.

In the circuit of Figure 9.3-3 we use trigger steering by means of bias applied to the base trigger diodes. The input negative pulses are applied to FFO, which scales in the normal manner. The system proceeds with carry pulses in the required straight binary counting sequence until state 1001 (= 9) is reached. Here the special decimal feature comes into play. The collector of Q_8 is now positive ($Y_3 = 1$). The OR gate controlling point X (near the base of Q_4) thus draws X positive also, with the result that the negative trigger pulse to Q_4 base is suppressed. The carry signal to FF1 is blocked, however, because of the suppression just described. The carry pulse from the collector of Q_2 is instead applied directly to the base of Q_7, causing the last flip-flop FF3 to return to 0 ($Y_3 = 0$). We thus go from 1001 directly back to 0000.

9.4 INTEGRATED-CIRCUIT BCD DECADE COUNTERS

An IC BCD decade counter is identical to a four-bit binary counter except for an internal circuit that resets all the flip-flops to 0 on the tenth count (binary 1010) rather than the 16th. Each four-bit counter is equivalent to one decimal digit and is separately decoded from the binary as a digit from 0 to 9. No four-bit binary numbers higher than 1001 exist. The decimal digit can be read by a seven-segment display after decoding. Each segment of the display can be independently illuminated, and the decoder/driver chooses the proper combinations of segments to form the desired digit, as determined by the BCD input. Illumination is provided by light-emitting diodes, neon bulbs, or incandescent lamps.

There are many decade counters available; some of them are shown in Figure 9.4-1. The TTL 7490A and 74L90 are two popular counters. Each of them contains four master-slave flip-flops and additional gating to provide a divide-by-two counter and a three-stage binary counter for which the count cycle length is divide-by-five. The 7490A and 74L90 have a gated zero reset and gated set-to-nine inputs for use in BCD nine's complement applications. To use the maximum count length (decade, divide-by-12, or four-bit binary) of these counters, the B input is connected to the Q_A input. The input count pulses are applied to input A. A symmetrical divide-by-ten count can be obtained from the 7490A or 74L90 counters by connecting the Q_D output to the A input and applying the input count to the B input, which gives a divide-by-ten square wave at output Q_A.

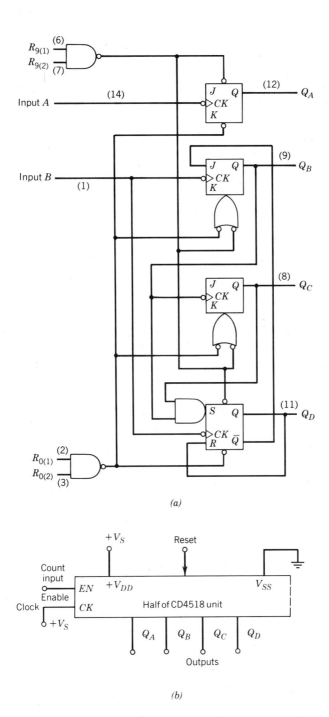

(a)

(b)

FIGURE 9.4-1 BCD decade counters: (*a*) TTL 7490A and 74L90; (*b*) half of CMOS CD4518 dual counter.

Two BCD counters per package are available in the CMOS CD4518, as shown in Figure 9.4-1*b*. Normally pulses to be counted are connected to the enable input because the flip-flops change on the enable falling edge. Counting is inhibited (turned off) by bringing the enable (strobe) input to 1. Clocking on the rising edge is made possible by interchanging the clock and enable inputs (inhibit on 0).

Synchronous Counters

An asynchronous or ripple counter, described previously, will exhibit a delay. The *carry propagation delay* is the time required for a counter to complete its response to an input pulse. The carry time of a ripple counter is longest when each stage is in the 1 state. In this situation, the next pulse must cause all previous flip-flops to change state. No particular binary will respond until the preceding stage has nominally completed its transition. The clock pulse effectively "ripples" through the chain. Thus the carry time will be of the order of magnitude of the sum of the propagation delay times of all the binaries. If the chain is long, the carry time may well be longer than the interval between input pulses. In such a case, it will not be possible to read the counter between pulses.

If the asynchronous operation of a counter is changed so that all flip-flops are clocked simultaneously (synchronously) by the input pulses, the propagation delay time may be reduced considerably. Another advantage of the synchronous counter is that no decoding spikes appear at the output since all flip-flops change state at the same time. Thus no strobe pulse is required when decoding a synchronous counter. In order to obtain these advantages, the clock must be connected to all stages, as shown in Figure 9.4-2, but each stage is inhibited unless the carry output of the previous stage occurs. Carry occurs just after the ninth pulse of a BCD counter, thus enabling the subsequent stage on the tenth count of the driving stage.

An IC 74160 synchronous decade counter is shown in Figure 9.4-3. It may be preset to any BCD count using the data inputs and an active low on the load input. An active low on the clear will reset the counter. The count enable inputs P and T should both be high for the counter to advance through its sequence of states in response to a positive transition on the CK input. The enable inputs in conjunction with the carry out (terminal count of 1001) provide for cascading several decade counters. A timing example for a 74160 is shown in Figure 9.4-4.

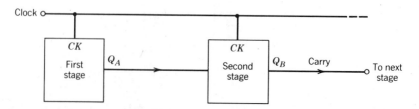

FIGURE 9.4-2 A synchronous counter.

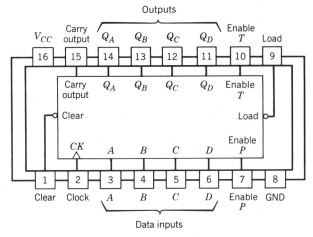

FIGURE 9.4-3 Block and pin diagram of the 74160 synchronous decade counter.

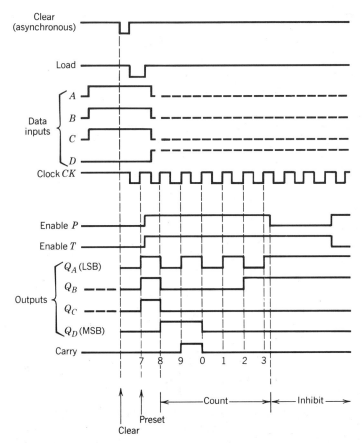

FIGURE 9.4-4 A timing diagram showing the IC 74160 counter being preset to BCD seven (0111).

9.5   LCD AND LED SEVEN-SEGMENT DISPLAYS

Major Characteristics of LCD and LED

Both the liquid-crystal display (LCD) and the light-emitting diode (LED) are commonly used for digital readouts. The LCD has the distinct advantage of having a lower power requirement than the LED. It is typically on the order of microwatts for the display, as compared to the same order of milliwatts for LEDs. The LCD requires an external or internal light source, whereas the LED will give off visible light when it is energized. The LCD is limited to the temperature range of about 0° to 60°C, whereas the LED is typically operated within the range of −55° to 100°C. Since LCDs can chemically degrade, their lifetime is limited to 10,000 + hours. The power requirement of LEDs is typically from 10 to 150 mW with a lifetime of 100,000 + hours. The LCD has a decay time of 150 ms or more; this is very slow compared to the 10-ns rise and fall times for the LED. It is so slow that the human eye can observe the fading out of segments switching off. At low temperatures the response time of LCDs is considerably increased. Perhaps the major disadvantage of LCDs is their slow decay time.

LCD Seven-Segment Displays

LCDs are commonly arranged in the same seven-segment numerical format as LED displays. The types of LCDs receiving the most interest today are the *dynamic-scattering* and *field-effect* units. Liquid-crystal cells may be either transmittive or reflective. The liquid-crystal material is an organic compound that exhibits the optical properties of a crystal. The light-scattering unit shown in Figure 9.5-1 usually uses nematic liquid crystal, which is layered between glass sheets and has transparent electrodes deposited on the inside faces. The unit thus formed is a liquid-crystal cell. When a voltage (the threshold level is usually between 6 and 20 V) is applied across the cell, charge carriers via the liquid disturb the molecular arrangement, resulting in the establishment of regions with different indices of refraction. The incident light is, consequently, reflected in different directions at the interface between regions of different indices of refraction; the result is that the scattered light has a frosted-glass appearance.

The construction of a field-effect or twisted nematic LCD is similar to that of the dynamic-scattering unit, except that two thin polarizing optical filters are placed at the inside surface of each glass sheet. The liquid crystal referred to as *twisted nematic* actually twists the incident light, and the passage of the light is not permitted when the cell is not energized. Hence when no voltage is applied, the viewer sees a uniformly dark pattern across the entire display. When a threshold voltage is applied (usually from 2 to 8 V), no twisting of the light occurs and a light area is seen from the display.

The total current flow through four small seven-segment displays is typically about 25 μA for dynamic-scattering cells and 300 μA for field-effect cells.

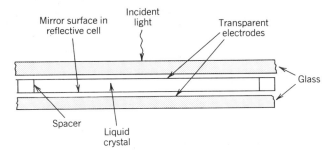

FIGURE 9.5-1   Construction of reflective dynamic-scattering liquid-crystal cell.

However the LCD requires an ac voltage supply, either in the form of a square wave or a sine wave. A dc current producing a plating of the cell electrodes could damage the device. Usually a dynamic-scattering LCD uses 30 V peak to peak, 60-Hz square wave, and a field-effect LCD uses 8 V peak to peak. The square wave drive method is illustrated referring to Figure 9.5-2. In the seven-segment LCD, the back plane is a terminal common to all cells and it is supplied with an ac voltage in square waveform. The square waves applied to the seven segments are either in phase or in antiphase with the back plane square wave. Each cell contains a segment with the common back plane. The waveform applied to a segment in phase with the back plane input results in no voltage developed across the cell. The square wave (e.g., 15 V peak) applied to a segment in antiphase with the back plane input results in an ac voltage (30 V peak to peak) developed across the cell.

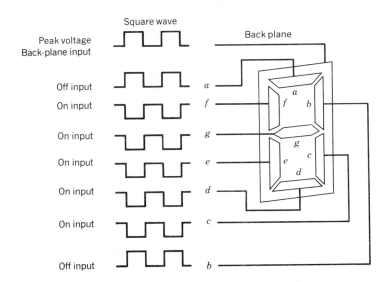

FIGURE 9.5-2   Square wave drive method for seven-segment LCD. In the case where cells *a* and *b* are off while others are on the result is an indication of number 6.

The reflective IEE series 1603-02 manufactured by Industrial Electronic Engineers, Inc., is a $3\frac{1}{2}$-decade liquid-crystal display with four floating decimals and an overflow plus or minus one ($\pm$). A colon is incorporated in the display for additional application advantages, such as clocks and so on. The series 1603-02 LCD consists of a layer of micron-thin liquid-crystal material confined between two sheets of glass, one sheet having a clear conductive electrode and the other a reflective coating etched in a segmented pattern to create a digital display. For optimum life, liquid crystals are operated on ac (40–100 Hz), which, coupled with a 15–30-V range, make them directly compatible with MOS logic. A $3\frac{1}{2}$-decade display means that the three right-hand units are complete seven-segment units while the fourth (left-hand) unit is only a single segment that indicates numeral 1 when energized. This unit is known as a *half unit*, and the complete display is then described as a $3\frac{1}{2}$-*decade display*. The maximum number that can be indicated by such a display is 1999.

LED Seven-Segment Displays

Visible light is emitted from an LED when it is properly forward biased. In any forward-biased *pn* junction, there is a charge carrier recombination, which takes place as electrons cross from the *n*-side and recombine with holes on the *p*-side. This combination requires that the energy possessed by the unbound free electron be transferred to another state. In silicon and germanium *pn* junctions, part of this energy will be given up in the form of appreciable heat and insignificant light. The semiconductor material commonly used for LED construction is either gallium arsenide phosphide (GaAsP) for red or yellow light emission, or gallium arsenide (GaAs) for green or red light emission. The process of giving off light by applying an electrical source of energy is referred to as electroluminescence. Since recombination occurs in the *p*-type material, the *p*-region becomes the surface of the LED. Single LEDs operate at voltage levels from 1.2 to 3.3 V, which makes them completely compatible with solid-state circuits.

The most popular LED arrangement is a seven-segment display used for numeric display of characters from 0 to 9. The LEDs in this type of display may have a common anode or common cathode, as illustrated in Figure 9.5-3a. The simplest and usually the most economical display contains only one LED per segment, as shown in Figure 9.5-3b. An external or built-in current-limiting resistor is required for each segment to limit the current to a safe value (typically 20 mA dc).

7447A BCD-to-Seven-Segment Decoder

Figure 9.5-4 shows a 7447A BCD-to-seven-segment decoder/driver connected to a common-anode LED seven-segment display (VLED). The 7447A is not designed for driving the common-cathode VLED. The 7447A decoder is actually a BCD-to-seven-segment code converter, as illustrated by its function table

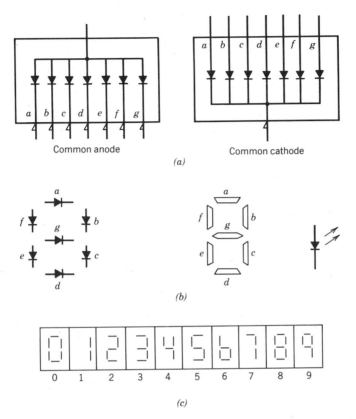

Common anode

Common cathode

(a)

(b)

(c)

FIGURE 9.5-3 LED seven-segment displays: (a) Common anode and common cathode connections; (b) LED segment designation and single-LED symbol; (c) resulting display for each numerical input.

(Table 9.5-1). Each LED in Figure 9.5-4 is turned on by a combination of low-level outputs from the 7447A, as illustrated by the display-output column in Table 9.5-1. To test all segments of a display, an active-low lamp-test (LT) input is provided on the 7447A. Some BCD-to-seven-segment decoders, including the 7447A, also have automatic leading- and/or trailing-edge zero-blanking control (ripple-blanking input and output—RBI and RBO). Their lamp-test may be performed at any time when the RBI/RBO node is at high level.

The 7447A decoder/driver has active low open-collector outputs (indicated by the presence of the low-level-indicator symbol on each output) and can operate up to 15 V and 40 mA. The value of resistor R depends on the supply voltage and the recommended LED operating voltage and current of the seven-segment display. The value of R for each segment is the same.

In the TTL NAND gate inverter of Figure 8.3-2, the output circuit made up of R_4, Q_4, D, and Q_3 is known as a *totem-pole output*; it has a higher fan-out and a faster switching speed than a single transistor. The main difference between

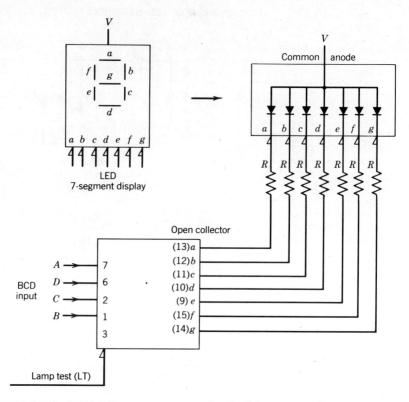

FIGURE 9.5-4 7447A BCD-to-seven-segment decoder/driver connected to a common-anode LED seven-segment display (VLED).

TABLE 9.5-1
7447A Function Table and Display Output

Decimal Digit	Input[a]					Output[a]							Display Output
	LT	D	C	B	A	a	b	c	d	e	f	g	
0	H	L	L	L	L	L	L	L	L	L	L	H	See Figure 9.5-3c
1	H	L	L	L	H	H	L	L	H	H	H	H	
2	H	L	L	H	L	L	L	H	L	L	H	L	
3	H	L	L	H	H	L	L	L	L	H	H	L	
4	H	L	H	L	L	H	L	L	H	H	L	L	
5	H	L	H	L	H	L	H	L	L	H	L	L	
6	H	L	H	H	L	H	H	L	L	L	L	L	
7	H	L	H	H	H	L	L	L	H	H	H	H	
8	H	H	L	L	L	L	L	L	L	L	L	L	
9	H	H	L	L	H	L	L	L	H	H	L	L	
LT	L	X	X	X	X	L	L	L	L	L	L	L	

[a] H = high level; L = low level; X = irrelevant.

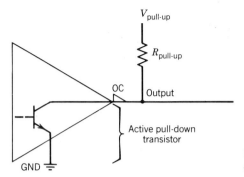

FIGURE 9.5-5 Inverter with an open-collector (OC) output.

an open-collector (OC) output and a totem-pole output is the absence of the active pull-up circuit (R_4-Q_4-D). When the active pull-down transistor in Figure 9.5-5 turns off, external resistor $R_{\text{pull-up}}$ pulls the output logic line to a high level ($H = V_{\text{pull-up}}$, without specifying a particular voltage). The open-collector output transistor (or active pull-down transistor) can be obtained with a maximum breakdown voltage of either 15 or 30 V, depending on the design. This allows $V_{\text{pull-up}}$ to be selected for interfacing with LEDs, lamps, relays, or other logic families. When open-collector output circuits are used in interfacing applications, they are generally referred to as *buffer* or *driver circuits*.

The 7448 and 7449 BCD-to-seven-segment decoder/drivers feature active-high outputs for driving common-cathode VLEDs or lamp buffers.

9.6 SCALER (COUNTER) CONTROL CIRCUIT

A fast waveshaping circuit for the scaler control is shown in Figure 9.6-1. It will receive inputs of either a negative or positive polarity. The inputs are direct coupled to a limit current that keeps any large input pulses from overloading the input transistors (Q_1 and Q_2). Both Q_1 and Q_2 are very fast (300 MHz or faster) transistors and are driven to saturation by the incoming pulse. The inverted output pulses from Q_1 and Q_2 are collector coupled to two NAND gates of a fourfold NAND-gate integrated circuit (e.g., MC846). The positive input section uses one section of the fourfold NAND gate as a simple inverter. The NAND function of the gates is supplied through the remote input jack. Any signal approaching ground through the remote input will activate the NAND gate, which will pass the signal from either the positive or negative input (whichever has an input signal at the time). The signal is then fed to an emitter follower that matches the control output to the input impedance of the first decade of the *n*-decade scaler.

In Figure 9.6-1, the input limit current passes via the four-diode bridge, which is an analog switching circuit, as discussed in Section 3.4. The direct-coupled inputs have an input impedance of about 1000 Ω. Input sensitivity levels are

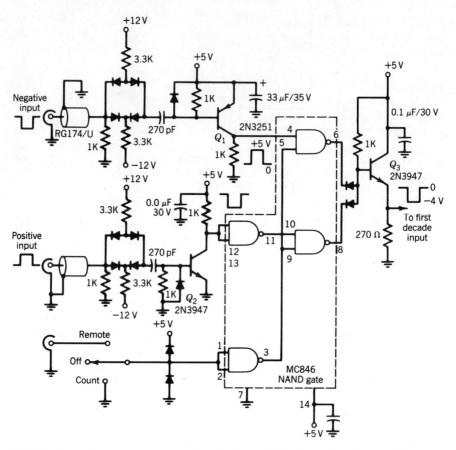

FIGURE 9.6-1 Scaler (counter) control circuit. (From H. H. Chiang, *Basic Nuclear Electronics*, Wiley, New York, 1969.)

approximately 1 V with fast rise times, and they decrease with increasing rise times. A nominal 3-V input requires a rise time of 1 μs or faster. Each decade-counting unit of the scaler contains a BCD counter (e.g., 7490A) and a BCD-to-seven-segment decoder (e.g., 7447A) or a BCD-to-decimal decoder (e.g., 74141). The scaler constructed like this may have a 10-MHz counting rate and a 50-ns double-pulse resolution.

9.7 REGISTERS

Introduction

A *register* is a set of flip-flops (*RS*, *D*, or *JK* devices) employed for the temporary storage of a digital word. Registers are most often involved in transferring information from one point to another. A register may receive information from one input unit, store this data until the control unit can finish another task, and then direct the register to transfer its contents to a specified location in memory. Registers used in computers or other digital systems are divided into two basic types: the *parallel register* and the *shift register*. In parallel registers all positions are generally filled simultaneously. Shift registers can be filled or emptied serially from one end of the register. In several cases, a shift register will combine parallel and serial operation. The shift register will form a ring counter if the Q output of the first flip-flop is connected to the shift input of the register (see Figure 9.7-2).

Parallel Registers

In the eight-bit register of Figure 9.7-1, the reset line is activated to clear the register before any data is entered. The input data lines and input data strobe line are suitably connected to the input terminals of all the input AND gates. As the input data lines are set, a strobe pulse appearing on the input data strobe line will set the data into the register due to the AND function. After the data word has been entered, it can be shifted to another section of the computer.

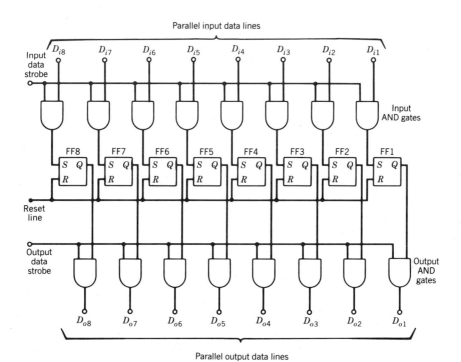

FIGURE 9.7-1 An eight-bit parallel register demonstrating the basic operation of parallel registers.

The output data strobe line and output Q of each flip-flop are connected to the two input terminals of each output AND gate. Activating the output data strobe line causes the data to appear on the output lines during the strobe. Typically a parallel register is employed to receive a word from memory and then to transfer this word to the control unit. Parallel registers can be rapidly filled or read even though they require a large number of gates.

Shift Registers

A shift register can be used in serial mode to receive or transmit serial information (data having bits that appear on a line sequentially in time). It can also operate in parallel-series combination to receive parallel information (each bit appears on a separate line at the same time) and transmit serially, or receive serially and transmit in parallel. A shift pulse applied to the serial-mode register moves all data one position to the right or left, depending upon the register design. A register designed to shift right is shown in Figure 9.7-2. It consists of six flip-flop stages (FF1–FF6) and so has six positions. If the LSB of the data is stored in FF1, it will also be presented to the output line. If we want five data

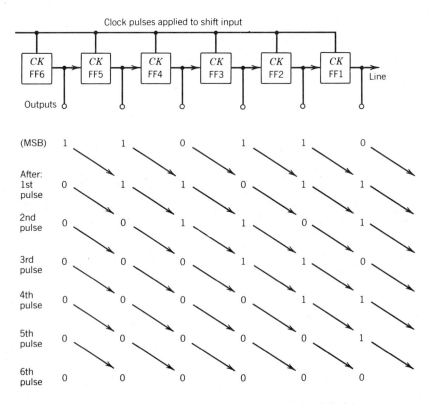

FIGURE 9.7-2 A six-position shift register designed to shift right.

bits to be transmitted to the line, the LSB must be stored in FF2 with a 0 stored in FF1. As these data are to be transmitted, six equally spaced clock pulses are applied to the shift input of the register. The first pulse causes all data to shift toward FF1, with the bit that was previously in this position shifting onto the output line. Assume that the code to be transmitted is 110110. If this information has been entered into the six positions of the register, each clock pulse will cause the data bits to move as shown in Figure 9.7-2. The operation of the shift register is based on the fact that a given stage will properly gate the next stage so that the clock pulse will set this next stage to the same state as that of the driving stage.

A shift register based on JK flip-flops is shown in Figure 9.7-3a. It is designed to shift on negative shift pulse transitions. The gates of FF8 should be connected such that $J8 = 0$ and $K8 = 1$ to cause 0s to fill the register when the code shifts into the line. The register may also be filled by connecting $J8$ to the incoming data line and $K8$ to the inverted data signal. The state of Q_8 before a shift transition gates FF7 so that Q_7 changes to this state as the transition occurs. The state of Q_7 before the transition determines the state of Q_6 after the transition, and so on down to the end of the register. Figure 9.7-3b shows a timing chart for the output signal of Figure 9.7-3a when 11011010 is initially contained by the eight-bit register. The leading 0 is not part of the data word but is inserted in the register to achieve equal spacing of the data bits. The code produced at the output of the register is NRZ (nonreturn to 0). In order to transmit RZ (return to 0) code, we can make the conversion with an AND gate, as shown in

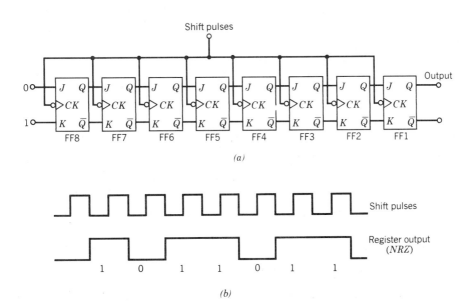

(a)

(b)

FIGURE 9.7-3 (*a*) Eight-position shift register based on JK flip-flops; (*b*) timing chart for the output of (*a*) when 11011010 is initially contained by the eight-bit register; the leading 0 is not part of the data word but is inserted in the register to achieve equal spacing of the data bits.

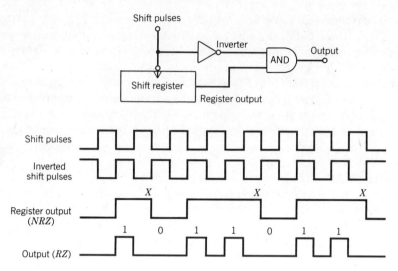

FIGURE 9.7-4 Conversion of NRZ to RZ code by ANDing the inverted shift pulses with the register output. If the register shifts on positive clock transitions, RZ code can be created by ANDing the shift pulses with the register output, eliminating the need of an inverter.

Figure 9.7-4. On the points marked X, one input is going positive while the other is dropping. Thus these points could cause slivers at the output of the AND gate. To avoid this problem we can choose a proper inverter for the inverted shift pulses, which are delayed more than the delay of the register output.

IC shift registers are commercially available. The 74164 (an eight-bit, parallel-out, serial shift register) and the 74165 (a parallel-load shift register) can be used in a digital communication system as the sending register and the receiving register, respectively. The 74194 is a four-bit bidirectional universal shift register that can be used to construct larger registers. It features parallel input and parallel output, or serial input and output (right or left shift).

9.8 DATA CONVERTERS

Analog configurations and digital circuits occupy mutually exclusive realms. The signals in the analog configurations are continuous and may assume any values within upper and lower limits, whereas the signals in digital circuits may assume only one of two discrete voltage levels—that is, either 0 or 1 of the two binary digits. A *data converter* is a circuit that examines a signal from one of these realms and then converts it to a proportional signal in the other. Thus a data converter is either an analog-to-digital (A/D) converter (ADC) or digital-to-analog (D/A) converter (DAC).

Analog-to-digital converters are numerous. Some of their types are the following:

1. *Counter ADC.* This ADC consists of a DAC, a comparator, a clock, a NAND gate, and a BCD counter. The inverting and noninverting terminals of the comparator accept the DAC output v_A and analog input v_i, respectively. The NAND gate accepts input pulses from the clock and comparator and sends its output pulse to the BCD counter. The counter output applied to the D/A converter is the digital output $Q_A Q_B Q_C Q_D$. The DAC allows comparison of the converter analog output v_A with the analog input v_i, which is assumed positive. Initially the counter is cleared by a convert pulse, so that $v_A = 0$, and the NAND gate allows clock pulses to pass. As time progresses, v_A increases in steps. At the point where v_A exceeds v_i, the comparator output switches negative and the pulses to the counter cease. At this point, the digital output is essentially equal to equivalent analog input. The number of bits can be increased although the time required grows quickly with the number of bits.

2. *Simultaneous ADC.* The simplest and fastest ADC is the simultaneous A/D converter. Several comparators are used to determine the input amplitude. The comparator outputs are connected to the inputs of the priority encoder. This circuit has been used primarily for four-bit ADCs.

3. *Successive approximation (SA) ADC.* The SA ADC (see Section 9.9) is similar to the counter except that the bits are tested in succession, that is, incremented in large steps starting with the most significant bit (MSB), rather than incrementing by the smallest step, as done when counting pulses. The SA ADC is much faster than counting pulses.

4. *Voltage-to-frequency (V/F) converter.* This is a voltage controlled oscillator (VCO) in which an input V_i is represented by an output frequency F. An ADC using a V/F converter may consist of a VCO and a binary counter between which a two-input AND gate is connected. The AND gate inputs are the VCO output and the gate pulse.

5. *Dual-slope ADC.* This ADC is a dual-slope integrator. Most digital voltmeters employ it as their ADC circuit (see Section 9.9).

A digital-to-analog converter (DAC) is used to convert binary digital information to analog form; for instance, it can convert the binary digital word 110001 to the decimal format 49, since

$$1 \times 32 + 1 \times 16 + 0 \times 8 + 0 \times 4 + 0 \times 2 + 1 \times 1 = 49,$$

and the DAC contains the weighting factors (2, 4, 8, 16, . . .), which can be summed or not depending upon the state of the input digit. Weighting factors can be either voltages or currents. Common DAC circuits will be discussed in Section 9.9. Frequency-to-voltage conversion is a form of DAC, in which an input frequency is converted to an output voltage.

9.9 ANALOG-TO-DIGITAL (A/D) CONVERTERS (ADCs)

Successive-Approximation ADCs

Figure 9.9-1 shows a successive-approximation A/D converter (SA ADC). A *successive approximation register* (SAR) includes the control logic, a shift register, and a set of output latches, one for each register section. The outputs of the latches drive a DAC. A start pulse sets the first bit of the shift register high, so the DAC will see the word 10000000_2, and consequently produces an output voltage equal to one-half of the full-scale output voltage ($V_{fs}/2$). If the input voltage V_i is greater than $V_{fs}/2$, then the $B1$ latch is set high. On the next clock pulse, register $B2$ is set high for trial 2. The output of the DAC is now $\frac{3}{4}$ scale. If, on any trial, it is found that V_i is smaller than V_A, then that bit is reset low. Thus, to test the bits in succession, a bit is switched on during a test phase. If the bit causes V_A to exceed V_i as detected by the comparator, it is switched off and kept off; otherwise it is left on. Sequentially the control logic tests the next most significant bit until the conversion is completed.

Here is an example, illustrating a three-bit SAR through a sample conversion. Assume that the full-scale voltage is 1 V and V_i is 0.625 V. Consider the timing during the positive-going clock-pulse times T_1 to T_4, as shown in Figure 9.9-2.

First trial during T_1: The start pulse is received; so register $B1$ goes high. Output word is now 100_2; so $V_A = 0.5$ V. Since V_A is less than V_i, latch $B1$ is set to 1; so at the end of the trial, the output word remains 100_2.

Second trial during T_2: On this trial (which starts on receiving the next clock pulse), register B_2 is set high, so that output word is 110_2. Voltage V_A is now

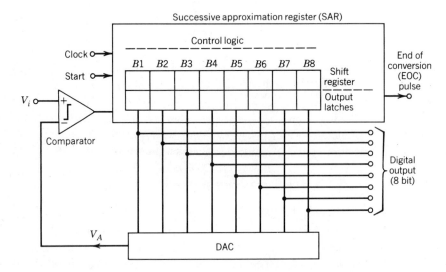

FIGURE 9.9-1 Block diagram of a successive-approximation A/D converter.

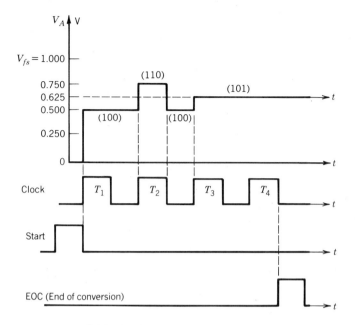

FIGURE 9.9-2   Timing chart for Figure 9.9-1.

0.75 V. Since V_i is less than V_A, the $B2$ latch is set to 0 and the output word reverts to 100_2.

Third trial during T_3: Register $B3$ is set high, making the output word 101_2. The value of V_A is now 0.625 V; so $V_i = V_A$. The $B4$ register is latched to 1, and the output word remains 101_2.

Overflow during T_4: During the fourth clock pulse, overflow occurs, telling the control logic to issue an end of conversion (EOC) pulse. In some cases, the overflow pulse is the EOC pulse.

Notice that in the example, we had a three-bit SAR and required four ($= n + 1$) clock pulses to complete the conversion. Hence an SA-type ADC takes ($n + 1$) clock pulses for a full-scale conversion. SA A/D converters are normally purchased as complete units or implemented by microprocessor software. The National Semiconductor ADC 0816 (a CMOS device) is an SA-type eight-bit ADC designed for use with a microprocessor. The input of this ADC can be connected directly to the analog signal, or it can be connected to a 16-channel analog multiplexer. In this situation the channel to be converted is chosen by a four-bit address presented to the chip.

Dual-Slope ADC

The dual-slope integrator is employed as an ADC in most digital voltmeters (DVMs), since the dual-slope DVMs or DMMs offer relative immunity to noise riding on the input voltage and relative immunity to error due to inaccuracy, long term drift in the clock frequency, and so on. The dual-slope DVM system is shown in Figure 9.9-3a, while associated waveforms are shown in Figure 9.9-3b. The op-amp A_1 and R_1-C_1 combination make up an integrator, where

$$I_f = -I_i, \quad I_i = \frac{E_{in}}{R_1}, \quad \text{and } I_f = C_1 \frac{dV_{o1}}{dt}.$$

Thus

$$\frac{C_1 \, dV_{o1}}{dt} = \frac{-E_{in}}{R_1}. \tag{9.9-1}$$

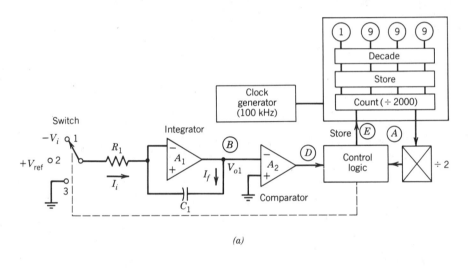

(a)

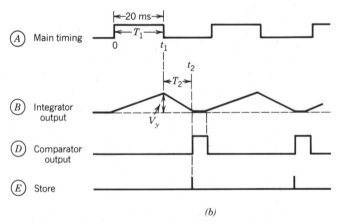

(b)

FIGURE 9.9-3 (a) System and (b) waveforms of the dual-slope DVM. The divide-by-2000 counter (for 1999 display) is further divided by two to drive the voltmeter. (From H. H. Chiang, *Electrical and Electronic Instrumentation*, Wiley, New York, 1984, p. 242.)

Integrating both sides,

$$\int \frac{C_1 \, d_{o1}}{dt} \, dt = - \int \frac{E_{in}}{R_1} \, dt;$$

$$C_1 V_{o1} = \int \frac{E_{in}}{R_1} \, dt.$$

Hence

$$V_{o1} = \frac{-1}{R_1 C_1} \int_0^t E_{in} \, dt. \tag{9.9-2}$$

Equation (9.9-2) is the transfer equation for the op-amp integrator circuit. The dual-slope DVM system operates as follows.

At the beginning of the measurement cycle, the capacitor C_1 in Figure 9.9-3a is fully discharged. The input to the integrator is connected to the negative input voltage $(-V_i)$ so that the capacitor C_1 begins to charge due to current $-V_i/R_1$. While the integrator output V_{o1} begins rising from zero, the counter starts to count the clock pulses from the 100-kHz clock generator. The charging is continued until the counter has counted 2000 (i.e., for 2K/100K or 20 ms). At the end of this period, the voltage V_y across the capacitor will be

$$V_y = \frac{-1}{R_1 C_1} \int_0^{t_1} (-V_i) dt = \frac{V_i T_1}{R_1 C_1}. \tag{9.9-3}$$

Then the control logic section switches the input of the integrator to the positive reference voltage V_{ref} so that the capacitor discharges under the influence of current V_{ref}/R_1. Since the reference voltage in the magnitude is larger than the voltage to be measured, the charge on the capacitor decreases more rapidly than it built up and at time t_2 it will be zero; that is,

$$0 = V_y - \frac{1}{R_1 C_1} \int_{t_1}^{t_2} V_{ref} \, dt = V_y - \frac{V_{ref} T_2}{R_1 C_1}$$

$$= \frac{V_i T_1}{R_1 C_1} - \frac{V_{ref} T_2}{R_1 C_1};$$

thus $V_i T_1 = V_{ref} T_2$, or

$$V_i = \frac{T_2}{T_1} V_{ref}. \tag{9.9-4}$$

The zero voltage condition is sensed by the comparator, which causes the control logic to switch the input of the integrator to zero volts (ground potential),

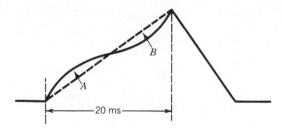

FIGURE 9.9-4 Integrator output when measuring a dc voltage with a superimposed 50- or 60-Hz signal. Area A cancels area B.

thus preventing any further change in the charge on the capacitor. At the same time, the control logic commands the counter to store the count. As indicated by Equation (9.9-4), the time displayed gives a direct measure of the input voltage in terms of the reference voltage. Hence the reference voltage can be chosen to give a suitable basic range for the digital voltmeter. For example, with a reference voltage of 2 V and $T_2 = T_1$, the basic range will be 2 V; however it will only be possible to display 1.999 V. The counter continues counting until it reaches the all zero state, at which point the measurement cycle is repeated. A measuring period of 20 ms is a good choice because it gives good rejection of 50- or 60-Hz interference. Figure 9.9-4 shows the integrator output when measuring a dc voltage with a superimposed 50- or 60-Hz signal. Since area A cancels area B, the interfering signal is effectively rejected.

We may see terms like "$3\frac{1}{2}$-digit" and "$4\frac{1}{2}$-digit" in advertisements for DVM/DMM products. These terms refer to the fact that the most significant digit can only be a 0 or a 1, while all other digits can be anything between 0 and 9. Such terminology indicates that the meter can read 100% over its basic range. For example, a $3\frac{1}{2}$-digit voltmeter will read 0–1999 mV, while its basic range is only 0–999 mV. If this range is exceeded, then the 1 lights up; otherwise it remains darkened.

A DVM usually has an input resistance larger than 10 MΩ and an accuracy better than $\pm 0.2\%$ of the reading.

9.10 DIGITAL-TO-ANALOG (D/A) CONVERTERS (DACs)

Binary-Weighted Resistor Ladder DAC Circuit

The D/A converter converts a digital code to an analog voltage that is proportional to the number represented by the code. A binary-weighted resistor ladder DAC circuit is shown in Figure 9.10-1. The resistors in the ladder are said to be *binary weighted* since their values are related to each other by powers of two. From first to last the ladder resistors in the chain have the sequential values R, $2R$, $4R$, $8R$, $16R$, ..., and $(2^{n-1})R$ if there are n resistors. The switches (B_1

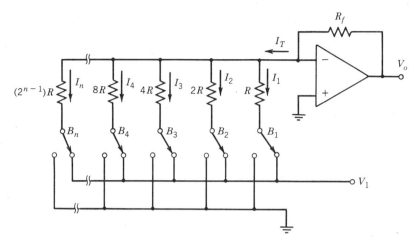

FIGURE 9.10-1   Binary weighted resistor ladder D/A converter circuit.

through B_n) represent the input bits of the digital word and would be transistor switches in actual practice. The switches are used to connect the input resistors to either voltage source V_1 or ground to represent binary states 1 and 0, respectively. The currents I_1 through I_n are given by

$$I_1 = \frac{V_1}{R},$$

$$I_2 = \frac{V_1}{2R},$$

$$I_3 = \frac{V_1}{4R},$$

$$\vdots$$

$$I_n = \frac{V_1}{2^{(n-1)}R}.$$

The total current I_T to the junction (inverting terminal) is expressed by the summation of currents I_1 through I_n:

$$I_T = \sum_{i=1}^{n} \frac{b_i V_1}{2^{(i-1)}R}, \tag{9.10-1}$$

where I_T = the total current in amperes (A)
V_1 = the reference potential representing digital one bit, expressed in volts (V)

R = the resistance of the lowest-value resistor in the ladder, expressed in ohms (Ω)

b_i = either 1 or 0, depending upon whether the input bit is 1 or 0 (e.g., for input 1011_2, $a_1 = 1$, $a_2 = 0$, $a_3 = 1$, $a_4 = 1$)

n = the number of bits, that is, the number of switches.

The output voltage is given by

$$V_o = -R_f \sum_{i=1}^{n} \frac{b_i V_1}{2^{(i-1)}R} = -\frac{V_1 R_f}{R} \sum_{i=1}^{n} \frac{b_i}{2^{(i-1)}} . \qquad (9.10\text{-}2)$$

The maximum conversion factor (CF_{max}) that would be used is

$$CF_{max} = \frac{V_{oM}}{N_M} = \frac{V_{oM}}{2^n - 1} , \qquad (9.10\text{-}3)$$

where V_{oM} is the maximum output voltage and N_M is the maximum number to be converted; this number generally equals $2^n - 1$, and n is the number of code bits. Typically the DAC accuracy is $\pm\frac{1}{2}$LSB, meaning that the error voltage (V_E) will never exceed one-half of the conversion factor; that is

$$V_E \leqslant \frac{V_o}{2N_M} = \frac{V_o}{2^{n+1} - 2} .$$

There are $2^n - 1$ distinguishable output voltage levels for an n-bit resolution of a DAC. Accuracy and resolution are the two specifications of major interest to the DAC user.

R-2R Resistor Ladder DAC Circuit

Figure 9.10-2 shows a basic DAC using a resistive (R-$2R$) ladder and an analog buffer amplifier. In practice, the digital input signals would come from a logic circuit (such as latch), but not through the mechanical switches as shown. In this basic DAC, each signal ($V_o - V_{n-1}$) is low (ground potential) or high ($= V_{ref}$) depending on the position of its respective switch. The analog output (V_a) of an R-$2R$ ladder is generally expressed by

$$V_a = \frac{V_0 + 2V_1 + 4V_2 + 8V_3 + 16V_4 + \cdots + 2^{(n-1)}V_{n-1}}{2^n}, \qquad (9.10\text{-}4)$$

where n is the number of digital inputs and V_0 represents the LSB.

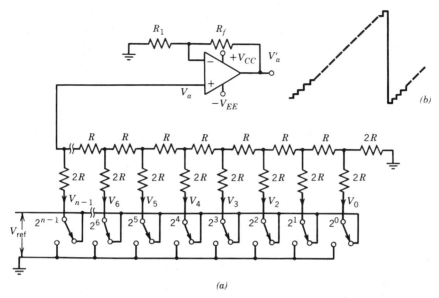

FIGURE 9.10-2 (a) R-2R resistor DAC circuit; (b) analog output voltage of a staircase waveform if the switches in (a) are replaced by logic signal outputs of an n-bit (e.g., eight-bit) binary counter.

REFERENCES

1. Bell, D. A., *Solid State Pulse Circuits*, 2nd ed., Reston Publishing Co., Reston, Virginia, 1981, Chaps. 12 and 13.

2. Taub, H., *Digital Circuits and Microprocessors*, McGraw-Hill, New York, 1982, Chap. 4.

3. Floyd, T. L., *Digital Logic Fundamentals*, 2nd ed., Charles E. Merrill, Columbus, Ohio, 1982, Chaps. 8 and 10.

4. Comer, D. J., *Electronic Design with Integrated Circuits*, Addison-Wesley, Reading, Massachusetts, 1981. Chaps. 2, 5, and 7.

5. Millman, J. and H. Taub, *Pulse, Digital, and Switching Waveforms*, McGraw-Hill, New York, 1965, Chap. 18.

6. Allocca, J. A. and A. Stuart, *Electronic Instrumentation*, Reston Publishing Co., Reston, Virginia, 1983, Chaps. 4 and 6.

7. Bouwens, A. J., *Digital Instrumentation*, McGraw-Hill, New York, 1984, Chaps. 6, 9, and 14.

8. *The Bipolar Digital Integrated Circuits Data Book for Design Engineers, Part One*, Texas Instruments, Canton, Massachusetts, 1981.

9. *COS/MOS Integrated Circuits Data Book*, RCA Solid-State Corp., Somerville, New Jersey, 1980.

QUESTIONS

9-1. Draw the complete circuit and the waveform chart for a four-flip-flop counting chain. Explain its counting process referring to the waveforms.

9-2. Draw the block diagram and the waveform chart for a down counter. Briefly explain its counting process.

9-3. Referring to Figure 9.3-1 explain how a scale-of-16 counter can be converted to a decade counter.

9-4. Draw the waveform chart for the BCD decade counter and explain its counting process.

9-5. What are the advantages of a synchronous counter over an asynchronous counter.

9-6. Compare the characteristics of the LCD and LED.

9-7. Briefly describe the LCD seven-segment display.

9-8. Briefly describe the LED seven-segment display.

9-9. Briefly explain the operation of the scaler control circuit shown in Figure 9.6-1.

9-10. What is the major difference between a parallel register and a shift register?

9-11. Draw a timing chart for the output signal of Figure 9.7-3 when 10010110 is initially contained by the eight-bit shift register.

9-12. Sketch the block diagram for a count-type ADC, and explain its operation.

9-13. Briefly explain how a successive approximation ADC works.

9-14. Why is the dual-slope ADC used in most digital voltmeters?

PROBLEMS

9-1. Sketch the waveforms appearing at points A, B, C, and D of Figure p9-1.

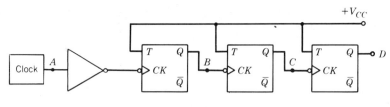

FIGURE P9-1

9-2. A 7490A is configured to operate as a divide-by-ten counter. Show the output waveform with respect to the clock.

9-3. Sketch the waveforms appearing at points A, B, and C of Figure p9-3.

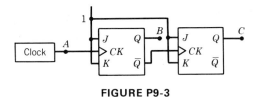

FIGURE P9-3

9-4. An LED is controlled by a transistor switch. The LED has a forward-voltage drop of 1.4 V, and the transistor has $h_{FE(min)} = 60$. The LED in the collector circuit is to have a minimum forward current of 18 mA. Design an appropriate circuit with $V_{CC} = 6$ V and $V_{in} = 6$ V.

Answer.   A resistor of 220 Ω in series with LED; a resistor of 18 kΩ in base circuit.

9-5. A four-bit (i.e., $n = 4$) D/A converter using a binary weighted resistor ladder has a reference source of 12 Vdc, and $R_f = R$. Find the output voltage V_o for the input word 1101_2.

Hint: For input 1101_2, $b_1 = 1, b_2 = 1, b_3 = 0, b_4 = 1$.

Answer. -7.38 V.

9-6. If in the circuit of Figure 9.10-2, $n = 8$, $V_{ref} = 10$ V, and the amplifier gain is 1, find the analog output V'_a with each of the following inputs: (a) 169_{10}; (b) 251_{10}, (c) $33_H (= 00110011_2)$.

Answer.   (a) 6.6 V; (b) 9.805 V; (c) 1.99 V.

10

LINEAR AND APPROXIMATE RAMP GENERATORS

10.1 INTRODUCTION

A sawtooth or rising-ramp triangular voltage used in a cathode-ray tube (CRT) causes the electron beam to sweep and flyback, thus producing a time base on the screen. Therefore the sawtooth or ramp generator is also called the *time base generator* or *sweep circuit*. The sweep portion of the output waveform from a ramp generator exhibits a linear variation of voltage with time. The ramp voltage can be applied to a resistor, thereby producing a current that increases linearly with time. The current time base generator is a current sweep circuit, which causes a linear time-varying current to flow instead through a coil. The coil of interest, called a *yoke*, is often used to produce a magnetic field that serves to cause deflection of the electron beam in a CRT. The yoke is mounted to the CRT near its electron gun and provides a magnetic field that is perpendicular to the deflection of the electron beam. Such magnetic deflection is primarily used in large-screen CRTs for television and radar displays. Time base generators can be employed for other applications, such as time measurements, time modulation, analog-to-time converters, and so on.

A simple sweep voltage generator can be constructed using a capacitor charged through a resistor in combination with a discharge transistor, as shown in Figure 1.4-1. The chief disadvantage of this simple circuit is its nonlinearity. To produce a linear sweep voltage, the capacitor charging current should be kept constant. This can be accomplished by instituting the charging resistor with a transistor constant current circuit.

The closely linear ramp voltage can be produced by the bootstrap, free-running, or Miller integrator circuits, which can be constructed using a transistor or an IC operational amplifier.

Approximate ramp voltages can be produced by relaxation oscillators, which can be constructed using negative-resistance devices such as unijunction transistors (UJT), Shockley or four-layer diodes, and programmable UJTs.

10.2 RAMP GENERATORS WITH CONSTANT CURRENT SOURCES

A true current source has an infinite impedance (zero admittance) since the current through it cannot be changed by voltage applied across it. Similarly, a bipolar transistor operated in its active region has extremely high impedance when a proper value of its forward-bias V_{EB} (or V_{BE}) is held constant, since the current through the transistor is essentially independent of the voltage across it (V_{EC} or V_{CE}). In this case, the transistor can be used as a constant current source. If the transistor constant current circuit is used instead of the charging resistance in the simple RC sweep arrangement of Figure 1.4-1, then it will become a linear ramp generator, as shown in Figure 10.2-1. Transistors Q_1 and Q_2 operate as a switch and a constant current source, respectively. The input pulse v_i is applied to the Q_1 base. When v_i is at ground level, Q_1 is off and C is quickly charged. When v_i is positive, Q_1 is switched on and C is rapidly discharged. During the positive input pulse, the output from Q_1 remains at voltage $V_{CE(sat)}$. Thus, with the positive input pulse train, the ramp generator produces the output sawtooth waves with linear ramps, as shown in Figure 10.2-1. Note that the

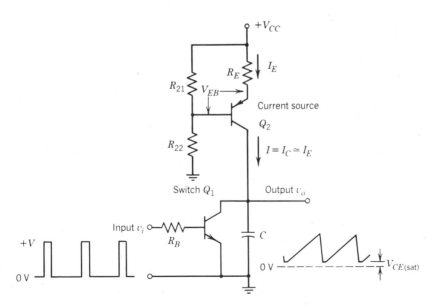

FIGURE 10.2-1   A linear ramp generator with a constant current transistor (Q_2) as the charging resistance.

bipolar transistor (BJT) Q_2 cannot be expected to function linearly if its collector-base voltage approaches the saturation level.

A *P*-channel FET with its source resistor and gate connected to $+V_{CC}$ can be used as a constant current source to replace the BJT (Q_2) constant current circuit in Figure 10.2-1. However the drain-source voltage V_{DS} is not permitted to fall below the maximum value of pinch-off voltage, since the FET will not function correctly in a linear circuit if its voltage V_{DS} falls below the pinch-off level. When it is desired to set the source current to a precise level, the source resistor should be made adjustable.

10.3   TRANSISTOR BOOTSTRAP RAMP GENERATOR

The transistor bootstrap ramp generator is shown in Figure 10.3-1. The discharge transistor Q_1 is direct coupled to the emitter follower Q_2. Capacitor C

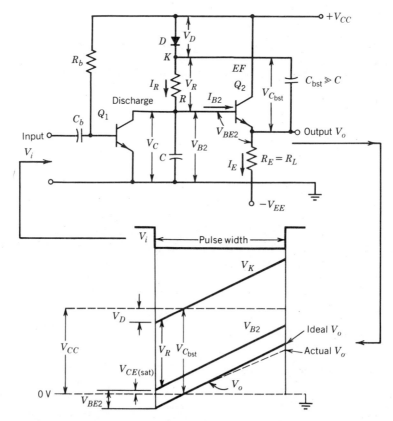

FIGURE 10.3-1   The circuit and waveforms of a transistor bootstrap ramp generator. Normally Q_1 saturates and Q_2 conducts.

of the basic RC network is connected across Q_1. The Q_2 emitter resistor R_E is connected to the negative supply $-V_{EE}$ to ensure that Q_2 remains conducting when its base voltage V_{B2} is close to ground. The bootstrapping capacitor C_{bst} is connected between the Q_2 emitter and the junction K of R and diode D, whose anode is connected to $+V_{CC}$. C_{bst} has a much larger capacitance than C. C_{bst} serves to maintain a constant voltage across R and so maintain the charging current constant. Normally (before an input signal is applied), Q_1 is on and its voltage is $V_{B2} = V_{CE(sat)} \simeq 0.2$ V. The Q_2 emitter is now at output voltage $V_o = V_{B2} - V_{BE2}$. At this time the diode cathode voltage is $V_K = V_{CC} - V_D$. The voltage across C_{bst} is $V_{Cbst} = V_K - V_o =$ constant. These voltages are indicated in the waveforms.

When an input negative pulse is applied, Q_1 is switched off and C starts to charge through R, causing V_{B2} and V_C to increase. As V_{B2} rises, V_o also rises, with only V_{BE2} below V_{B2}. When V_o rises, the larger capacitor C_{bst} retains its charge, and so V_K goes above the level of V_{CC}, causing diode D to be reverse biased. The constant voltage V_{Cbst} maintains voltage V_R constant. Consequently the charging current I_R is kept constant and C charges linearly, providing a linear output ramp. During the ramp time, D is reverse biased and charging current I_R is furnished by C_{bst}. If C_{bst} is very large and I_R is small, then C_{bst} will discharge by only a very small amount. As the negative input pulse is removed, C is discharged quickly by Q_1 and output V_o drops to its initial level. Also V_K drops, causing D to become forward biased. At this time a current pulse flows through D instead of the small charge lost from C_{bst}. The generator is then ready to produce another output ramp.

The advantages of the bootstrap circuit are its very linear output ramp and the output ramp amplitude approaching the supply voltage level. The actual output has some nonlinearity. The ramp is said to have 1% nonlinearity if the difference between the actual output and the ideal output is 1% of the output peak voltage. Nonlinearity sources are the Q_2 base current I_{B2} and the slight discharge of C_{bst} that occurs during the ramp time.

10.4   INTEGRATED CIRCUIT BOOTSTRAP RAMP GENERATORS
Bootstrap Ramp Generator Using an IC Op-Amp

If the emitter follower Q_2 in the circuit of Figure 10.3-1 is replaced by an op-amp connected as a voltage follower, then the circuit becomes an IC op-amp ramp generator, as shown in Figure 10.4-1. This IC generator has the following features:

1. The load resistor R_L is grounded rather than connected to $-V_{EE}$.
2. The output ramp starts at $V_{CE(sat)}$ instead of $V_{CE(sat)} - V_{BE2}$.
3. The low input current to the op-amp has an almost negligible effect on the charging current to capacitor C.

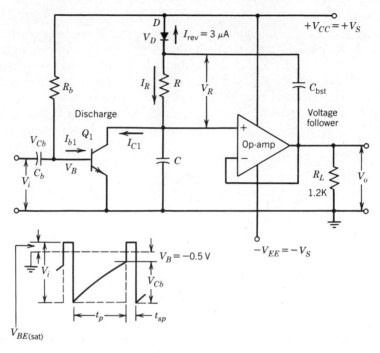

FIGURE 10.4-1 Bootstrap ramp generator using an IC op-amp as the voltage follower. Input V_i is a negative pulse; its pulse width (t_p) is typically larger than the space width (t_{sp}).

4. The maximum input bias current I_{MIB} is 500 nA for the op-amp 741, whereas the reverse saturation current of diode D is typically 3 μA, which is much more significant than the value of I_{MIB}. Thus the reverse leakage current of diode D can be the starting point for the IC bootstrap generator design.

5. When the IC ramp generator is utilized as part of an oscilloscope with an input capacitance of 30 pF, the value of capacitor C chosen must be about 1000 times greater than 30 pF so that C will not be affected by the stray capacitance.

When an IC op-amp is employed as a voltage follower, the inverting input terminal is connected to the output, as shown in Figure 10.4-1. The output from an op-amp is the amplified voltage difference between the two input terminals. Assume that an input of 1 V is applied at the noninverting terminal of the voltage follower. When the output increases positively, the voltage at the inverting terminal also increases positively. When the inverting input equals the noninverting input or 1 V, there is no longer an input signal, and the output voltage ceases to increase. Thus the voltage follower has a voltage gain of 1 with a high input impedance and a low output impedance. The output voltage closely follows the input. Actually there is a small voltage difference between the input terminals of a voltage follower. For a 741 with a gain of 200,000 and an output of 10 V, the input difference is typically 50 μV. Thus the output voltage is only 50 μV behind the input voltage. This is an improvement upon the transistor emitter follower, for its output V_o is typically 0.7 V behind input V_i.

Example 10.4-1. Design a bootstrap time-base generator using an IC op-amp as the voltage follower. The specifications for the circuit are listed below:

Load resistance $R_L = 1.2$ kΩ.
When diode D is reverse biased, $I_{rev} = 3$ μA.
1% nonlinearity due to I_{rev}.
1% nonlinearity due to C_{bst} discharge.
Ramp time $t = 1.5$ ms.
Output amplitude over the ramp time is $V_p = 7$ V.
$+V_{CC} = 16$ V; $V_{D1} = 0.7$ V; $V_{CE(sat)} = 0.2$ V; $V_{BE(sat)} = 0.8$ V.
A negative-going input pulse has an amplitude of 4 V.
Interval between input pulses is 0.1 ms.
$h_{FE(min)}$ of $Q_1 = 100$.

Solution.

(a) Since 1% nonlinearity due to I_{rev} is permitted, the charging current through R is

$$I_R = 100 \times I_{rev} = 100 \times 3 \ \mu\text{A} = 300 \ \mu\text{A}.$$

$$C = \frac{I_R t}{\Delta V} = \frac{I_R \times (\text{Ramp time})}{V_p} = \frac{300 \ \mu\text{A} \times 1.5 \ \text{ms}}{7 \ \text{V}}$$

$$= 0.064 \ \mu\text{F}.$$

Use 0.06 μF standard value.

(b) $V_R = V_{CC} - V_D - V_{CE(sat)} = 16$ V $- 0.7$ V $- 0.2$ V $= 15.1$ V.

$$R = \frac{V_R}{I_R} = \frac{15.1 \ \text{V}}{300 \ \mu\text{A}} = 50.3 \ \text{k}\Omega.$$

Use 47 kΩ standard value.

(c) For 1% nonlinearity due to C_{bst} discharge:
$\Delta V_{Cbst} = 1\%$ of initial V_{Cbst}.
$V_{Cbst} \simeq V_{CC} = 16$ V.

$$\Delta V_{Cbst} = \frac{16 \ \text{V}}{100} = 0.16 \ \text{V}.$$

C_{bst} discharge current $= I_R = 300 \ \mu A$.

$$C_{bst} = \frac{I_R t}{\Delta V_{Cbst}} = \frac{300 \ \mu A \times 1.5 \ ms}{0.16 \ V} = 2.8125 \ \mu F.$$

Use 2.7 μF standard value.

(d) Since discharge time of $C = \frac{1}{10}$ of the charge time,

$$I_{C1(min)} \ of \ Q_1 = 10 \times I_R = 10 \times 300 \ \mu A = 3 \ mA.$$

$$I_{b1} = \frac{I_{C1(min)}}{h_{FE(min)}} = \frac{3 \ mA}{100} = 30 \ \mu A.$$

$$R_b = \frac{V_{CC} - V_{BE(sat)}}{I_B} = \frac{16 \ V - 0.8 \ V}{30 \ \mu A} = 506 \ k\Omega.$$

Use 470 $k\Omega$ standard value.

(e) For Q_1 to remain biased off at the end of the input negative pulse, let $V_B = -0.5 \ V$.

$$V_{Cb} = V_i - V_{BE(sat)} - V_B = 4 - 0.8 - 0.5 = 2.7 \ V.$$

When Q_1 is off, the charging current for C_b is

$$I = I_b = \frac{V_{CC} - V_i}{R_b} = \frac{16 \ V - (-4 \ V)}{470 \ k} = 42.55 \ \mu A.$$

Hence

$$C_b = \frac{It}{V_{Cb}} = \frac{42.55 \ \mu A \times 1.5 \ ms}{2.7 \ V} = 0.0236 \ \mu F.$$

Use 0.02 μF standard value.
The total voltage applied to the $R_b C_b$ circuit is

$$7.4 \times V_{Cb} \quad or \quad 20 \ V.$$

Since V_{Cb} is only $1/7.4$ of 20 V, the charging current I_b for C_b is nearly constant when Q_1 is off.

Free-Running IC Bootstrap Ramp Generator

In the IC bootstrap ramp generator of Figure 10.4-1, the inverter Q_1 is an *npn* transistor used to discharge capacitor C. If the *npn* stage is replaced by a *pnp* stage and an op-amp Schmitt trigger is connected between the output and input terminals for feedback action, then the sweep circuit becomes a free-running bootstrap ramp generator, as shown in Figure 10.4-2. When the Schmitt output is negative, *pnp* transistor Q_1 discharges capacitor C. When the Schmitt output is positive, diode D_1 will protect the emitter-base junction of Q_1 if the reverse bias is excessive. Diode D_2 makes the Schmitt trigger have a lower-trigger level (LTL) close to ground. When the Schmitt output is negative, D_2 is reverse biased and only the diode reverse leakage current I_{rev} flows. The LTL now becomes $V_2 = -I_{rev}R_2$. The very low current I_{rev} through a selected resistance R_2 of a few kilohms will result in an amount of only millivolts from ground.

During the time when the Schmitt output is positive, Q_1 remains off and C charges, thereby producing a positive-going ramp output. As the ramp amplitude reaches the upper-trigger level (UTL) of the Schmitt circuit, the Schmitt output becomes negative. Then the base of Q_1 produces current I_{B1}, biasing the transistor on and discharging capacitor C. When the capacitor voltage drops rapidly, the ramp output also drops quickly. This fall continues until the LTL is reached. Note that diode D_2 makes the Schmitt LTL close to ground. Therefore, as the ramp output drops to ground level, the Schmitt output goes positive again, switching Q_1 off and permitting the next ramp to start.

If a negative synchronous pulse is applied to the sync input terminal, the UTL of the Schmitt trigger is reduced, and so the Schmitt output becomes negative, causing the ramp to go to zero. The ramp length and the output frequency can

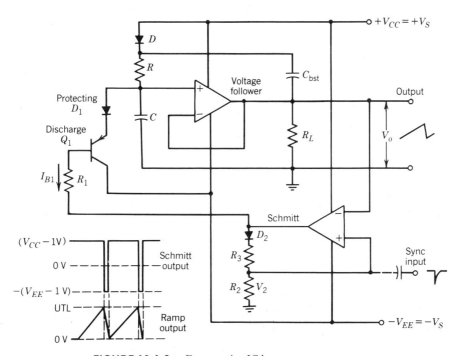

FIGURE 10.4-2 Free-running IC bootstrap ramp generator.

be controlled by varying resistance R. In order to control the ramp amplitude, the R_2-R_3 voltage divider can be replaced by a R_2-R_p-R_3 series combination, with the moving tap of potentiometer R_p connected to the Schmitt noninverting terminal. R_p affords the adjustment of the Schmitt UTL, thus providing the output amplitude control.

10.5 MILLER INTEGRATOR USING IC OPERATIONAL AMPLIFIER(S)

Miller Effect

Let V_i, V_o, and A_v be the amplifier input voltage, output voltage, and voltage gain, respectively. If the noninverting input is grounded and a capacitor C is connected between the inverting input and output terminals, then the voltage on the input side of C increases by V_i while that on the output side of C decreases by A_vV_i as V_i becomes positive. Thus the total capacitor voltage change is

$$\Delta V_i = V_i + A_vV_i = (1 + A_v)V_i, \tag{10.5-1}$$

and the charge supplied to C is

$$Q = C\Delta V_i = (1 + A_v)CV_i. \tag{10.5-2}$$

Equation (10.5-2) indicates that the input has supplied a charge to a capacitance of $(1 + A_v)C$. The result that capacitance C is magnified by a factor of $(1 + A_v)$ is referred to as the *Miller effect*.

Miller Integrator Ramp Generators

A Miller integrator shown in Figure 10.5-1 uses the Miller effect to generate a linear ramp. The charging current of capacitor C is supplied from a square-wave or pulse input. In order to ensure that the small bias currents cause equal voltage drops at each input terminal, the noninverting terminal is grounded via a resistance R_2 equal to the resistance R_1 at the inverting terminal. Since the voltage difference between the two input terminals is only about 50 μV at most, the inverting input terminal is always very close to ground level; thus it is often called a *virtual ground*. Therefore $V_i/R_1 = I_1$ is the constant input current. Since I_1 is much larger than the input bias current, all of I_1 effectively flows through capacitor C. A positive input voltage produces a flow of I_1 into C, resulting in a negative output voltage. On the other hand, a negative input voltage produces a flow of I_1 out of C, resulting in a positive output voltage. Since I_1 is a constant current, C is charged linearly, resulting in a linear output ramp, either negative or positive. Thus the output waveform will be triangular if the input voltage is a square wave.

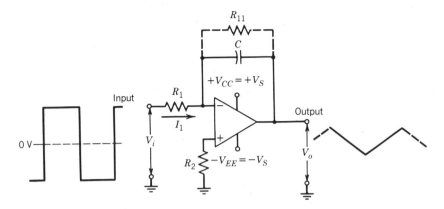

FIGURE 10.5-1   Miller integrator ramp generator with its input and output waveforms.

Remember that the inverting input terminal is a virtual ground. Suppose that the input is zero but is above 20 μV away from ground. If the voltage gain is 20,000, then the output will be $+4$ V or -4 V drifted from its zero level. The output voltage drift will produce a charge on capacitor C, giving the output V_o an offset. That is why V_o is not symmetrical above and below ground. The output drift can be minimized if the dc voltage gain is cut down using a high value resistor R_{11} ($\simeq 10R_1$, typically) connected between the output and the inverting input terminals; but the presence of R_{11} will affect the performance of the integrator at low frequencies. Thus capacitance C should be properly selected. As a lower limit,

$$\frac{1}{2\pi f C} = \frac{R_{11}}{10}.$$

Therefore the lowest operating frequency of the Miller integrator is

$$f = \frac{10}{2\pi R_{11} C}. \tag{10.5-3}$$

The op-amp has an input bias current ranging from tens to hundreds nA, typically. The input current I_1 should be much larger than the input bias current.

Resistors R_{11} in the circuit of Figure 10.5-1 can be replaced by an N-channel FET with its gate and source connected to the input terminal and the inverting input terminal, respectively. To ensure that the FET is biased off when an input pulse is present, the input pulse should have a negative amplitude greater than the FET pinch-off voltage. If the input consists of negative pulses, one of which has a pulse width (t_p) much larger than the space width (t_{sp}), then the output during $t_p + t_{sp}$ is a sawtooth waveform in which one ramp is much steeper than the other.

Example 10.5-1. Design a Miller integrator ramp generator to produce a triangular waveform with a peak-to-peak amplitude of 5 V. The input is a ± 12-V square wave with a frequency of 500 Hz. Use an IC op-amp with a maximum input bias current $I_B = 500$ nA and a supply $V_{CC} = 18$ V. Determine the lowest operating frequency of the integrator.

Solution. Choose $I_1 = 1$ mA ($\gg 500$ nA $= I_B$); then

$$R_1 = \frac{V_i}{I_1} = \frac{12 \text{ V}}{1 \text{ mA}} = 12 \text{ k}\Omega.$$

Let $R_{11} = 10R_1 = 120$ kΩ; then

$$R_2 = R_{11} \| R_1 \simeq 12 \text{ k}\Omega.$$

Each ramp time equals one-half of the period of the input:

$$t = \frac{1}{2f} = \frac{1}{2 \times 500 \text{ Hz}} = 1 \text{ ms}.$$

Ramp amplitude $\Delta V = 5$ V.

$$C = \frac{It}{\Delta V} = \frac{1 \text{ mA} \times 1 \text{ ms}}{5 \text{ V}} = 0.2 \text{ }\mu\text{F}.$$

From Equation (10.5-3), the lowest operating frequency is

$$f = \frac{10}{6.28 \times 120 \text{ k}\Omega \times 0.2 \text{ }\mu\text{F}} = 66 \text{ Hz}.$$

10.6 UJT RELAXATION OSCILLATOR AS APPROXIMATE RAMP GENERATOR

Unijunction Transistor (UJT)

The unijunction transistor (UJT) is a three-terminal device, as shown in Figure 10.6-1. A slab of lightly doped n-type silicon material has two base contacts (B_1 and B_2) attached to both ends of one surface and an aluminum rod alloyed to the opposite surface. The emitter pn junction of the device is formed at the boundary of the aluminum rod and the n-type silicon slab, Vith the p-type region attached along (about halfway) the slab. The pn junction forms a reverse-biased emitter if the emitter voltage V_E is less than the voltage V_{RB1} at the attachment point A. Let R_{B1} be the resistance from A to B_1 and R_{B2} the resistance

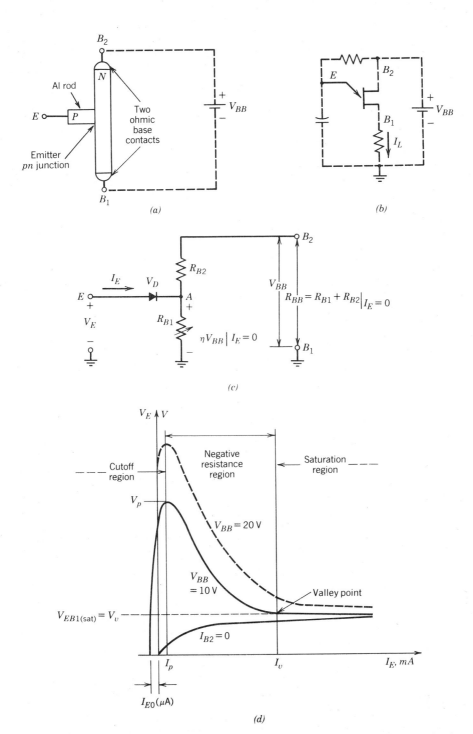

FIGURE 10.6-1 · UJT: (*a*) Construction; (*b*) circuit representation; (*c*) equivalent circuit; (*d*) static emitter characteristics.

from A to B_2. When the pn junction is reverse biased, the emitter current I_E is zero and the voltage V_{RB1} is given by

$$V_{RB1} = \frac{R_{B1}V_{BB}}{R_{B1} + R_{B2}} = \eta V_{BB}\Big|_{I_E = 0}, \qquad (10.6\text{-}1)$$

where

$$\eta = \frac{R_{B1}}{R_{B1} + R_{B2}}\Big|_{I_E = 0}.$$

The letter η (eta) is termed the *intrinsic stand-off ratio* ($\simeq 0.5$ to 0.8). In fact the slab acts as a resistive voltage divider. When the applied emitter voltage V_E is greater than ηV_{BB} by the forward-voltage drop of the diode V_D ($= 0.35$ to 0.7 V), the junction diode will fire. The emitter firing potential (peak-point voltage) is given by

$$V_p = \eta V_{BB} + V_D. \qquad (10.6\text{-}2)$$

UJT Relaxation Oscillator

Figure 10.6-2a with the switch in position 1 shows the simple circuit of a UJT relaxation oscillator, where C_E is charged through resistor R_E toward supply voltage V. As long as the capacitor voltage V_E is below the emitter firing potential V_p, the UJT emitter lead appears as an open circuit. When V_E approaches V_p, the UJT switches on and a large emitter current I_E flows. This causes capacitor C to discharge rapidly. When the capacitor voltage falls to the emitter saturation level, the UJT switches off, allowing C to begin to charge again.

From the waveform shown in Figure 10.6-2b, we see that the peak-point voltage V_p is given by

$$V_p = (V - V_v)(1 - e^{-T/RC}).$$

Since $V \gg V_v$,

$$V_p \simeq V(1 - e^{-T/RC}).$$

Hence

$$T \simeq RC \ln \frac{V}{V - V_p} = RC \ln \frac{1}{1 - V_p/V},$$

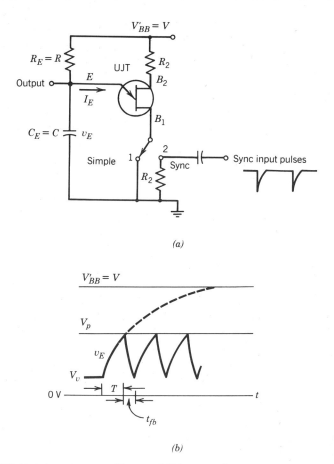

(a)

(b)

FIGURE 10.6-2   *(a)* UJT oscillator and *(b)* its approximate output ramp waveform.

or

$$T = R_E C_E \ln \frac{1}{1 - \eta}, \tag{10.6-3}$$

where $R_E = R$, $C_E = C$, and $\eta = V_p/V$. (Since $V_D \ll V_p$, $V_p = V_D + \eta V_{BB} = \eta V_{BB} \simeq \eta V$.) The time set is approximately proportional to the time constant $R_E C_E$. The oscillator frequency $f \simeq 1/T$ if the flyback time t_{fb} is negligible.

If the switch in Figure 10.6-2*a* is turned to position 2, then the resistor R_2 in series with UJT terminal B_1 allows synchronizing input negative pulses to be applied. As an input pulse pulls B_1 negative, voltage V_{EB1} is raised to the level at which the UJT fires, As the UJT fires, C discharges rapidly until the UJT switches off. The output v_E with an approximate rising ramp is produced across capacitor C.

The following table shows typical data representing major characteristics of 2N3980 and 2N4947 UJT devices at 25°C:

	2N3980	2N4947
I_p, µA	2	2
$V_{EB1(sat)}$, V	3 (max)	3 (max)
I_v, mA	1 (min)	4 (min)
η (average)	0.75	0.60
$I_{E(max)}$, mA	50	50
$V_{EB2(rev)}$, V	− 30 (max)	− 30 (max)
Use	General purpose	HF oscillator

Example 10.6-1. Design a simple sawtooth waveform generator using a 2N4947 UJT. The V_{BB} supply is 25 V, and output frequency is to be 10 kHz. Determine the amplitude of the output waveform.

Solution. Figure 10.6-2 with the switch in position 1 shows a simple sawtooth-waveform generator. From Equation (10.6-2), the UJT emitter firing voltage $V_p = \eta V_{BB} + V_D$. Capacitor C charges from $V_{EB1(sat)}$ to V_p. The data listed previously for the 2N4947 gives the following specifications:
$V_v = V_{EB1(sat)} = 3$ V max, $I_p = 2\,\mu A$, $I_v = 4$ mA, $\eta = 0.6$ average, and $Vp = 0.6 \times 25 + 0.7 = 15.7$ V.
Hence, for capacitor C,

V_{app} = supply voltage = V_{BB} = 25 V,
V_{ini} = initial charge voltage = $V_{EB1(sat)}$ = 3 V,
V_{fin} = final charge voltage = V_p = 15.7 V.

The range of R is calculated as follows:

$$R_{max} = \frac{V_{BB} - V_p}{I_p} = \frac{25\text{ V} - 15.7\text{ V}}{2\ \mu A} = 4.65\ \text{M}\Omega.$$

$$R_{min} = \frac{V_{BB} - V_{EB(sat)}}{I_V} = \frac{25\text{ V} - 3\text{ V}}{4\text{ mA}} = 5.5\ \text{k}\Omega.$$

Thus R should be in the range 5.5 kΩ to 4.65 MΩ. Let R have a value of 100 kΩ. The voltage of capacitor C is

$$v_C = V_{app} - (V_{app} - V_{ini})e^{-t/RC}.$$

Then

$$C = \frac{1/f}{R \ln \left(\dfrac{V_{\text{app}} - V_{\text{ini}}}{V_{\text{app}} - v_C} \right)} = \frac{1/10 \text{ kHz}}{100 \text{ k}\Omega \ln \left(\dfrac{25 \text{ V} - 3 \text{ V}}{25 \text{ V} - 15.7 \text{ V}} \right)}$$

$$= 1161 \text{ pF}.$$

Use 1200 pF standard value.
Output amplitude $= V_p - V_{EB1(\text{sat})} = 15.7 - 3 = 12.7$ V.

10.7 OTHER RELAXATION OSCILLATORS AS APPROXIMATE RAMP GENERATORS

Shockley Diode Relaxation Oscillator

The Shockley diode shown in Figure 10.7-1a is a four-layer *pnpn* diode with only two external terminals. When anode *A* is biased positively with respect to cathode *K*, junctions j_1 and j_3 are forward biased while junction j_2 is reverse biased. Hence, at small forward-bias voltages between *A* and *K* only a very low leakage current flows. When the forward bias is raised to the break-down voltage of junction j_2, a large forward current flows. To help understand the basic operation of the device, the four-layer *pnpn* structure is split into *pnp* transistor (Q_1) and *npn* transistor (Q_2) structures, as shown in Figure 10.7-1b. From the two-transistor equivalent circuit shown in Figure 10.7-1c, we see that when current I_{E1} flows into the Q_1 emitter, I_{C1} ($\simeq I_{E1}$) flows into the Q_2 base. Similarly, Q_2 has $I_{E2} \simeq I_{C2}$ and I_{C2} ($= I_{B1}$) flows out of the Q_1 base. The result is that both transistors are in saturation and the total anode-to-cathode voltage V_{AK} is about 0.9 V. The characteristics of the Shockley diode are exactly the same as those encountered for the silicon-controlled rectifier (SCR) with its gate current equal to zero ($I_{\text{gate}} = 0$). When V_F is small, only a low leakage current flows. When the breakover voltage $V_{(BR)F}$ (or switching voltage V_s) is reached, junction j_2 breaks down, the two transistors switch into saturation, and the device voltage quickly drops to a low V_F, presenting the negative resistance characteristic. The device reverse characteristics are similar to those of a reverse-biased conventional diode, except that the junctions j_1 and j_3 should break down before the Shockley diode goes into reverse breakdown.

A low pass *RC* network with its *C* shunted by a Shockley diode operates as a relaxation oscillator, as shown in Figure 10.7-2. Capacitor *C* is charged through *R* until the diode switching voltage is reached. Then *D* quickly switches to the low level on voltage V_F, and so *C* rapidly discharges via *D* until its current falls below I_H, the holding current. I_H is the minimum level that can maintain the

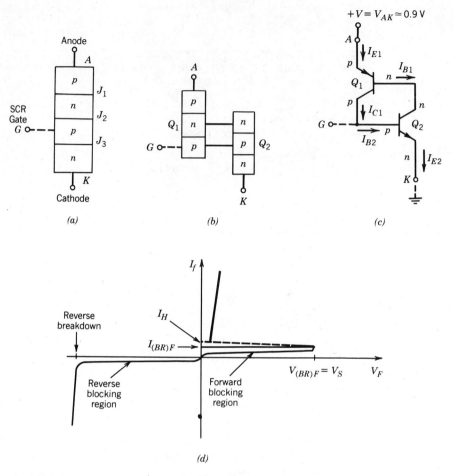

FIGURE 10.7-1 Shockley (four-layer) diode: (*a*) construction; (*b*) construction modified to show equivalent circuit; (*c*) two-transistor equivalent circuit; (*d*) characteristics.

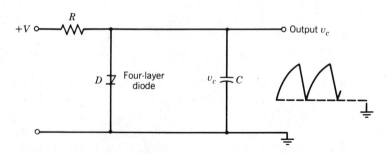

FIGURE 10.7-2 Four-layer diode relaxation oscillator.

diode conduction. Once D ceases to conduct, current flows into C again until it is charged up to $V_{(BR)F}$. The cycle is repetitive, producing an output waveform with the approximate rising ramp. The diode holding current I_H and forward breakover current $I_{(BR)F}$ restrict resistance R. Resistor R should not be so small that its current is greater than I_H. Nor can it be so large that its current is less than $I_{(BR)F}$. Therefore R is chosen between these limitations.

PUT Relaxation Oscillator

The programmable unijunction transistor (PUT) is a *pnpn* silicon switch very similar to the silicon-controlled switch (SCS). The SCS is essentially a mini-ature SCR with leads attached to all four semiconductor layers. Its four ter-minals are the anode, cathode, cathode gate, and anode gate (SCR has no anode gate). The main differences between PUT and SCS are that the PUT anode gate operates at a very low current level, usually less than 1 μA, and only the anode gate is available for controlling the triggering of the PUT. The PUT is so named since it is functionally similar to the UJT. The major difference between these two devices is that the PUT firing voltage V_p is adjustable (programmable) by the external resistors (R_1 and R_2 in Figure 10.7-3), whereas the UJT emitter

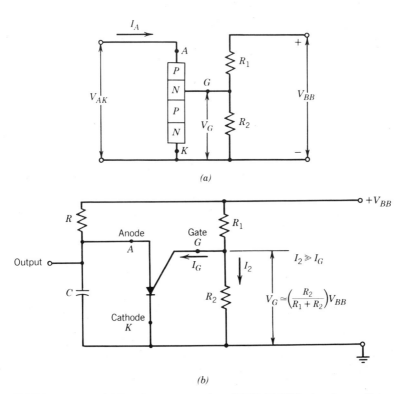

(a)

(b)

FIGURE 10.7-3   (a) Four-layer construction of PUT; (b) PUT relaxation oscillator.

fire voltage V_p is not adjustable. The PUT anode takes the place of the UJT emitter. Both the UJT and PUT are devices that switch to a high conductance, or on state, when voltage V_p or V_T is reached and switch back to a low conductance, or off state, when brought back to a lower current.

As shown in Figure 10.7-3, the gate of the PUT is connected to the junction of R_1 and R_2. The gate voltage is

$$V_G = \frac{V_{BB}R_2}{R_1 + R_2}.$$

The PUT will trigger on if input V_{AK} makes the anode positive with respect to the gate. When this occurs, V_{AK} rapidly drops to a low level and the PUT conducts heavily until the current through the device becomes too low to sustain conduction. Figure 10.7-3b shows a relaxation oscillator using the PUT. The fire voltage is

$$V_p = V_G + pn \text{ junction voltage drop} = V_G + 0.7 \text{ V}.$$

The following table shows typical data representing major characteristics of 2N6027 and 2N6028 PUT devices (40 V, 375 mW) at 25°C:

	2N6027	2N6028
I_p, μA	1.25	0.08
V_v = offset, V	0.70 (R_G = 1 M)	0.50 (R_G = 1 M)
I_V, μA	18	18
Leakage I_{GAO}, nA	1.0	1.0
V_F, V	0.8	0.8
V_{o-p}, V	11	11
I_G, μA	5	5

REFERENCES

1. Bell, D. A., *Solid-State Pulse Circuits*, 2nd ed., Reston Publishing Company, Reston, Virginia, 1981, Chap. 7.
2. Tocci, R. J., *Fundamentals of Pulse and Digital Circuits*, 2nd ed., Charles E. Merrill, Columbus, Ohio 1977, Chap. 12.
3. Pettit, J. M. and M. M. McWhorter, *Electronic Switching, Timing, and Pulse Circuits*, 2nd ed., McGraw-Hill, New York, 1970, Chap. 6.
4. Millman, J. and H. Taub, *Pulse, Digital, and Switching Waveforms*, McGraw-Hill, New York, 1965, Chaps. 12–15.
5. Boylestad, R. and L. Nashelsky, *Electronic Devices and Circuit Theory*, 3rd ed., Prentice-Hall, Englewood Cliffs, New Jersey, 1982, Chap. 11.

6. Chiang, H. H., *Electrical and Electronic Instrumentation*, John Wiley & Sons, New York, 1984, Chap. 13.

7. Clay, F. P. and M. S. Eaton, Two-Chip VCO Linearly Controls Ramp's Amplitude and Frequency, *Electronics*, November 30, 1981, p. 113.

QUESTIONS

10-1. Draw the circuit and waveforms of a simple RC sawtooth generator. Explain how the circuit operates and how it is limited.

10-2. Draw the circuit and waveforms of a ramp generator containing a constant current source. Explain how the circuit operates.

10-3. Draw an FET constant current circuit. Explain its operation.

10-4. Draw the circuit and waveforms of a transistor bootstrap ramp generator. Explain how the circuit operates.

10-5. Draw the circuit of a bootstrap ramp generator using an IC op-amp. Explain how the circuit operates.

10-6. Draw the circuit of a free-running IC bootstrap ramp generator. Show the waveforms and explain the operation and controls of the circuit.

10-7. Explain the Miller effect.

10-8. Draw a Miller integrator circuit using an IC op-amp. Show the waveforms and explain the circuit operation.

10-9. Draw a Miller integrator circuit using an IC op-amp and an FET.

10-10. Draw the UJT and relaxation oscillator circuit. Show the waveforms and explain the circuit operation.

10-11. Describe the four-layer diode characteristics using the related diagrams. Draw the circuit of a four-layer diode relaxation oscillator. Explain the circuit operation.

10-12. Draw the PUT relaxation oscillator circuit and explain its operation.

PROBLEMS

10-1. (a) A low-pass RC circuit with its capacitor C is shunted by a grounded-emitter transistor, whose base is connected to the supply voltage $+V_{CC}$ through the biasing resistor R_b. The input negative-going pulse is applied to the base via a capacitor C_b so that the output voltage across C is an approximate rising ramp. Sketch the circuit and its possible input and output waveforms.

(b) Design the circuit as a simple RC ramp generator to produce an output of 5 V peak. The supply voltage is 25 V, and the load R_L to be connected at the output is 390 kΩ. The ramp is to be triggered by a negative-going pulse with an amplitude of 4 V, pulse width of 2 ms, and time interval between pulses of 0.2 ms. Take $h_{FE(min)} = 60$ for transistor Q_1.

Note.   (1)   Choose the minimum capacitor charging current

$$I \gg I_{L(max)} = \frac{V_p}{R_L}.$$

Let $I = 100I_{L(max)}$ at peak output V_p.
(2)   Final capacitor voltage $v_C = V_p = V_{CC}(1 - e^{-t/RC})$.
(3)   (current) × (time) = CV_p = constant;

$$\frac{\text{discharge time}}{\text{charge time}} = \frac{0.2 \text{ mA}}{2 \text{ mA}} = \frac{1}{10};$$

collector current $I_{C1} \simeq 10I$.
(4)   At the end of the negative input pulse, let $V_B = -0.5$ V for Q_1 to remain off; thus the blocking capacitor voltage is $\Delta V_{Cb} = 4$ V $- 0.8$ V $- 0.5$ V.

Answer. $R = 15K$; $C = 0.56 \,\mu F$; $R_b = 120 \,k\Omega$; $C_b = 0.12 \,\mu F$.

10-2.   Design an RC ramp generator to provide an output of 4 V peak. The supply voltage is 18 V, and the maximum load current $I_{L(max)}$ from the output is 40 μA. The input trigger pulse has an amplitude of -5 V, a pulse width of 1 ms, and a space width of 0.1 ms. Take $h_{FE(min)} = 40$ for the transistor in the same circuit as that described in Problem 10-1a.

Answer. $R_L = 100 \,k\Omega$; $R = 3.3 \,k\Omega$; $C = 1.2 \,\mu F$; $R_b = 18 \,k\Omega$; $C_b = 0.2 \,\mu F$.

10-3.   (a) Draw the circuit of a constant current ramp generator, which is formed when the direct-coupled input of transistor Q_1 in Figure 10.2-1 is replaced by the capacitor (C_b)-coupled input with a resistor R_b connected between the base and $+V_{CC}$.

(b) Using the Q_2 constant current circuit in the ramp generator described in part a, modify the simple ramp generator designed in Problem 10-2 to provide a linear ramp output.

Hint.   Let $V_{CE2} = 3$ V minimum and $I_{R21} \simeq I_E$.
Answer. $R_E = 2.2 \,k\Omega$; $R_{21} = 2.2 \,k\Omega$; $R_{22} = 1.2 \,k\Omega$.

10-4.   A transistor bootstrap ramp generator is to give an output of 6 V peak, with a time period of 2 ms. The load resistor is to be 1.5 kΩ and the ramp is to be linear within 2%. The input pulse has an amplitude of -4 V, a width of 2 ms, and a space width of 0.5 ms. Design a suitable circuit using transistors with $h_{FE(min)} = 70$ and a supply voltage of ± 16 V.

Hint.   1% nonlinearity due to ΔI_{B2}; 1% nonlinearity due to C_{bst} discharge.
Answer. $C = 1.8 \,\mu F$; $R = 2.7 \,k\Omega$; $C_{bst} = 18 \,\mu F$.

10-5. Design a bootstrap ramp generator using a 741 op-amp to produce an output of 8 V peak over a time period of 1.5 ms. The load resistor is to be 1 kΩ, and the ramp is to be linear within 3%. The supply voltage is to be ±15 V. The input pulse has an amplitude of −5 V, a width of 1.5 ms, and a space width of 0.3 ms. Take the transistor $h_{FE(min)} = 70$. The reverse-biased diode current $I_{rev} = 3\,\mu A$.

Hint. 1.5% nonlinearity due to I_{rev}; 1.5% nonlinearity due to C_{bst} discharge.

Answer. $C = 0.039\,\mu F$; $R = 68\,k\Omega$; $C_{bst} = 1.2\,\mu F$; $R_b = 1\,M\Omega$; $C_b = 8200$ pF.

10-6. The 741 op-amp used in a Miller integrator has an input bias current $I_B = 500\,nA$, maximum. The supply voltage is to be ±15 V. Design a suitable circuit for the integrator to produce a triangular waveform output with a peak-to-peak amplitude of 5 V. The input is a ±12-V square wave with a frequency of 1 kHz. Determine the lowest operating frequency for the integrator.

Answer. Let $I_1 = 1\,mA$, $R_{11} = 10R_1$, $R_1 = 12\,k\Omega$, $R_2 \simeq 12\,k\Omega$, $R_{11} = 120\,k\Omega$, $C = 0.1\,\mu F$; 132.7 Hz.

10-7. The circuit of Figure 10.6-2 with the switch in position 1 is to use a 2N3980 UJT. The V_{BB} supply is 20 V, and the output frequency is to be 2 kHz. Design a suitable circuit and determine the output amplitude.

Answer. Let $R = 100\,k\Omega$, $C = 3300\,pF$, and output amplitude = 12.7 V.

10-8. Design a relaxation oscillator using a 2N4947 UJT. The V_{BB} supply is 15 V, and the output frequency is to be 4 kHz. Calculate the amplitude of the output waveform.

10-9. Design a relaxation oscillator using a 2N6027 PUT. The supply voltage is to be 18 V, and the output amplitude is to be 6 V at 2 kHz.

Answer. Let $I_2 = 100I_G$, $R_2 = 10\,k\Omega$, $R_1 = 22\,k\Omega$, and $R = 961\,k\Omega − 9.34$ MΩ. Let $R = 1\,M\Omega$ and $C = 1300\,pF$.

10-10. Design a relaxation oscillator using a 2N6028 PUT. The supply voltage is to be 20 V and the output amplitude is to be 8 V at 4 kHz.

11

MODULATION, DEMODULATION, AND TIME DIVISION MULTIPLEXING

11.1 INTRODUCTION TO MODULATION

Conventional Modulation

The process of placing information onto an information carrier is called *modulation*, while the converse process is referred to as *demodulation*. A radio transmitter transmits the modulated radio waves via the antenna to the destinations where the appropriate radio receivers receive the modulated radio frequency. The receivers reproduce the original signals (audio, video, or data) through detection (demodulation) and amplification. There are three types of conventional modulation used in the transmitters: amplitude modulation (AM); frequency modulation (FM) (see Section 1.5); and phase modulation (PM). They are shown in Figures 11.1-1a, 11.1-1b, and 11.1-1c, respectively.

In amplitude modulation, the amplitude of a sinusoidal carrier wave, whose frequency and phase are fixed, is varied in proportion to a given signal. This alters the given signal by translating its frequency components to higher (sideband) frequencies. The use of AM may be advantageous whenever a shift in the frequency components of a given signal is desired. The modulating signal of frequency f_m mixing with carrier frequency f_c by amplitude modulation will produce the lower sideband (LSB) $f_c - f_m$ and the upper sideband (USB) $f_c + f_m$. Information contained in the two sidebands are the same; no information is contained in the carrier. Hence we only need to amplify the LSB or

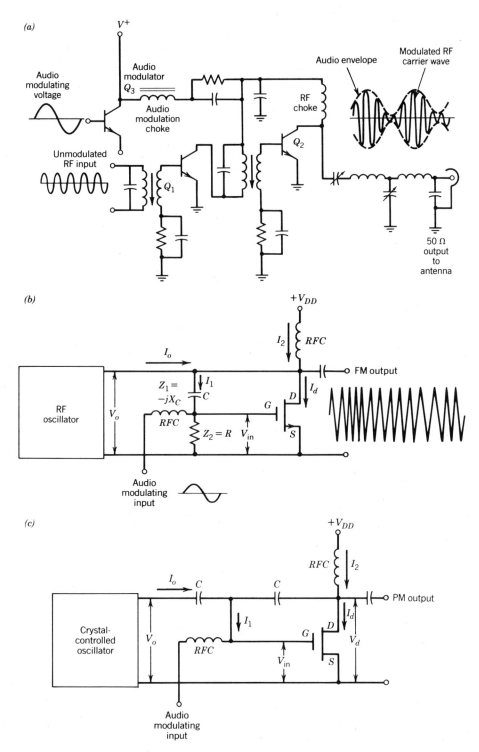

FIGURE 11.1-1 (*a*) AM in a transistor collector circuit, (*b*) FM in an FET varactor circuit, and (*c*) PM in an FET varactor circuit.

USB in order to save power and increase the channel to transmit much more data. But the single-sideband demodulation requires a higher precision receiver, since the carrier must be reinserted in order to take the f_m signal from the detector. In this case, the stability of the local oscillator needs to be very high; we can use the phase locked loop (PLL) or quartz crystal to increase the stability. Figure 11.1-1a is an example of a collector modulation method for AM; both transistors Q_1 and Q_2 are modulated.

In the FM circuit of Figure 11.1-1b, currents I_1 and I_2 are negligible at oscillator frequency ω_o, so that the oscillator output current I_o is approximately equal to the drain current I_d. Thus

$$I_o = I_d = g_m V_{\text{in}}, \qquad (11.1\text{-}1)$$

where

$$V_{\text{in}} = \frac{Z_2}{Z_1 + Z_2} V_o, \qquad (11.1\text{-}2)$$

in which $Z_1 \gg Z_2$. The small signal admittance presented to the oscillator is

$$Y = \frac{I_o}{V_o} = g_m \frac{Z_2}{Z_1 + Z_2} = g_m \frac{R}{1/j\omega C + R}, \qquad (11.1\text{-}3)$$

or

$$Y = j\omega_o C_{\text{eq}} \simeq j g_m \omega_o C R.$$

Hence

$$C_{\text{eq}} = g_m R C. \qquad (11.1\text{-}4)$$

Therefore the modulating voltage varies transconductance g_m, which varies equivalent capacitance C_{eq}, which, in turn, varies oscillator frequency ω_o.

In Figure 11.1-1c, the unmodulated oscillator voltage is V_o and the modulated voltage is V_d. The phase relationship between V_d and V_o is derived as follows:

$$I_o = (V_o - V_d) \times j\frac{\omega_o C}{2}; \qquad (11.1\text{-}5)$$

$$V_{\text{in}} \simeq \frac{V_o + V_d}{2}. \qquad (11.1\text{-}6)$$

I_1 and I_2 are negligible at ω_o. Substituting Equation (11.1-6) in Equation (11.1-1) and equaling to Equation (11.1-5), we obtain

$$g_m \frac{V_o + V_d}{2} = (V_o - V_d) \times j \frac{\omega_o C}{2} \, ;$$

from which

$$V_d = -V_o \frac{g_m - j\omega_o C}{g_m + j\omega_o C} = V_o \left[\frac{(\omega_o C)^2 - g_m^2}{g_m^2 + (\omega_o C)^2} + j \frac{2\omega_o C g_m}{g_m^2 + (\omega_o C)^2} \right], \quad (11.1\text{-}7)$$

where $|V_d| = |V_o|$ since the modulus of the complex term in the numerator is equal to that in the denominator. Thus the phase of V_d with respect to V_o is equal to the phase angle given by

$$\phi = \arctan \frac{2\omega_o C g_m}{(\omega_o C)^2 - g_m^2} . \quad (11.1\text{-}8)$$

Make $(\omega_o C)^2 \gg g_m^2$. Then

$$\phi \simeq \arctan \frac{2g_m}{\omega_o C} . \quad (11.1\text{-}9)$$

Keep the deviation $\phi < 0.1$ rad. Then

$$\phi \simeq \frac{2g_m}{\omega_o C} . \quad (11.1\text{-}10)$$

Thus ϕ can be changed by varying g_m with the modulating voltage. Since any change in phase is equivalent to a change in frequency, the phase modulation is an equivalent FM or indirect FM.

Pulse Modulation

We know that amplitude, frequency, and phase are the three variable components of a wave. If we convert the modulating signal into a pulse train, then there are many more factors we can change to obtain different types of pulse modulation, such as pulse amplitude modulation (PAM), pulse duration modulation (PDM), pulse position modulation (PPM), and pulse code modulation (PCM). These factors are shown in Figure 11.1-2. Several separate pulse-modulated signals can be transmitted or recorded on one channel by time division multiplexing (TDM). In PAM, the amplitude of each pulse is proportional to the instantaneous amplitude of the modulating signal v_m. In PDM, the duration of each pulse is proportional to v_m. In PPM, the position in time of the pulses is made to vary in proportion to the amplitude of signal v_m; each PPM pulse occurs just at the end of a corresponding PDM pulse. In

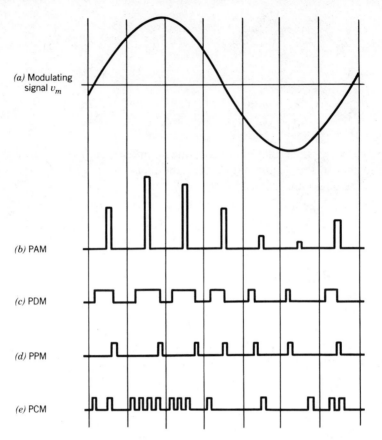

FIGURE 11.1-2 A low-frequency modulating signal and four types of pulse modulation: pulse amplitude modulation (PAM), pulse duration modulation (PDM), pulse position modulation (PPM), and pulse code modulation (PCM).

PCM, each amplitude sample of signal v_m is converted to a binary number. For all four modulation methods, the sampling frequency is at least about ten times the highest signal frequency.

11.2 AMPLITUDE MODULATION

Expressions for Amplitude-Modulated Waveform

The amplitude-modulated or AM waveform can be expressed in general terms as

$$v = V_c \cos \omega_c t + V_{mp} \cos \omega_m t \cos \omega_c t$$

$$= V_c \left[1 + \frac{V_{mp}}{V_c} \cos \omega_m t \right] \cos \omega_c t$$

or

$$v = V_c[1 + m \cos \omega_m t] \cos \omega_c t, \qquad (11.2\text{-}1)$$

where

$$m = \text{modulation index} = \frac{V_{mp}}{V_c}. \qquad (11.2\text{-}2)$$

When modulating amplitude V_{mp} equals carrier amplitude V_c, 100% modulation is present. Since

$$\cos \omega_c t \cos \omega_m t =. \tfrac{1}{2}[\cos(\omega_c - \omega_m)t + \cos(\omega_c + \omega_m)t],$$

Equation (11.2-1) can be rewritten as

$$v = V_c \cos \omega_c t + \frac{m V_c}{2} \cos(\omega_c - \omega_m)t + \frac{m V_c}{2} \cos(\omega_c + \omega_m)t, \qquad (11.2\text{-}3)$$

where the radian frequencies ω_c, $\omega_c - \omega_m$, and $\omega_c + \omega_m$ are of the carrier, LSB and USB, respectively. The sidebands each contain one-fourth the power of the carrier, giving a total power of

$$P_T = P_C + P_{LSB} + P_{USB} = P_C(1 + \tfrac{1}{4} + \tfrac{1}{4}) = \tfrac{3}{2}P_C. \qquad (11.2\text{-}4)$$

Single-Sideband Generation

The simplest method of single-sideband generation uses filtering to eliminate one sideband from a suppressed carrier when the carrier frequency is as low as several tens of kHz. With higher carrier frequencies, two stages of modulation, as shown in Figure 11.2-1, can be used to ease the filtering problem. In the first stage, the upper sideband results from modulating a 100-kHz carrier by a signal varying from 300 to 3400 Hz. Using a balanced modulator results in a lower sideband extending from 96.6 to 99.7 kHz. The first bandpass filter with a passband extending from 100.3 to 103.4 kHz follows the first modulator. This filter should provide 40-dB attenuation within 600 Hz ($= 100.3 - 99.7$ kHz) at 100 kHz, a 0.6% frequency change. The second filter should provide 40-dB attenuation between the frequencies 9.8997 and 10.1003 MHz, or within 200.6 kHz at 10 MHz, a 2% frequency change. The balanced modulator is actually a multiplier circuit forming the product of the carrier and modulating signals. Its output signal consists of two sidebands (USB and LSB) without a carrier and so is called a *suppressed carrier signal*.

Integrated Circuit Modulators

Figure 11.2-2 shows the MC1596/MC1496 integrated circuit modulator, along with the external components required to construct a balanced modulator. A

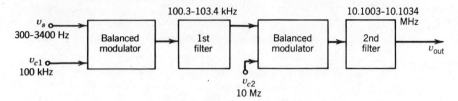

FIGURE 11.2-1 Upper-sideband generation using two-step method.

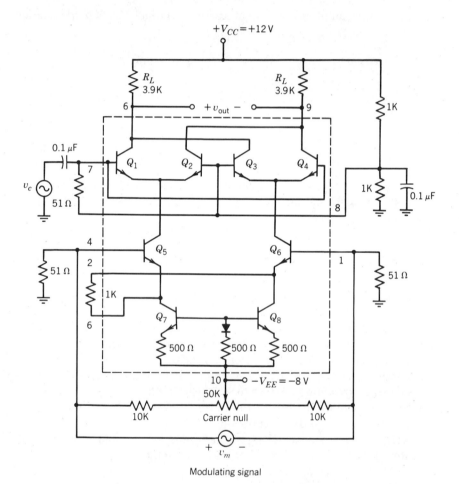

FIGURE 11.2-2 A balanced modulator containing a typical integrated circuit and the required external components.

single-end output can be attained by following the double-ended outputs with a difference amplifier. Transistors Q_7 and Q_8 are current sources to set the bias currents of Q_5 and Q_6 approximately equal. If a positive-going modulating signal is applied, current through Q_5 will increase while current through Q_6 will decrease. The circuit of Figure 11.2-2 can become an amplitude modulator containing the carrier signal by lowering the 10-kΩ resistors to a much smaller value (680 to 820 Ω), since the 50-kΩ potentiometer can be allowed to have more control over the bias currents of Q_5 and Q_6; in this case the two dc currents are adjusted to be different so that a carrier component is added to the sidebands.

The MC1596/1496 will function as a phase detector. When both inputs are at the same frequency, this IC unit will deliver an output that is a function of the phase difference between the input signals.

11.3   AM DETECTION

Diode Detector

The information transmitted by an AM wave is carried by the peak amplitude of the waveform. This information should be recovered by a peak-detecting circuit. Since basic detection is half-wave rectification, a diode in series with an RC combination makes up a simple AM detector. The video detector shown in Figure 12.5-1 is a typical AM demodulator used to recover the video information transmitted by the TV composite waveform.

In the simple diode detector of Figure 11.3-1a, the output voltage v_o contains a high frequency ripple voltage, a much lower frequency voltage that is proportional to the amplitude of the peaks of the AM waveform, and a dc voltage equal to the amplitude of the unmodulated carrier signal. The time constant of the $R_1 C_1$ combination is selected to be much larger than the period of the carrier and much smaller than the period of the modulating signal. One of the reasonable values is a time constant that is the geometric mean of the two periods divided by 2π, or

$$R_1 C_1 = \sqrt{\frac{1}{(2\pi f_m)(2\pi f_c)}}. \tag{11.3-1}$$

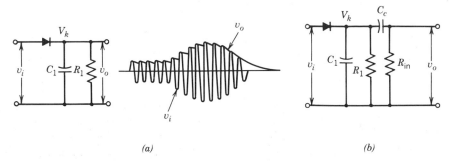

(a)   *(b)*

FIGURE 11.3-1 (a) A simple AM diode detector with its waveform; (b) a practical diode detector circuit, in which R_{in} is the input impedance to an amplifying stage.

Since the R_1C_1 time constant is very large, the discharge between cycles of the carriers is negligible; however the charge time constant is very small, so capacitor C_1 rapidly charges during each positive half-cycle of the carrier.

In the practical diode detector of Figure 11.3-1b, capacitor C_c removes the dc component of the demodulated signal from the output voltage v_o. When an unmodulated carrier is applied to the input, voltage V_k is a positive peak signal to the amplitude of the carrier, and so output voltage v_o equals zero. The input current from the source is positive. The diode cannot pass negative current. The current with unmodulated input is

$$I_{in} = \frac{V_c}{R_1},$$

where V_c is the amplitude of the unmodulated carrier. If the carrier is 100% modulated, the input current can be expressed as

$$I_{in} = \frac{V_c}{R_1} + \frac{V_c}{R_{eq}} \cos \omega_m t, \qquad (11.3\text{-}2)$$

where $R_{eq} = R_1 R_{in}/(R_1 + R_{in})$. When $\cos \omega_m t$ approaches -1, current I_{in} becomes negative and so the output waveform distorts, except that the amplifier input impedance R_{in} is much larger than R_1. To avoid the distortion resulting from the negative current, the modulation index should be limited such that

$$\frac{V_c}{R_1} = \frac{mV_c}{R_{eq}}$$

or

$$m = \frac{R_{in}}{R_1 + R_{in}}. \qquad (11.3\text{-}3)$$

When R_{in} is much larger than R_1, R_{eq} approaches R_1, m approaches unity, and thus no distortion occurs.

Precision-Rectifier AM Demodulator

If the amplitude of the peaks is in the millivolt range, a simple diode rectifier cannot accurately demodulate the signal. A precision rectifier, as shown in Figure 11.3-2a, will serve for signals falling into the range of tenths of volts. The inverting input terminal is a virtual ground. Diode D_1 is forward connected between the inverting terminal and the output end of the op-amp. Diode D_2 is forward connected from the D_1 cathode. The feedback resistor R_f is connected between the inverting terminal and the D_2 cathode. D_1 and D_2

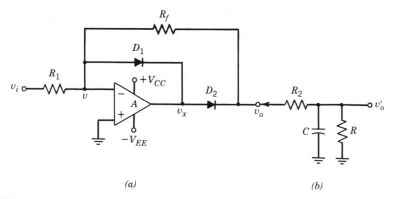

(a)   *(b)*

FIGURE 11.3-2 An AM detector containing (a) a precision rectifier and (b) a charge storage network.

are silicon diodes; they have a constant voltage drop of $V_D = 0.6$ V or more when forward biased. The input signal v_i is applied to the inverting terminal through resistor R_1. When v_i goes positive, D_1 becomes forward biased and voltage v_x at the $D_1 - D_2$ junction is clamped at $-V_D$. Then D_2 is reverse biased, and thus no current flows through R_f; the result is that the output voltage (v_o) is equal to zero. When v_i goes negative, D_1 becomes reverse biased and D_2 is forward biased. The output voltage is given by

$$v_o = v_x - V_D = -vA - V_D.$$

Voltage v at the inverting terminal is, using superposition theorem,

$$v = v_i \frac{R_f}{R_1 + R_f} + v_o \frac{R_1}{R_1 + R_f} ;$$

then v_o becomes

$$v_o = \frac{-A}{R_1 + R_f} (v_i R_f + R_1 v_o) - V_D.$$

Solving for v_o gives

$$v_o = - \frac{R_f v_i}{R_f/A + R_1(A + 1)/A} - \frac{(1 + R_f/R_1)V_D}{R_f/R_1 + (A + 1)}.$$

Typically the open-loop voltage gain $A = 200{,}000$. Since A is so very large, the output voltage can be approximated by

$$v_o \simeq - \frac{R_f}{R_1} v_i - \frac{(1 + R_f/R_1)R_D}{A + 1} \simeq - \frac{R_f}{R_1} v_i. \qquad (11.3\text{-}4)$$

The precision rectifier operates for half-wave rectification. It gives zero output for positive-going inputs and a positive output proportional to the magnitude of the input for negative-going inputs. The error involved in rectification amounts to microvolts and is given by

$$\frac{V_D(R_1 + R_f)}{(A + 1)R_1}.$$

When the precision rectifier is used to detect the peaks or envelope of an amplitude-modulated wave, the simple charge storage network shown in Figure 11.3-2b can be added as the output section of the envelope detector.

11.4 FREQUENCY MODULATION

Expressions for Frequency-Modulated Waveform

Frequency modulation (FM) occurs when the carrier frequency is caused to change in accordance with a modulating signal. For most FM communication systems, the frequency deviation is related linearly to the amplitude of the modulating signal. The expression for the radian frequency of an FM signal with sinusoidal modulation is

$$\omega_i = 2\pi f_i = 2\pi f_c + 2\pi f_d \cos \omega_m t, \tag{11.4-1}$$

where $f_d (= \Delta f)$ is the frequency deviation proportional to amplitude V_{mp} of the modulating signal. The FM waveform is expressed as

$$v = v_p \cos[\int \omega_i(t)dt] = V_p \cos(\omega_c t + m_f \sin \omega_m t), \tag{11.4-2}$$

where $m_f = \omega_d/\omega_m = f_d/f_m$ = index of frequency modulation.

Phase Modulator

The phase modulator is used to form the basis for some frequency modulation systems. A small-deviation phase modulator is shown in Figure 11.4-1. The carrier with a 90°-phase shift is added to a much smaller AM waveform that has the same carrier frequency. The resultant waveform is the vector sum of the two components, as shown in Figure 11.4-1b. When the AM waveform varies in amplitude, the output voltage v_c varies in phase and slightly in amplitude. If the phase deviation is limited to approximately $30°(\pm 15°)$, a reasonably linear variation of phase with modulating voltage results. The limiter is used to keep the amplitude variations of the output waveform at a minimum level. The expression for the output voltage is then

$$v_o = C \cos(\omega_c t + \phi_o + m_p \sin \omega_m t), \tag{11.4-3}$$

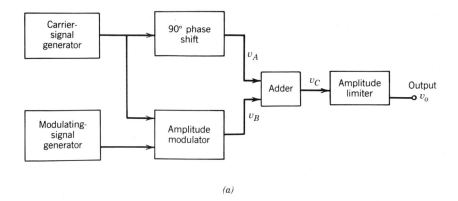

(a)

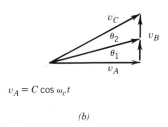

$$v_A = C \cos \omega_c t$$

(b)

FIGURE 11.4-1 A small-deviation phase modulator: (a) block diagram; (b) vector diagram.

which corresponds to a phase-modulated wave. The quantity m_p is termed the *phase modulation index*, and the constant ϕ_o is the phase in the absence of modulation. Neglecting ϕ_o, the form of Equation (11.4-3) is the same as that of Equation (11.4-2):

$$v_o = C \cos(\omega_c t + m_p \sin \omega_m t). \qquad (11.4\text{-}4)$$

The relationship between phase and frequency modulation can be derived mathematically, starting with the expression for a phase-modulated waveform given by Equation (11.4-4). The frequency of this signal is not constant since the total instantaneous phase or angle

$$\theta = \omega_c t + \phi_o + m_p \sin \omega_m t$$

does not vary linearly with time. The instantaneous radian frequency is found from

$$\omega = \frac{d\theta}{dt} = \omega_c + m_p \omega_m \cos \omega_m t. \qquad (11.4\text{-}5)$$

Comparison of this relation with Equation (11.4-1) shows that the maximum frequency deviation produced by phase modulation is

$$(f_d)_p = m_p f_m. \tag{11.4-6}$$

The quantity $(f_d)_p$ is proportional both to the amplitude and to the frequency of the modulating signal. In contrast, the frequency deviation of an FM wave is independent of the modulating frequency. When the phase of a sine wave is modulated, the frequency is also modulated. The instantaneous frequency change is proportional to the derivative of the phase-modulating signal. Consequently one method of obtaining an FM signal with a frequency of

$$\omega = \omega_c + A(t)$$

is to use $\int A(t)dt$ as the modulating signal of a phase modulator. For instance, if the desired FM signal frequency is

$$\omega = \omega_c + 2\pi k_f M \cos \omega_m t,$$

then the phase-modulating signal should be

$$\frac{2\pi k_f M}{\omega_m} \sin \omega_m t.$$

We can say that modulation of phase is merely a way of obtaining a frequency-modulated wave in which the frequency deviation will be proportional to the modulating frequency.

Phase Modulator Converted to Frequency Modulator

The system of Figure 11.4-1 can be converted from a phase modulator to a frequency modulator simply by inserting an integrator between the modulating signal generator and the amplitude modulator. Instead of applying the carrier signal to the system input; a subharmonic might be applied. The modulation can then be performed to produce a low-level, comparatively low-frequency FM signal. A harmonic of this signal is then amplified and limited by the class-C output amplifier. Greater indexes of modulation can be obtained by cascading phase modulators.

FM systems can be made utilizing heterodyning in connection with multipliers, and they can be constructed using the voltage-controlled oscillator (VCO). The FM system can also be built employing the varactor diode or transistor (BJT or FET) varactor as part of the oscillator tuned circuit (see Figures 1.5-3 and 11.1-1*b*).

Advantages of FM over AM

Compared to amplitude modulation, frequency modulation has certain advantages: (1) the signal-to-noise ratio can be increased without increasing transmitted power (but at the expense of an increase in frequency bandwidth required); (2) certain forms of interference at the receiver are more easily suppressed; and (3) the modulation process can take place at a low-level power stage in the transmitter, thus avoiding the need for large amounts of modulating power.

11.5   FM DEMODULATION USING A PHASE DETECTOR

The phase detector shown in Figure 11.5-1 consists of a balanced modulator and a low-pass filter. The modulator produces an output voltage given by

$$v = KV_1V_2 \cos \omega_c t \cos(\omega_c t + \theta);$$

$$v = \frac{KV_1V_2}{2}[\cos(2\omega_c t + \theta) + \cos \theta]. \qquad (11.5\text{-}1)$$

The low-pass filter eliminates the component of frequency $2\omega_c$, giving an output voltage of

$$v_o = \frac{KV_1V_2}{2} \cos \theta. \qquad (11.5\text{-}2)$$

This equation indicates that the output voltage varies quickly with phase difference.

The FM detector shown in Figure 11.5-2 uses a phase detector circuit. The amplitude limiter reduces variations in amplitude that would affect the output signal. The series RLC tuned circuit is a bandpass filter that resonates at the carrier frequency. The transfer function of this filter is given by

$$\frac{v_2}{v_1} = \frac{\omega_o^2}{\omega_o^2 - \omega^2 + j\omega R/L}, \qquad (11.5\text{-}3)$$

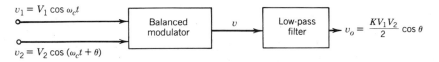

$v_1 = V_1 \cos \omega_c t$

$v_2 = V_2 \cos (\omega_c t + \theta)$

Balanced modulator → v → Low-pass filter → $v_o = \frac{KV_1V_2}{2} \cos \theta$

FIGURE 11.5-1 A phase detector whose output varies with phase difference.

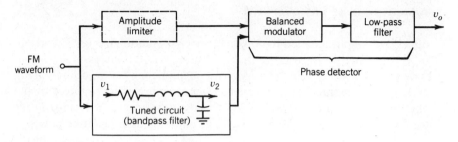

FIGURE 11.5-2 An FM detector using a phase detector whose two input signals have a phase difference.

where $\omega_o = 1/\sqrt{LC}$. The impedance of the circuit is

$$Z = R\left[1 + j\left(\frac{\omega L}{R} - \frac{1}{\omega CR}\right)\right],$$

from which we obtain

$$\tan(-\theta) = \frac{1}{-(\omega L/R - 1/\omega CR)} = \frac{1}{1/\omega CR - \omega L/R}$$

$$Q_o = \frac{\omega_o L}{R} = \frac{1}{\omega_o CR} = \frac{1}{R}\sqrt{\frac{L}{C}}.$$

$$\frac{1}{\omega CR} - \frac{\omega L}{R} = \left(\frac{1}{\omega_o CR}\frac{\omega_o}{\omega} - \frac{\omega_o L}{R}\frac{\omega}{\omega_o}\right) = Q_o\left(\frac{\omega_o}{\omega} - \frac{\omega}{\omega_o}\right)$$

$$= \frac{1}{R}\sqrt{\frac{L}{C}}\left(\frac{\omega_o}{\omega} - \frac{\omega}{\omega_o}\right);$$

$$\tan(-\theta) = \frac{\omega R}{\omega\sqrt{L/C}(\omega_o/\omega - \omega/\omega_o)} = \frac{\omega R}{\sqrt{L/C}(\omega_o - \omega^2/\omega_o)}$$

$$= \frac{\omega R}{\omega_o\sqrt{L}\sqrt{L}(\omega_o - \omega^2/\omega_o)} = \frac{\omega R}{L(\omega_o^2 - \omega^2)}.$$

Hence the phase shift of the tuned circuit is

$$\theta = -\arctan\frac{\omega R}{L(\omega_o^2 - \omega^2)}. \qquad (11.5-4)$$

The resonant frequency ω_o equals carrier ω_c. We are interested in the case where ω is near ω_c. The series expansion of arctan x is

$$\tan^{-1}x = \frac{\pi}{2} - \frac{1}{x} + \frac{1}{3x^3} - \frac{1}{5x^5} + \frac{1}{7x^7} - \cdots, \qquad x > 1.$$

We can apply the first two terms of the expansion for

$$\tan^{-1}x = \frac{\pi}{2} - \frac{1}{x}$$

to give

$$\theta \simeq -\frac{\pi}{2} + \frac{L(\omega_o^2 - \omega^2)}{\omega R}. \tag{11.5-5}$$

Let

$$\omega = \omega_o \pm \Delta\omega,$$

where $\Delta\omega$ is the frequency deviation. Then Equation (11.5-5) becomes

$$\theta \simeq -\frac{\pi}{2} + \frac{L}{\omega R}(\pm 2\omega_o\Delta\omega - \Delta\omega^2). \tag{11.5-6}$$

Since $\omega_o \gg \Delta\omega$ and $\omega \simeq \omega_o = \omega_c$, Equation (11.5-6) can be written as

$$\theta \simeq -\frac{\pi}{2} \pm \frac{2\Delta\omega}{\omega_o}\frac{\omega_o L}{R} = -\frac{\pi}{2} \pm \frac{2\Delta\omega Q_o}{\omega_o}. \tag{11.5-7}$$

Equation (11.5-7) represents the phase difference of the two input signals to the balanced modulator. From Equation (11.5-2), we see that the low-pass filter output will be

$$v_o = K_1 \cos\left(-\frac{\pi}{2} \pm \frac{2\Delta\omega Q_o}{\omega_c}\right) = K_1 \sin\left(\frac{\pm 2\Delta\omega Q_o}{\omega_c}\right) \simeq K_2 \frac{\Delta f}{f_c}. \tag{11.5-8}$$

Since the frequency deviation $\Delta f \ll f_c$, output v_o is small; however, because v_o varies linearly with Δf, the output represents the information contained in the modulating signal of the FM waveform.

11.6 PHASE-LOCKED LOOP FM DEMODULATION

Basic Phase-Locked Loop (PLL) Circuit

The basic phase-locked loop or PLL is a feedback configuration, as shown in Figure 11.6-1. It consists of a phase comparator, a low-pass filter, an amplifier,

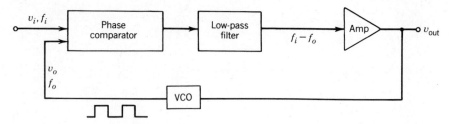

FIGURE 11.6-1 Block diagram of basic PLL circuit.

and a voltage-controlled oscillator (VCO). A feedback signal from the output drives the VCO to generate a free-running frequency f_o equal to the input frequency f_i, while the phase difference between the two signals to the comparator equals a constant value. In this case the loop is operating in lock. The fixed phase difference results in a fixed dc voltage to the VCO. The VCO then provides output of a fixed-amplitude square-wave signal at the frequency of the input. Changes in the input-signal frequency will result in changes in the dc voltage to the VCO. The phase comparator is generally based on a multiplier circuit or balanced modulator that forms the product of two input signals. Thus the comparator output contains frequency components at the sum and difference of the signals compared. The low-pass filter passes only the lower-frequency component of the signal. The PLL circuit can be used in FM detectors, AM detectors, TV receivers, frequency synthesizers, and satellite signal-tracking systems. If the VCO free-running frequency f_o is near the carrier frequency f_c, the PLL circuit can be used as an FM detector. When f_o and f_c are equal and the PLL achieves lock, the demodulated output voltage for an unmodulated carrier equals zero. The instantaneous radian frequency of the FM signal is given by Equation (11.4-5):

$$\omega = \omega_c + m_p \omega_m \cos \omega_m t.$$

If the change in frequency is sinusoidal, then $s = j\omega$. If this frequency is so low that the transfer function of the PLL becomes

$$\frac{V_{out}(s)}{\Delta\omega(s)} = \frac{1}{G},$$

then the ac output voltage is given by

$$v_{out}(t) = \frac{\Delta\omega(t)}{G}, \tag{11.6-1}$$

where G is the feedback factor, expressed in rad/s/V. The frequency deviation about the carrier is given by

$$\Delta\omega(t) = m_p \omega_m \cos \omega_m t.$$

Thus

$$v_{\text{out}}(t) = \frac{m_p \omega_m}{G} \cos \omega_m t. \qquad (11.6\text{-}2)$$

This equation indicates that the ac output voltage is proportional to the modulating signal.

565 PLL FM Detector

A PLL IC unit 565 is shown in Figure 11.6-2. It contains a phase detector, an amplifier, and a VCO. They are partially connected internally. R_1 and C_1 externally connected are used to determine the free-running frequency of the VCO with zero control voltage input. The VCO free-running frequency is given by

$$f_o \simeq \frac{0.3}{R_1 C_1}. \qquad (11.6\text{-}3)$$

The VCO output is connected to an input of the phase detector, while the FM signal is connected to the other input. The output from the phase detector is amplified by a stage with an output resistance of 3.6 kΩ. Capacitor C_2 serves as a low-pass filter whose bandwidth is

$$\omega_2 = \frac{1}{3600 C_2}. \qquad (11.6\text{-}4)$$

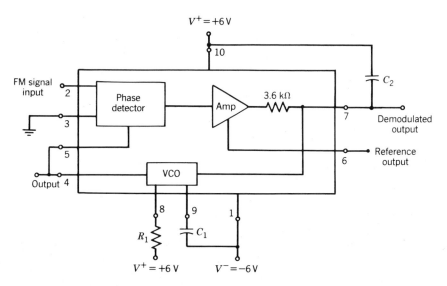

FIGURE 11.6-2 565 PLL unit connected as FM demodulator.

The filtered signal is the demodulated output, which is fed back to the VCO. The capture range of the PLL is the frequency range centered about the VCO free-running frequency, over which the loop can attain lock with the input signal. Once the PLL has achieved capture, it can maintain lock with the input signal over a somewhat wider frequency range, which is termed the *lock range*. The lock range is given by

$$f_L = \pm \frac{8f_o}{V^+}.$$

(11.6-5)

The capture range is specified as

$$f_{c'} = \pm \frac{1}{2\pi} \sqrt{2\pi f_L \omega_2} = \pm \frac{1}{2\pi} \sqrt{\frac{2\pi f_L}{3600 C_2}}.$$

(11.6-6)

Assume supply voltage $V^+ = 6$ V, $R_1 = 10$ kΩ, $C_1 = 220$ pF, and $C_2 = 330$ pF; then $f_o = 136.36$ kHz, $f_L = \pm 181.8$ kHz, and $f_{c'} = 156.1$ kHz. An FM input signal within the lock range of 181.8 kHz will result in the demodulated output voltage varying around its dc voltage level set with input signal at f_o. The output voltage is a function of the input-signal frequency within the range from $f_o - f_L/2 = 45.45$ kHz to $f_o + f_L/2 = 227.27$ kHz. Typically, a $+5$-V output voltage set at 136.36 kHz varies between $+5.3$ V at 45.45 kHz and $+4.7$ V at 227.27 kHz. As a result, the dc voltage at the output is linearly related to the input-signal frequency within the lock range $f_L = 181.81$ kHz around the center frequency $f_o = 136.36$ kHz.

The frequency range of the 565 PLL extends from 0.001 Hz to 500 kHz and, therefore, cannot be used directly for commercial FM detection due to its IF $= 10.7$ MHz. For a 400-kHz carrier with a 5-kHz frequency deviation, $C_1 = 150$ pF if we choose $R_1 = 10$ kΩ within the limitation 2 k$\Omega \leqslant R_1 \leqslant 20$ kΩ. The low-pass filter bandwidth should be selected to pass a 5-kHz signal; thus f_2 can be set at 6 kHz. Therefore $C_2 = 1/3.6 \times 10^3 \times 2\pi \times 6 \times 10^3 = 7368$ pF.

Other FM detectors, such as the conventional discriminator (Figure 12.3-2) and the ratio detector (Figure 12.9-2), will be discussed in Section 12.3 and Section 12.9.

11.7 MODULATION AND DEMODULATION TECHNIQUES FOR PAM, PDM, AND PPM

PAM Modulation and Detection

In pulse-amplitude modulation (PAM), the greatest pulse represents the largest positive signal amplitude sample, while the smallest pulse represents the largest negative sample. If a radio frequency (RF) carrier is pulse-amplitude modulated instead of simply being amplitude modulated, much less power is needed for the transmission of information since the transmitter is actually switched off between pulses.

A PAM system based on the one-shot multivibrator is shown in Figure 11.7-1. The astable clock triggers the one-shot at a sampling frequency. The modulating signal superimposed on a positive dc supply voltage is applied to the output through the load resistor of the multivibrator (see Figure 1.6-1).

Another chief advantage of pulse modulation is that the signals can be recovered simply by clipping off the superimposed noise.

The pulse-amplitude modulator shown in Figure 11.7-2 consists of two operational amplifiers and two electronic switches. Carrier v_c applied to the inverter causes the voltage-controlled switches S_1 and S_2 to operate in opposite states. When S_1 is off and S_2 is on, the output voltage is 0 V. When S_1 is on and S_2 is off, the output voltage is

$$v_o = v_m \left(-\frac{R_2}{R_1} \right)\left(-\frac{R_5}{R_3 + R_4} \right),$$

or

$$v_o = \frac{R_2 R_5}{R_1(R_3 + R_4)} v_m. \tag{11.7-1}$$

This equation indicates that the amplitude of v_o is proportional to the modulating signal v_m.

Pulse-amplitude modulation can be demodulated by a simple low-pass RC filter, as illustrated in Figure 1.6-2.

PDM Modulation and Detection

In pulse duration modulation or PDM (which is also called pulse-width modulation or PWM), the pulses have a constant amplitude and a variable time duration that is proportional to the amplitude of the signal samples. A PDM system

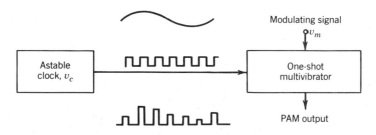

FIGURE 11.7-1 A PAM system based on a one-shot (monostable) multivibrator.

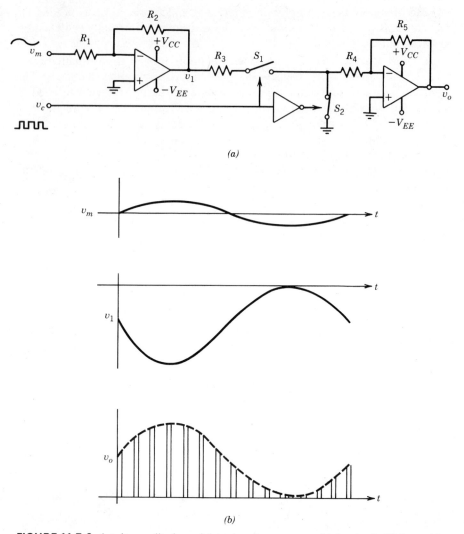

(a)

(b)

FIGURE 11.7-2 A pulse-amplitude modulator based on op-amps: (*a*) the circuit; (*b*) the positive output waveform resulting from a bias circuit added to cause v_1 to have the proper negative bias.

is shown in Figure 11.7-3. It consists of a voltage comparator whose inverting and noninverting terminals are connected to a free-running ramp generator output and a modulating signal source, respectively. The noninverting terminal is also connected to the junction of the R_1-R_2 voltage divider, across which is a supply voltage $V_{CC} - (-V_{EE})$. Thus the potential at that terminal is a dc level with the ac signal superimposed. When the ramp is at its zero level, the inverting input terminal voltage is below the potential at the noninverting input terminal, and thus the comparator output is at its extreme positive voltage level. When the ramp voltage becomes equal to the level at the noninverting terminal, the

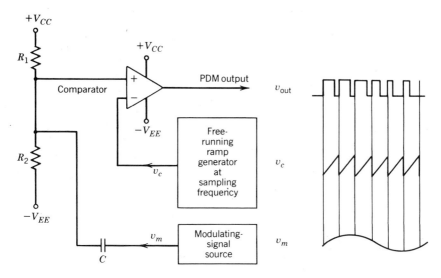

FIGURE 11.7-3 A PDM system consisting of a voltage comparator whose inverting and noninverting terminals are connected to a free-running ramp generator output and a modulating signal source, respectively.

comparator output switches quickly from its extreme positive voltage level to its extreme negative level. As the ramp returns to zero, the inverting input voltage is once again below the noninverting terminal voltage and the comparator output returns to its extreme positive level. Each output pulse begins at the instant the ramp returns to zero and ends when the ramp coincides with the signal voltage. At the instant that the signal voltage is at its highest level, the ramp takes the longest time to reach the same voltage as that at the noninverting terminal; at the instant that it is at its lowest level, the ramp takes the shortest time to reach this voltage. Hence the width of the output pulses is maximum at the instant of highest signal voltage and minimum at the instant of lowest signal voltage.

The PDM demodulating system shown in Figure 11.7-4 consists of an integrator circuit and a low-pass filter. The integrator receiving the PDM wave produces a ramp-type output. The ramp peak is proportional to the pulse duration. The ramp peaks represent the amplitude samples of the original signal. When the ramp voltage is applied to the low-pass filter, it passes the recovered low frequency signal to the output.

PPM Modulation and Detection

The PPM system shown in Figure 11.7-5 is a PDM system with the addition of a one-shot multivibrator, which is triggered by the trailing edges of the PDM pulses. The PPM output is a series of constant-amplitude constant-width

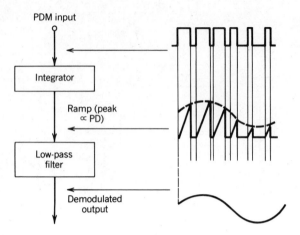

FIGURE 11.7-4 A PDM demodulating system containing an integrator and a low-pass filter.

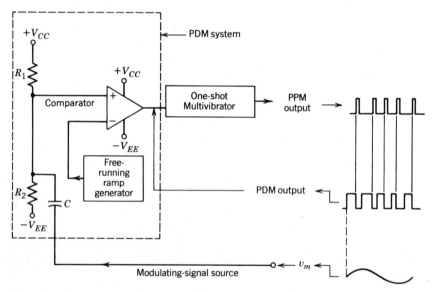

FIGURE 11.7-5 A PPM modulating system containing a PDM system and a one-shot multivibrator.

pulses, but the position in time of the pulses varies in proportion to the amplitude of the modulating signal. Thus the delay of the output pulses from the beginning of the period becomes greater when the modulating signal increases.

The PPM demodulating system shown in Figure 11.7-6 consists of an astable clock, an RS flip-flop, and a PDM demodulating system. For PPM detection, a PDM waveform is first produced by triggering the flip-flop. The output of the flip-flop is a PDM waveform that is detected by the PDM demodulator. The

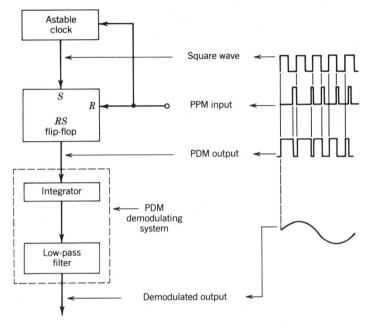

FIGURE 11.7-6 A PPM demodulating system, consisting of an astable clock, an *RS* flip-flop, and a PDM demodulating system.

flip-flop is triggered into its set state by the leading edges of a square wave that should be synchronized with the original signal source. Synchronization is necessary so that the leading edges of the square wave coincide with the leading edges of the PDM wave that was used to generate the PPM pulses. The flip-flop is reset by the leading edge of the PPM pulses. Synchronization is one of the most difficult requirements for PPM demodulation. Some alternations are available.

11.8   TIME DIVISION MULTIPLEXING

Multiplexing

Multiplexing deals with the transmission of multiple-data (information) channels over a single line or radio link; the information channels are thus said to be multiplexed onto one channel. There are two multiplexing structures: frequency division multiplexing (FDM) and time division multiplexing (TDM). FDM is generally associated with CW (continuous wave) modulation, whereas the discontinuous nature of pulse modulation is generally associated with TDM. In both instances a number of (or *M*) data channels are interleaved in frequency or time onto a single channel, as illustrated in Figure 11.8-1. FDM is

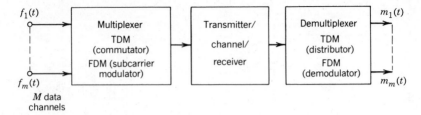

FIGURE 11.8-1 The process for multiplexing and demultiplexing M information channels onto a single transmission channel.

produced by the superposition (addition) of M waveforms, isolated in frequency, which then overlap in time. TDM is produced by the superposition (addition) of M waveforms, isolated in time (nonoverlapping pulses), which then overlap in frequency. The time multiplexing of M information channels onto a single transmission channel consists of interleaving the pulse modulation from each channel in time. This process is termed *commutation*. Upon arrival at the receiver, the interleaved channels are decommutated or distributed. The distributor inputs are taken to be baseband TDM pulses.

TDM System for Multiplexing Five PAM Signals

PAM samples can be generated using a very small duty cycle. If only one PAM wave were sent over a channel, much of the transmission time available would go unused. This time can be employed by sandwiching the pulse trains of other PAM signals into the unused time intervals between pulses. This process is time division multiplexing or TDM. Figure 11.8-2*a* illustrates the TDM of five PAM signals—each with the same sampling rate. Only two of the five waves are shown.

Figure 11.8-2*b* illustrates a five-channel time-devision-multiplexed system. An electronic switch called a *commutator* connects the output of each PAM channel modulator to the communication channel input in turn, dwelling on each contact only for the duration time of one pulse. The channel should have sufficient bandwidth to handle these pulses, which now occur at five times the sampling rate of one channel. At the receiver end, another switch, rotating in synchronism with the sending commutator, decommutates the pulse trains and connects them to the proper demodulators. One channel of the modulator group is commonly reserved for the sending of synchronizing signals along with the information signals in order to prevent the receiver from getting out of step with the transmitter. The pulses from the PAM modulators are made shorter than the allotted pulse-duration time so that a guard time (t_g) exists between each pair of pulses to make intersymbol crosstalk negligible. When a pulse is passed through a low-pass filter or limited channel, its corners are rounded off, its amplitude is reduced, and it is smeared by unequal time delays into the adjacent time slot: The pulse tail endures long after the nominal pulse period and can interfere with a following pulse; this effect is referred to as *intersymbol crosstalk*.

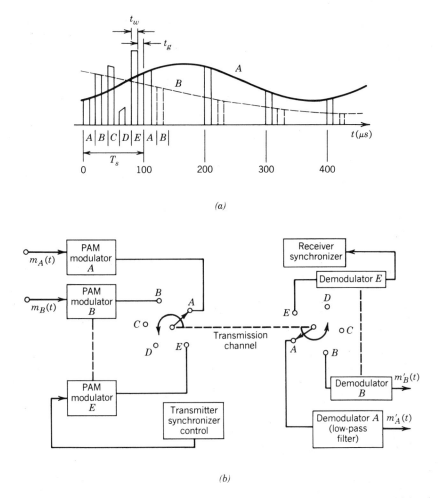

(a)

(b)

FIGURE 11.8-2 (a) A five-channel TDM PAM signal; (b) a TDM system used to multiplex the five signals.

TDM of Three PDM Channels

The typical TDM waveforms of three PDM channels are shown in Figure 11.8-3. Channel A is a series of PDM pulses obtained from sampling signal A. Wave A is a 2.5-kHz signal with a period of 400 μs. Since A is sampled ten times in every cycle, the samples occur at 40-μs intervals. Let the maximum pulse width be less than 10 μs. Then a 30-μs space is left between pulses. If signals B and C are sampled at the same rate, the samples of channel B and channel C can be included in the 30-μs interval. As shown in Figure 11.8-3a–c, the first pulse in channel A begins at t_0, the second pulse in channel B occurs 40 μs after t_1, and the first pulse in channel C begins at t_2, which is 20 μs after t_0 and 10 μs after t_1.

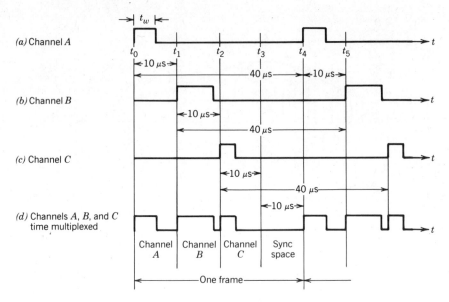

FIGURE 11.8-3 Typical TDM waveforms of three PDM channels.

As shown in Figure 12.8-3*d*, three channels of information are contained in the waveform, which can be recorded on a single magnetic track, transmitted on a radio frequency, or otherwise processed. The series of three or more information pulses together with the synchronizing space is commonly known as *one frame of the TDM waveform.* Synchronization is necessary in the demodulation process.

PPM and PCM can also be processed in time-multiplexed form.

TDM Coding and Decoding Systems

In a TDM coding system, a ring counter is used to select the signals to be sampled in the correct repetitive sequence. The ring counter is a digital circuit made up of a series of flip-flops in which the contents are continuously re-circulated. The ring counter triggered by input pulses switches through a number of states equal to one more than the number of TDM channels: that is, the number of channels plus the synchronizing (sync) space. For a five-channel system, the ring counter should have six states equal to five channels plus the sync space. A TDM system is also called a *TDM coding system.* The system for separating the waveform into individual channels is called a *TDM decoding system.* Time-multiplexed signals should be decoded before demodulation.

Figure 11.8-4 shows the block diagram and waveforms for a three-channel TDM and PDM system. The clock output toggles the ring counter at the desired frequency. The counter outputs switch the sampling gates on and off in the correct sequence. When one output from the counter is high, all others

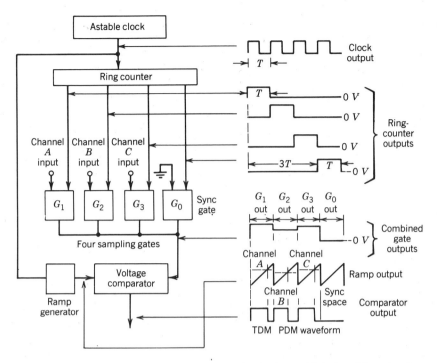

FIGURE 11.8-4 Block diagram and waveforms for the time multiplexing of three signals.

are low. Hence one sampling gate is on while all others are off. The clock frequency and number of channels determine the time periods during which a sampling gate is in the on and off states. For the three-channel system, each gate is off for three time periods of the clock. The input of the synchronizing gate G_0 is grounded, so that its output is 0 V during the sync space. The signals to be sampled—A, B, and C— are applied to the input terminals of sampling gates G_1, G_2, and G_3, respectively. The outputs of gates G_0 to G_3 are common. When each gate switches on in turn, amplitude samples of the signals are time multiplexed into a single waveform. The combined output of the four gates is a PAM waveform with no spaces between pulses and with a sync space at the end of each set of samples. The PAM waveform from the gates is converted to PDM by applying it to one input of a voltage comparator when the other input terminal has a ramp generator output applied to it. The voltage comparator output becomes high when the ramp output becomes zero. The comparator output goes to zero when the ramp amplitude equals the amplitude being sampled. During the sync space, the comparator output does not switch positively when the ramp goes to zero. The system output is a TDM PDM waveform with intervening sync spaces.

The decoding system for three-channel time-multiplexed PDM signals is shown in Figure 11.8-5. It consists of an integrator (ramp generator), a Schmitt trigger, a ring counter, and three AND gates. The time-division-multiplexed

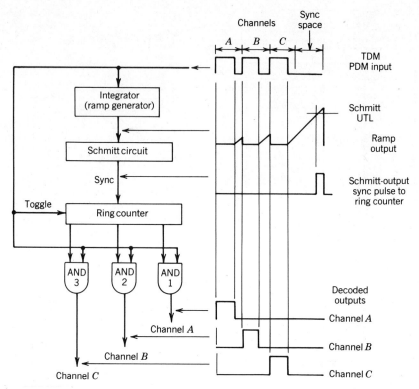

FIGURE 11.8-5 Block diagram and waveforms of a decoding system for three-channel time-multiplexed PDM signals.

PDM waveform is applied simultaneously to the integrator, the toggle input of the ring counter, and one input on each of the AND gates. The integrator and Schmitt trigger serve to detect the sync space in the TDM waveform. During the time when the input pulses are applied, the ramp output remains at zero level. When the pulses are absent, the ramp output rises linearly; then it quickly returns to zero again at the beginning of the next input pulse. The integrator generates a series of small ramps during the space time and a larger-than-usual output during the sync space. When the ramp output reaches the upper-trigger level of the Schmitt circuit, the Schmitt output pulse sets the ring counter to its synchronized state. The next input pulse from channel A toggles the ring counter to its channel A state. In the channel A state, the counter gives a positive output to AND gate 1 and a zero output to other gates. At this time, a channel A pulse is applied to one input of all AND gates and only gate 1 has a positive voltage at both input terminals; this results in a positive output only from gate 1. The output pulse from gate 1 starts at the beginning of channel A and finishes at the end of the channel A PDM pulse. At the beginning of channel B, the counter toggles to its channel B state. Hence AND gate 2 now has positive levels at both input terminals, while the other gates have zero levels applied from the counter.

Therefore the channel *B* PDM pulse is produced at the gate 2 output. The operation for channel *C* is similar to that for channel *B*. Each channel from the decoding system can be demodulated to recover the original modulating signals.

11.9   PCM MODULATION AND DEMODULATION

Introduction

In pulse code modulation (PCM), each amplitude sample pulse of a modulating signal is converted to a binary number. Since a certain amount of uncertainty is introduced by noise in transmitting AM signals, a PCM signal, rather than an AM signal, is transmitted. Recovery of the transmitted information does not depend on the height, width, or energy constant of the individual pulses, but only on their presence or absence. Because it is relatively easy to recover pulses under these conditions, even in the presence of large amounts of noise and distortion, PCM systems tend to be immune to interference and noise. Regeneration of the pulses on route is also relatively easy, resulting in a system that produces excellent results for long-distance communication. Figure 12.9-1 illustrates the sampling and coding processes and the resultant PCM signal.

Quantization

In PCM, a coded number is transmitted for each level sampled in the modulating signal. If the exact number corresponding to the exact voltage were to be transmitted for every sample, an infinitely large number of different code symbols would be needed. *Quantization* has the effect of reducing this infinite number of levels to a relatively small number, which can be coded without difficulty. In the quantization process, the total range of the modulating signal is divided up into a number of small subranges. The number will be in the range from 8 to 128. A number that is an integer power of 2 is usually selected due to the ease of generating binary codes. A new signal is generated by producing, for each sample, a voltage level corresponding to the midpoint level of the subrange in which the sample falls. Thus, if a range of 0 to 4 V were divided into four 1-V subranges and the signal were sampled when it was 2.8 V, the quantizer would put out a voltage of 2.5 V and hold that level until the next sampling time; 2.5 V corresponds to the midpoint of the third range, and 2.8 V − 2.5 V, or 0.3 V is the quantization error. The result is a stepped waveform that follows the contour of the original modulating signal, with each step synchronized to the sampling period. The process for quantizing a signal is illustrated in Figure 11.9-2.

The difference between the quantized staircase waveform and the original waveform amounts to noise added to the signal by the quantizing circuit. The

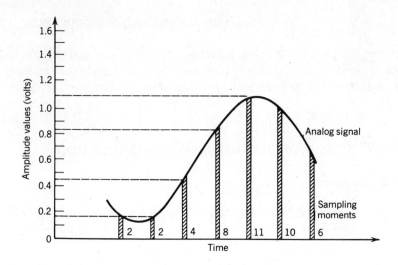

(a)

Amplitude value	Binary-coded equivalent	Pulse-code-modulated signal
0	0000	
1	0001	
2	0010	
3	0011	
4	0100	
5	0101	
6	0110	
7	0111	
8	1000	
9	1001	
10	1010	
11	1011	
12	1100	
13	1101	
14	1110	
15	1111	

(b)

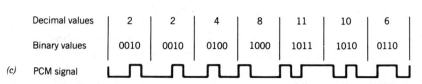

Decimal values	2	2	4	8	11	10	6
Binary values	0010	0010	0100	1000	1011	1010	0110

(c)   PCM signal

FIGURE 11.9-1 The sampling and coding processes and the resultant PCM signal: (a) sampling; (b) coding reference table; (c) PCM signal. If the analog signal is sampled and then coded using the table, the transmitted pulse-code-modulated signal becomes the waveform shown in (c).

366

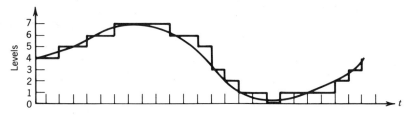

FIGURE 11.9-2 Process for quantizing a signal.

mean square quantization noise voltage has a value of

$$E_{nq}^2 = \frac{S^2}{12},$$ (11.9-1)

where S is the voltage of each step, or the subrange voltage span. Consequently the number of quantization levels should be kept high in order to maintain the quantization noise below some acceptable limit; this limit is given by the power signal-to-noise ratio—that is, the ratio of average signal power to average noise power. For a sinusoidal signal that occupies the full range, the mean square signal voltage has a value of

$$E_s^2 = \frac{1}{2} E_{peak}^2 = \frac{1}{2} \left(\frac{MS}{2} \right)^2 = \frac{(MS)^2}{8},$$ (11.9-2)

where M is the number of steps and S is the step height voltage. The signal-to-noise ratio is given by

$$\frac{\text{signal}}{\text{noise}} = \frac{E_s^2}{E_{nq}^2} = \frac{(MS)^2}{8} \times \frac{12}{S^2} = \frac{3}{2} M^2.$$ (11.9-3)

The signal-to-noise ratio derived previously was related to a maximum amplitude signal. In practice, the signal may be much smaller than this—as much as 30 dB less. When the number of steps is given, the number of binary bits required is found from the expression

$$I = \log_2 M.$$ (11.9-4)

When the case $6 < I < 7$ is met, the next higher integer (7) can be selected.

Companding and PCM Encoding

Because the noise level is dependent on the step size, which is a constant, a much lower signal-to-noise (S/N) ratio will result with small signals than with large ones. The process of companding is utilized to overcome this degradation of the S/N ratio. *Companding* is a compound process of volume compression before transmission combined with volume expansion after transmission. The compressor amplifier amplifies low-level signals more than it does high-level signals, thus compressing the input voltage range into a smaller span. The steps transmitted have equal amplitudes, but they are equivalent to using smaller steps in the low-level signals and larger steps in the high-level signals. The result is lower-amplitude quantization noise during the low-level periods. Companding itself does not produce any distortion in the recovered signal.

The encoding process generates a binary code number corresponding to the quantization level number to be transmitted for each sampling interval. Ordinary binary coding is most often used, where the binary number corresponding to the decimal number of the level in question is transmitted. This binary number will contain a train of 1 and 0 pulses, with a total of $\log_2 N$ pulses in

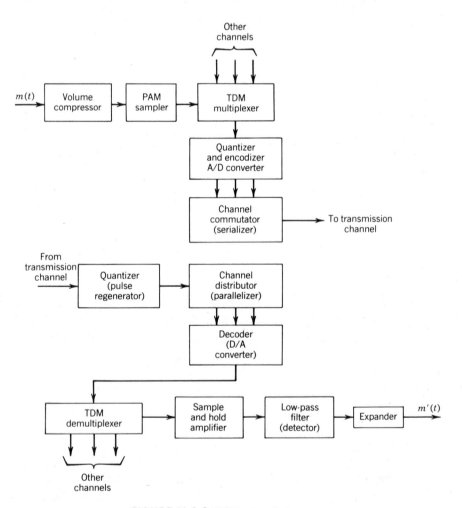

FIGURE 11.9-3 PCM transmission system.

each number (N is the number of levels in the full range). The PCM encoding process corresponds exactly to the process of analog-to-digital (A/D) conversion.

PCM Transmission System

Since PCM signals are essentially time-division-multiplexed pulses, modulation and demodulation techniques for PCM are similar to TDM methods. The block diagram of a PCM transmission system is shown in Figure 11.9-3. The modulating signal $m(t)$ is applied to the input of the volume compressor. A PAM sampler generates a PAM signal from the compressed signal, which is then multiplexed with signals from other input channels. An analog-to-digital converter performs the two functions of quantization and encoding, producing a binary-coded number for each channel-sampling period. A channel commutator transmits the code bits in serial fashion.

At the receiver, a two-level quantizer (pulse regenerator) reshapes the input pulses and eliminates most of the transmission noise. A channel distributor (parallelizer) decommutates the pulses and passes the bits in parallel groups to a digital-to-analog (D/A) converter for decoding. Another distributor (a TDM demultiplexer) demultiplexes the several PAM signals and routes them to the appropriate output channels. Each channel has a sample-and-hold amplifier, which maintains the pulse level for the duration of the sampling period, re-creating the staircase waveform approximation of the compressed signal. A low-pass filter is used to reduce the quantization noise, and, finally, an expander removes the amplitude distortion, which was intentionally introduced in the compression of the signal, to yield the output signal $m'(t)$.

11.10   DIGITAL MODULATOR FOR SIMULTANEOUS TRANSMISSION OF THREE SIGNALS

General Description

The digital modulator shown in Figure 11.10-1 allows the simultaneous transmission of three analog signals on the same carrier. The heart of the modulator circuit is the astable multivibrator (Q_2-Q_3), which generates the required carrier frequency signal—a periodic pulse train—whose amplitude, frequency, and duty cycle can be simultaneously controlled by the three information-bearing signals earmarked for transmission. The resistors normally used for charging the timing capacitors in a multivibrator can be replaced by two identical constant-current sources. A desired modulating signal from the FM input terminal is used to vary the initial voltage on the timing capacitors ($C_1 = C_2$) in a linear fashion; this produces a linear variation of the period and thus of the frequency of the modulator output. The circuit also includes a difference amplifier (Q_6-Q_7) with a potential divider bias so that the currents from the two

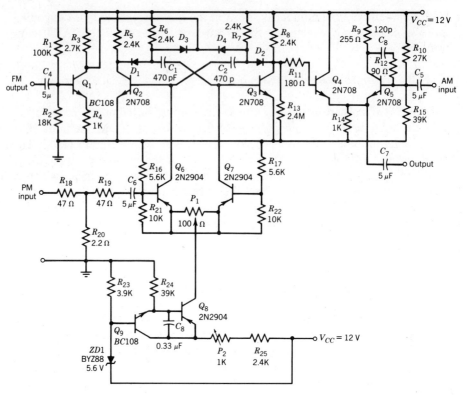

FIGURE 11.10-1 Digital modulator circuit. (From T. I. Raji, and A. I. Akinwande, *IEEE Transactions on Instrumentation and Measurement*, Vol. IM-31, No. 2, June 1982, pp. 140–142.)

current sources can be readily adjusted to produce a constant sum value while their difference equals the value of the modulating signal; this arrangement ensures that the multivibrator frequency remains constant while the duty cycle and thus the pulse duration varies in accordance with the modulating signal. The astable multivibrator circuit, when used with constant current sources, acts basically as a pulse or square-wave generator. Combining such a generator with the Q_4-Q_5 emitter follower circuit results in a pulse-amplitude modulator.

Operating Principle

Transistors Q_2, Q_3, Q_4, and Q_5 are high-frequency *npn* switching transistors. Q_6, Q_7, and Q_8 are *pnp* switching transistors. The complementary pair Q_8-Q_9 is used as constant current sources. The carrier frequency can be varied by adjusting the current furnished by the current source, using the preset potentiometer P_2 in the Q_8 emitter. Potentiometer P_1 is a preset resistor used to equalize the capacitor charging currents in Q_6 and Q_7 under no-input-signal conditions; this results in a symmetrical multivibrator output.

The slant normally observed in the trailing edge of a conventional astable multivibrator output waveform is caused by the discharging of the multivibrator capacitors through the associated collector resistors; it can be eliminated by the modified astable multivibrator (Q_2-Q_3) in Figure 11.10-1. Diodes D_1 and D_2 disconnect capacitors C_1 and C_2, respectively, to allow a full V_{CC} swing at the collectors of Q_2 and Q_3, while diodes D_3 and D_4 provide necessary isolation between capacitors C_1, C_2, and the modulating signal. If transistor Q_2 is driven off, its collector voltage immediately rises by V_{CC} such that transistor Q_3 goes into saturation. The saturation base current of Q_3 passes through C_1 and R_6 rather than R_5. Because the base current no longer passes through R_5, the Q_3 collector waveform will have the desired vertical trailing edges. The slant in the leading edge can also be minimized by making the load resistor R_{13} fairly large.

The desired modulating signal from the FM input is inserted at the junction of diodes D_3 and D_4 through the buffer stage Q_1. The input signal is used to vary the initial voltage on capacitor C_1 and C_2 for purposes of frequency modulation. The stage Q_1 must provide the dc bias necessary to ensure that the negative excursion of the modulating signal does not keep the multivibrator permanently in saturation. The symmetry or phase of the multivibrator output is varied by adjusting the relative currents into C_1 and C_2 so that one is larger than the other. The signal at the phase modulation (PM) input readily achieves this objective; the capacitor receiving the larger current will recharge faster, thus altering the symmetry of the multivibrator output. However, this alteration has no effect on the multivibrator frequency since the total current from the constant current source does not change. The T network at the PM input provides the required attenuation of the modulating signal so that the input to the difference amplifier stays below the maximum permissible level for linear operation.   '

Circuit Performance, Advantages, and Uses

The digital modulator circuit of Figure 11.10-1 should be compact to ensure good overall circuit performance and to minimize undershoots and undesired damped high-frequency oscillations. The circuit was tested for a nominal carrier frequency f_o of 100 kHz and a supply voltage V_{CC} of 12 Vdc. The performance of the circuit was checked by applying various waveforms—a sine wave, a saw-tooth wave, and a square wave—simultaneously at the three input terminals of the circuit, with one at each terminal. Each signal level was set between 4 and 10 V peak to peak and at a frequency between 4 and 8 kHz, to allow easy viewing of output-signal variations on an oscilloscope. The performance of the circuit with specified component values compares favorably with theoretical expectations.

In the fields of voice and data communications, telemetry, computer interfacing, broadcasting, and allied areas, the need often arises for the simultaneous transmission of several analog signals from a common source to a common destination, or perhaps to different destinations. The transmission objective may be achieved by employing conventional multiplexing techniques and circuits; they inevitably involve the use of a large number of components, which results in a consequent loss of simplicity, a probable loss of reliability, and an undesirably high cost. The digital modulator circuit of Figure 11.10-1 is capable of accomplishing the desired objective simply, reliably, and inexpensively. The circuit can also be employed for the recording of analog signals on magnetic tapes, using PDM–TDM or PPM–TDM techniques.

11.11 MODULATION AND DETECTION IN OPTICAL FIBER SYSTEMS

Introduction

Optical fibers are dielectric waveguide structures that are used to confine and guide light. The simplest waveguide structure consists of a dielectric rod; the cladding is air. Since the change in refractive index is abrupt, such a fiber is known as a *step index* fiber. Suitable materials for the rod are dielectrics transparent in the visible and near-infrared region of the electromagnetic spectrum (0.5–1.5-μm wavelength); thus the choice is restricted to glasses or fused silica. Plastic may also be used, but, in general, higher losses result.

The energy from the light source is the information carrier. When the light generated is at a single wavelength or frequency and maintains a uniform phase front, it is known as a *coherent source*. Such a light source is said to have *temporal and spatial coherence* and is analogous to a radio-wave oscillator. However the light-wave oscillator operates at an extremely high frequency (such as 10^{14} Hz), which is much higher than that of a radio-wave oscillator (typically 10^6 Hz). However most light sources are incoherent: they generate a spectrum of light at different wavelengths; the phase front of the light is not uniform.

LEDs are important light sources for fiber systems. The lasers with the characteristics approaching the ideal for fiber systems are semiconductor lasers. The term *laser* is an acronym for "light amplification by stimulated emission of radiation." It describes the process of light emission and amplification through the process of stimulated emission. As energy is added to an atomic structure, the electron absorbs the energy and moves to higher energy states. As it drops back to its original energy state, a radiative transition takes place and a photon is emitted. There is a time constant involved for the decay process. This process results in random emission of radiation and is termed *spontaneous emission*. Under special circumstances the electrons at a higher energy state could be trapped in that state for a time period longer than the time constant of the decay process. Under this condition, the population of electrons in the higher energy state build up, and a single transition would cause an additional transition to take place in such a way that the two emissions would be additive. This could induce more transitions, resulting in the emission of an amplified version of the first emission. This phenomenon is known as *stimulated emission*. If the emitted photons are made to travel again through the region where stimulated emission can take place (the active region), the amplification will continue until the gain balances the losses and a sustained output results. The emitted radiation is coherent in time and space. In fact, the amplifier has been transformed through feedback into an oscillator. We can say that a laser is a light source whose output involves the process of stimulated emission. A GaAlAs laser is a semiconductor laser emitting in the wavelength region from about 0.8 to over 0.9 μm. If the GaAlAs laser is not equipped with the Fabry-Perot cavity, stimulated emission can be suppressed in favor of spontaneous emission. By properly tailoring the device structure, GaAlAs light-emitting diodes can be fabricated. In LEDs, only spontaneous emission takes place. Properties of LEDs and laser diodes are listed in Table 11.11-1.

The optical fiber communication system is shown in Figure 11.11-1. It is composed of a transmitter, a channel, and a receiver for information transfer between remotely located points. Messages may be waveforms in time, digital symbols, and so on. In order to cast the message into a form suitable for propagation, it is transformed by modulating an optical carrier before conveying it over the fiber channel. The optical fiber is coupled to the transmitter by input coupler and to the receiver by output coupler. A photodetector in the receiver transfers the instantaneous optical power, as it reaches the receiver, into an electrical signal, which is then processed through amplification, filtering, equalizing, and so on.

Modulation

A single-frequency carrier source is sometimes described as a *line source* in the optical spectroscopy sense. The information signal can be imposed on the

TABLE 11.11-1
Properties of LEDs and Laser Diodes Used in Optical Communications

Item	LEDs		Laser Diodes	
	GaAlAs	InGaAsP	GaAlAs	InGaAsP
Wavelength (μm)	0.75–0.9	1.1–1.6	0.75–0.9	1.1–1.6
Frequency spectrum width (Å)	300–400	500–1000	< 1–20	< 1–20
Optical coupling power (mW)	0.10–0.80	0.05–0.10	1–10	1–5
Modulation bandwidth (MHz)	150	50–150	> 1000	> 1000
Life time (hr)	10^8	10^8	10^6	10^5–10^6
Variation of threshold current with temperature (%/°C)			0.6–1	1.2–2.0

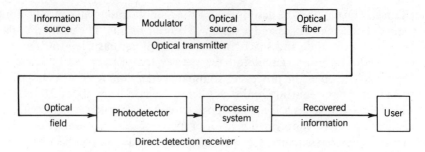

FIGURE 11.11-1 Block diagram of optical fiber communication system.

single-frequency carrier as an amplitude modulation, frequency modulation, or phase modulation. The carriers at optical wavelengths range from relatively incoherent emitters, such as LEDs, to relatively coherent emitters, such as semiconductor lasers. For the LED, the frequency spread is about $\pm 10^{13}$ Hz from a center frequency of 3.5×10^{14} Hz (at a center wavelength of 0.85 μm). For a multimode laser, the spread is about $\pm 0.4 \times 10^{12}$ Hz, and for a single longitudinal mode laser, the spread is plus or minus tens of megahertz. It is easily demonstrated that the higher the carrier frequency, the greater its potential to carry a wide band of information. Thus communication channels at optical frequencies have enormous bandwidth. In fact, the bandwidth offered with the use of an optical carrier is more likely to be limited by the baseband electronics and modulator-demodulator designs rather than by the carrying capacity of the optical carrier. Direct frequency or phase modulation is improper at optical wavelengths, except under exceptional circumstances when special arrangement is made to enable frequency or phase to be detected. The modulation processes that can easily be applied to an optical carrier essentially produce intensity variation of a noiselike carrier.

For an analog information signal, either an amplitude-, frequency-, or phase-modulated signal can be imposed onto a subcarrier. This frequently facilitates the design of filters for the separation of different channels of information carried out on an FDM (frequency-division multiplexing) basis. It can avoid cross-modulation products. It can also improve noise performance. The subcarrier with the information intensity modulates the optical carrier. At the receiver, the subcarrier is first intensity demodulated, and the information on the subcarrier is extracted by detectors. The analog signal can also be converted first to a digital form. The digital signal then modulates the optical carrier directly; alternatively it is first processed into FSK (frequency-shift keying) or PSK (phase-shift keying) format in order to provide a certain discrimination after detection and permit error to be controlled more easily.

For a semiconductor optical source, intensity variations can be directly produced by the change of the drive current. For linear modulation, the source is biased at a convenient point to permit an increase or a decrease of output intensity corresponding to the signal-level changes over a linear-characteristics region. For pulse signals of the binary type, the device can initially be biased at a point to produce small or no output, thus lowering the total power consumption.

Noise, Distortion, and Signal-to-Noise Ratio Considerations

The most dominant noise is the noise contributed by the detection process, while distortions could arise at the stages of modulation, transmission, and detection. The noise associated with the transmitter can also be significant when a laser source is employed. For an analog signal, the required signal-to-noise ratio (S/N) is rather great. For a telephone signal, the noise is required to be below audibility when a soft-spoken person is speaking. For a television signal, a 42-dB S/N ratio of peak-to-peak signal and root-mean-square (RMS) noise is generally acceptable to an audience. For a music channel, a 50-dB S/N ratio may still offend a hi-fi connoisseur's ears. This type of S/N ratio requirement is difficult to meet in an optical fiber system, where the transmitted power is about 0 dBm (around 1 mW) and the noise level of the receiver is of the order of 6 pW/Hz. One way to reduce the S/N ratio requirement for an analog signal is to process the signal in frequency-modulated format on a subcarrier. The S/N ratio improvement is related to the index of modulation ($\Delta f/f_m = f_d/f_m$), which is essentially a bandwidth-expansion factor. The improvement factor is $(\Delta f/f_m)^2$. The signal is placed on a subcarrier, and the bandwidth can then be several times larger. This means that the optical intensity modulation is at a higher frequency. The highest permissible frequency is set by the frequency response of the light source and the photodetector. The S/N ratio required for a digital signal is much lower. For a 10^{-9}-bit error rate, an S/N ratio of about 20 dB is sufficient. This value of S/N ratio is easily accomplished in an optical-fiber system.

Detection (Demodulation)

A photodetector with its associated electronic circuit serves to accomplish signal detection at the receiver. The input signal v_{in} is in the form of intensity variation of the optical carrier. Through the demodulation process, v_{in} is converted to its electrical form, which then restores the signal with a minimum addition of noise and distortion to its original state through appropriate amplification. Photodetectors for application in communications must have a high quantum efficiency at the proper spectral region, adequate frequency response, low dark current, and low signal-dependent noise. There are three types of photodetectors: the photoemissive type (e.g., a photomultiplier), the photoconductive type (e.g., CdS), and the photovoltaic type (e.g., photodiodes). The photovoltaic type has the best overall performance and is the most suitable type for optical-fiber-system application. Photodetectors convert the incoming photons to electrons. The photon-to-electronic conversion efficiency is referred to as its *quantum efficiency*. In the absence of signals, the residual electron emission is termed the *dark current*. Receiver performance is governed by the selection of the type of photodetector as well as by the design of the associated electronic circuit. A receiver is designed to achieve a high sensitivity, which allows a large attenuation between the transmitter and the receiver for a given transmitter power output and a given required S/N ratio. In a practical receiver, the electronics amplification stages following the detector contribute thermal noise. When the thermal noise is the dominant noise, the receiver design must minimize this noise so as to approach the performance limit set by the quantum limit. The finite probability of having no electrons emitted over the time with a given photon input is an error since the presence of a photon must constitute a 1. This error is referred to as the *quantum limit*.

Photovoltaic Devices (Photodiodes) as Photodetectors

A semiconductor *pn* junction diode circuit (including a load resistor) will generate a current proportional linearly to the incident photon energy if it exceeds the bandgap between conduction band and valence band. In this case, a photon absorption at the *pn* junction gives rise to the excitation of an electron into the conduction band, thereby forming a hole in the valence band. If a reverse bias is applied to the *pn* junction, the transit time can be made very small. At a constant bias, the current is a linear function of incident optical energy. A positive-intrinsic-negative (PIN) photodiode is created with its junction separated by an intrinsic region in order to improve frequency response.

An efficient PIN photodiode for high-frequency operation is made as small as practical to match the size of the optical beam spot that it is designed to detect, with its *pn* junction close to the surface so as to minimize the amount of optical absorption before the *pn* junction. The depth of the junction is made large enough for photon absorption to be complete.

An avalanche photodiode (APD) operates under high-field conditions. It can have secondary emission as the current flows across the junction, resulting

TABLE 11.11-2

Properties of Photodetectors Used in Optical Communications

Item	InGaAs PIN	Ge APD	InGaAs APD	Si APD
Quantum efficiency	0.8	0.8	0.8	0.8
Responding time (ps)	60	100	100	100
Capacitance (pF)	< 0.5	< 1	< 0.5	< 1
Dark current (nA)	1–5	50–500	1–5	1
Leakage current (nA)	—	50–200	1–5	1
Hole-electron ionization ratio	—	0.7–1.0	0.3–0.5	0.02

in a noiseless gain just like that in the photomultiplier. This process is called an *avalanche action*. For an APD to be created, the material should have a high uniformity and a relatively low ionization potential in order to avoid the possibility of load breakdown and plasma formation caused by local field intensity.

Material systems for photodiodes include germanium and silicon; InGaAs and InGaAsP are also possible photodetector materials. For photodiodes with an area of the order of 10^{-2} mm^2, a cutoff frequency of the order of 1 GHz is readily obtained. APDs can respond to gigahertz oscillations and maintain a gain in the order of 100. Semiconductor photodiodes are ideal for optical-fiber applications because of their high quantum efficiency, broad bandwidth, and fiber-compatible size. The dark current for some devices is not negligible. Properties of photodetectors are listed in Table 11.11-2.

11.12 DATA PROCESSING IN MULTICHANNEL BIOTELEMETRY SYSTEMS

Introduction

A multichannel biotelemetry system includes a transmitting section and a receiving section. The various data variables are combined or multiplexed into one transmitted signal. At the receiving end, the received and detected signal is separated (or demultiplexed) to restore the individual variables for observation and recording. Frequency division telemetry separates the biologic variables on the basis of the frequencies of subcarrier oscillators (SCOs) contained in the multiplexer. The SCOs have center (unmodulated) frequencies in the audio range from 100 Hz to 20 kHz. Each SCO has a different center frequency (f_c) and is typically frequency modulated up to $\pm 7.5\%$ from f_c by a biologic variable. Frequency deviation bands and center frequencies are selected to prevent overlapping and interchannel crosstalk. The individual frequency-modulated SCO outputs are summed into a composite audio signal and transmitted. At the receiving end, the individual SCO signals are recovered from the composite audio by bandpass filtering, and they are then demodulated to recover the physiologic information from each filtered frequency-modulated audio signal. The abbreviation FM–FM indicates the type of encoding and the type of modulation of the radio carrier, respectively. FM–FM biotelemeters are popular in a variety of restraint-free monitoring studies, such as the telemetries of ECG (electrocardiography), EEG (electroencephalogy), blood pressure, pH, and so on. FM–FM biotelemetry systems offer a relatively inexpensive and uncomplicated means for telemetric monitoring (for either external or implanted studies) when up to four data channels are required. Figure 11.12-1 shows the block diagram of a three-data-channel FM–FM telemetry system.

A time division biotelemetry system sequentially samples the biologic data inputs from transducers (i.e., divided in time), and then it converts the analog

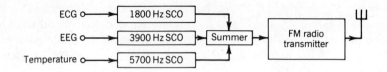

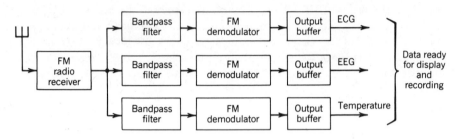

FIGURE 11.12-1 Block diagram of a multichannel FM-FM biotelemetry system, including a transmitting section and a receiving section. Although FM–AM can be used, FM–FM systems are more popular.

samples to digital form prior to encoding. This system not only permits sending a relatively large number of data channels but also provides improved resolution, accuracy, and reliability.

Encoding Techniques

PAM, PDM, PPM, PFM, and PCM are commonly used encoding techniques, each of which utilizes a different pulse parameter for modulation. Encoded pulses are transmitted serially, from the first channel to the last, in groups termed *frames*. A frame is preceded by and followed by synchronization (sync) information so that the receiver/decoder knows which channel is which when it sequentially demodulates the encoded serial pulses. Serial PAM can be derived directly from the multiplexer (MUX) output, since each pulse height at the output corresponds to the amplitude of one biologic signal at that instant. PDM is easily derived from PAM by comparison with a ramp of known slope; PCM can be derived from PDM by binary counting of the internally generated high frequency clock pulses over each pulse's duration time and then by serially transmitting the binary count for each data channel. In each case, sync information must be included; it can be transmitted in the form of higher pulse amplitude than normal data pulses, negative with respect to data pulses, missing pulses, extra pulses, and pulse code words. All time-division situations need the following sampling consideration. The sampling rate must be at least twice the highest frequency content biologic data channel. For example, if a three-channel biotelemeter is used to send ECG, temperature, and respiration rate, the signal with the broadest bandwidth (i.e., the highest frequency content) is the ECG at approximately 150 Hz. The sampling rate then must be at least 300 Hz. With three data channels, plus another devoted to synchronization, the clock frequency must be $300 \times 4 = 1200$ Hz. Thus from the sampling rate, the telemeter's clock frequency can be found.

PDM Techniques

The most popular time-division systems are PDM and PPM; they both perform better in noisy environments than PAM does, and their signal-to-noise ratio (S/N) is the same as FM. One technique for developing PDM uses one-shot multivibrators with a thermistor to study the equilibrium body temperature of free-swimming fish. An astable multivibrator provides the clocking rate signal to trigger the one-shot. The duration of the one-shot output pulse is proportional to the fish's temperature because of the thermistor incorporated into the timing circuit of the one-shot.

Another technique that permits telemetry of both resistance- and bio-potential-derived physiologic signals is illustrated in Figures 11.12-2a and 11.12-2b. The conditioned ECG, temperature, and blood-pressure signals are input to a six-channel multiplexer. The fourth input channel is grounded to provide a zero reference that permits differential cancellation at the decoder of any common mode ambient temperature and battery voltage drifts. The fifth channel causes the generation of a sync gap (or missing pulse) through a sync switch. The clock increments the binary counter, which gives three-bit addresses to the multiplexer. The sixth channel resets the counter to the first channel. Hence the first through fourth channels are sequentially sampled and appear one at a time at the common output of the multiplexer in PAM form. These PAM pulses represent instantaneous amplitude values of the input signals and are then amplified and applied to the comparator noninverting input. The signal at the inverting input is a linear ramp against which the amplitude of each amplified pulse is compared. When the ramp input rises to the other input, the positive comparator output changes from logic one to logic zero. The comparator output results in PDM form, since the time from the beginning of a clock pulse to the time the comparator output changes is a function of the amplitude of the input signals, as indicated in Figure 11.12-2b. The ramp is produced by an op-amp integrator that is automatically reset at the beginning of each clock pulse by shorting the feedback capacitor (C_{fb}) via a solid state switch. At the end (beginning) of each frame, the sync switch automatically connects the comparator to the negative supply voltage ($-V_S$) and no pulse (sync gap) appears at the output. The generated PDM is then applied to the FM radio transmitter.

PDM Decoding Techniques

Figure 11.12-3 shows the functional composition of a multichannel PDM decoder with an FM radio receiver at the input. This diagram illustrates the

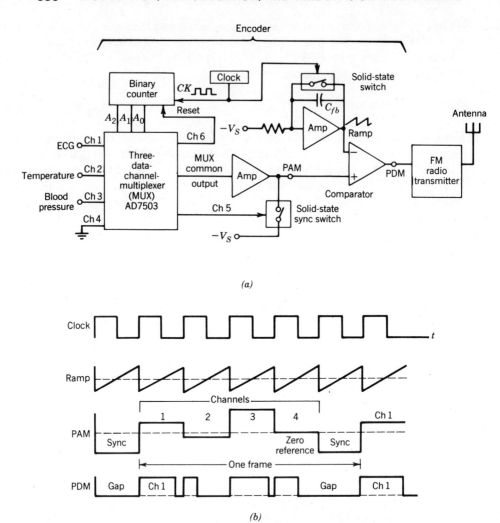

FIGURE 11.12-2 A six-channel PDM biomedical telemeter including an encoder and an FM radio transmitter: (a) functional diagram; (b) encoder waveforms. A monolithic CMOS, 8-channel analog multiplexer such as AD7503, low enables (Analog Devices, Inc.) can be employed as a 6-channel device while channels 7 and 8 are unused. Channel 1 (or 2) is ON when A_2, A_1, A_0, and enable are 0 (or 0), 0 (or 0), 0 (or 1), and 0 (or 0), respectively.

main operations involved in restoring the biologic data from the PDM signal. The level detector provides noise-free and squared-up PDM pulses that are directed to a one-shot multivibrator, a binary counter, and a sync gap detector. Because the leading edges of the PDM pulses occur at the encoder clock rate, the multivibrator can provide short width reset pulses at that rate to the decoder ramp generator. The clock rate sawtooth waveform is applied to the demulti-plexer (DMUX) common input. At the same time, the level detector output is

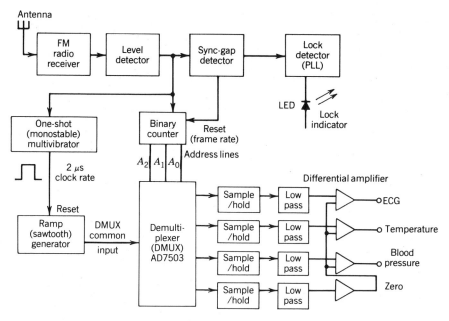

FIGURE 11.12-3 Functional diagram of a possible multichannel PDM decoder with an FM radio receiver at the input.

applied to the counter that provides data channel addresses to the DMUX. Thus PDM to PAM conversion can be done one channel at a time with the DMUX. For each channel, a linear ramp and a duration-modulated pulse commence at the same time; the ramp rises and appears at an addressed DMUX output until it is turned off when the DMUX is advanced to the next channel. This output is similar to a piece of a sawtooth whose area is proportional to the pulse duration. This sample is stored in the hold capacitor of a sample-and-hold circuit (sample/hold), subjected to low-pass filtering to remove residual clock frequency components, and then presented as output via a differential amplifier as the restored biologic waveform. Any common mode drift at the biotelemeter can be minimized through cancellation by applying the restored zero reference to all differential data amplifiers.

For the PDM data to be decoded in the proper order, the sync gap must be detected and the missing pulse restored. The regenerated missing pulse is then available to reset the counter to the first channel at the end (or beginning) of each frame of data. The frame rate equals the sampling rate per channel. The missing pulse can be used to operate a phase locked loop (PLL) circuit that can turn on an LED indicator when the decoder output is correct with the system in lock. Moreover the clock information can be used to drive chart recorder pens to a preset position if the telemeter must drop out. A CMOS microprocessor-based PDM decoder is being developed. It is expected to be low power, small, and portable.

Sample-and-Hold Circuit

The sample-and-hold circuit shown in Figure 11.12-4 is composed of an n-channel FET and a hold capacitor C between two voltage (unity) followers A_1 and A_2, which provide high input impedance and low output impedance for matching impedance purposes. The ramp to be sampled and the control pulse-train are applied to the A_1 noninverting terminal and the FET gate, respectively. The gate pulse train causes the FET to switch on and off, so that the circuit samples the instantaneous amplitude of a signal, and then holds the output voltage constant until the next sampling instant. The FET equivalent on resistance is $R_{D(on)}$. During the sampling time ($t_1, t_3, t_5, \ldots$), the FET is on

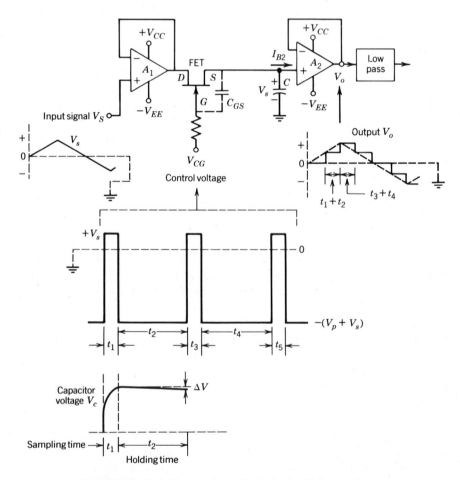

FIGURE 11.12-4 The sample-and-hold circuit with its waveforms.

and C is charged via $R_{D(on)}$. A capacitor voltage follows the exponential law

$$v_C = V(1 - e^{-t/RC}), \qquad (2.2\text{-}3)$$

provided that the initial charged voltage is zero. If the sampling time is

$$t_1(= t = 7RC) = 7R_{D(on)}C, \qquad (11.12\text{-}1)$$

the capacitor is charged to 0.999 of the input sampling voltage $V_s(= V)$, and the result is a 0.1% error in the sampled amplitude. This error is acceptable. During the holding time $(t_2, t_4, \ldots)$, C is partially discharged by the bias current I_{B2} flowing into A_2 and by the negligible FET source-gate leakage current $(\ll I_{B2})$. When the control voltage v_{CG} on the FET gate goes to its lowest level, the gate-source capacitance C_{GS} is charged to

$$-V_{GS(max)} = -(V_s + V_{CG}), \qquad (11.12\text{-}2)$$

thus decreasing the output voltage V_o. Note that the control voltage v_{CG} has a positive peak $+V_s$ and a negative peak $-(V_p + V_s)$, where V_p is the pinch-off voltage of the FET and V_s is the sampled signal voltage amplitude.

The C value of the hold capacitor can be computed from the given values of I_{B2}, t_2, V_s, and the error due to C discharge. The sampling time t_1 can be computed if C and $R_{D(on)}$ are known. The negative peak of the control voltage is calculated using the expression

$$-V_{CG} = -[V_{p(max)} + V_{s(peak)}]. \qquad (11.12\text{-}3)$$

The charge on C_{GS} is

$$Q = C_{GS}V_{GS(max)}. \qquad (11.12\text{-}4)$$

When Q is removed from C, the output decrement is

$$\Delta V_o = \frac{Q}{C}, \qquad (11.12\text{-}5)$$

and the error due to C_{GS} is

$$\% \text{ error due to } C_{GS} = \frac{\Delta V_o}{V_s} \times 100\%.$$

Example 11.12-1. A sample-and-hold circuit utilizes two 741 operational amplifiers and a 2N4856 FET whose $R_{D(on)} = 25\,\Omega$, $C_{GS} = 18$ pF, and $-V_{p(max)} = -10$ V. For simplicity, assume that the signal amplitude to be sampled is $V_s = \pm 0.8$ V. The holding time is 400 μs. Calculate (a) the hold-capacitor value C, (b) the minimum sampling time t_2, and (c) the % error due to C_{GS}.

Solution.

(a) The 741 has $I_{B(max)} = 500$ nA (see Table 5.7-1). Permit $I_{B2(max)}$ to discharge C by 0.1% during holding time t_2. Then the capacitor voltage decrement is

$$\Delta V = 0.1\% \times 0.8 \text{ V} = 0.8 \text{ mV};$$

$$C = \frac{I_{B2}t_2}{\Delta V} = \frac{(500 \text{ nA})(400 \text{ } \mu s)}{0.8 \text{ mV}} = 0.25 \text{ } \mu F.$$

(b) Permit 0.1% error in V_o due to the sampling time t_1. From Equation (11.12-1),

$$t_1 = 7R_{D(on)}C = 7 \times 25 \text{ } \Omega \times 0.25 \text{ } \mu F = 43.75 \text{ } \mu s.$$

(c) The negative peak of the control voltage is

$$- V_{CG} = -[V_{p(max)} + V_{s(peak)}]$$

$$= -(10 + 0.8) \text{ V} = -10.8 \text{ V}.$$

$$V_{GS(max)} = V_{CG} + V_s = 10.8 + 0.8 = 11.6 \text{ V}.$$

Charge on C_{GS} is

$$Q = C_{GS}V_{GS(max)} = (18 \text{ pF})(11.6 \text{ V}) = 208.8 \text{ pC}.$$

Since Q is removed from C, the output decrement is

$$\Delta V_o = \frac{Q}{C} = \frac{208.8 \text{ pC}}{0.25 \text{ } \mu F} = 835.2 \text{ } \mu V.$$

Thus

$$\% \text{ error due to } C_{GS} = \frac{\Delta V_o}{V_s} \times 100\%$$

$$= \frac{835.2 \text{ } \mu V}{0.8 \text{ V}} \times 100\% \simeq 0.104\%.$$

Monolithic Sample and Hold Circuits

The AD583 (Analog-Devices, Inc.) shown in Figure 11.12-5 is a monolithic sample-and-hold circuit consisting of a high-performance op-amp (input buffer) in series with an ultralow leakage analog switch and unity follower. An external holding capacitor (C_H), connected to the switch output, completes the sample-and-hold function.

With the analog switch closed, the AD583 functions like a standard op-amp; any feedback network may be connected between the output (pin 7) and the inverting input (pin 1) to control gain and frequency reponse. With the switch open the capacitor holds the output at its previous level.

The current output from the input amplifier charges the capacitor through the switch. The capacitor is unloaded by a unity follower. The output is fed back to the inverting input (as in an op-amp follower configuration), and thus, in sample, the charge on the capacitor is compelled to follow the input. In hold, the input amplifier no longer drives the capacitor; it retains its charge, unloaded by the output follower.

During the sample-to-hold, hold, and hold-to-sample states, the dynamic nature of the mode-switching introduces a number of specifications. The most important of these are defined as follows.

Acquisition time is the time required by the device to reach its final value within $\pm 0.1\%$ after the sample command has been given. This contains switch delay time, slewing time, and settling time and is the minimum sample time required to obtain a given accuracy.

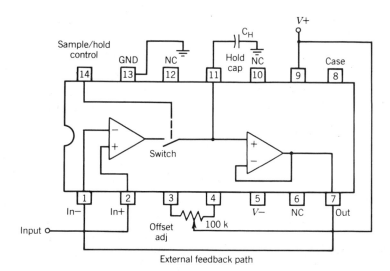

FIGURE 11.12-5   Functional diagram of the AD583 monolithic sample-and-hold circuit.

Charge transfer is the small charge transferred to the holding capacitor from the interelectrode capacitance of the switch when the unit is switched to the sample mode. Sample-to-hold offset error is directly proportional to this charge, where

$$\text{Offset error (V)} = \frac{\text{Charge (pC)}}{C_H \text{ (pF)}}. \qquad (11.12\text{-}6)$$

The aperture time is the time required after the hold command until the switch is fully open. This delays the effective sample timing with rapidly changing input signals. High slew rate and low aperture time permit sampling of rapidly changing signals.

Leakage currents from the holding capacitor during the hold mode cause the output voltage to drift. Drift rate (or droop rate) is calculated from drift current values using the formula

$$\frac{\Delta V}{\Delta T} \text{ (V/s)} = \frac{I \text{ (pA)}}{C_H \text{ (pF)}}. \qquad (11.12\text{-}7)$$

The AD582 is a low-cost IC sample and hold amplifier consisting of a high-performance op-amp, a low-leakage analog switch, and a JFET integrating amplifier. An external holding capacitor usually used as the feedback element of the integrator, completes the sample-and-hold function.

REFERENCES

1. Stremler, F. G., *Introduction to Communication Systems*, 2nd ed., Addison-Wesley, Reading, Massachusetts, 1982, Chaps. 5–7.

2. Roddy, D. and J. Coolen, *Electronic Communications*, 2nd ed., Reston Publishing Company, Reston, Virginia, 1981, Chaps. 8–11, 17.

3. Gregg, W. D., *Analog and Digital Communication*, Wiley, New York, 1978, Chaps. 7–12.

4. Comer, D. J., *Electronic Design with Integrated Circuits*, Addison-Wesley, Reading, Massachusetts, 1981, Chaps. 7, 8.

5. Owen, F. F. E., *PCM and Digital Transmission Systems*, McGraw-Hill, New York, 1982, Chaps. 1–7.

6. Raji, T. I. and A. I. Akinwande, Digital Modulator Permits the Simultaneous Transmission of Three Signals, *IEEE Transactions on Instrumentation and Measurement*, Vol. IM-31, No. 2, June 1982, pp. 17–24.

7. Kao, C. K., *Optical Fiber Systems: Technology, Design, and Applications,* McGraw-Hill Book Company, New York, 1982, Chap. 6.

8. Technical Staff of CSELT (Centro Studi e Laboratori Telecommunication), Torino, Italy, *Optical Fibre Communication*, McGraw-Hill, New York, 1981, Part IV, Chap. 1 and Part V, Chap. 2.

9. Chen, F. S., Modulators for Optical Communications, *Proceedings of the IEEE*, Vol. 58, No. 10, pp. 1440–1457 (1970).

10. Jeutter, D. C., Overview of Biomedical Telemetry Techniques, *IEEE Engineering in Medicine and Biology Magazine*, March 1983, pp. 17–24.

11. *Linear Integrated Circuits Data Book*, pp. 6-138 to 6-141 (PLL565), Motorola, Inc., Phoenix, Arizona, 1979.

12. *Linear (ICs) Data Book*, pp. 9-38 to 9-42 (PLL565), National Semiconductor Corp., Santa Clara, California, 1980.

13. *The European Consumer Selection* (Linear ICs), Motorola, Inc., pp. 9-61 to 9-70, 1977.

14. *Data-Acquisition Databook*, Volume I: Integrated Circuits, Analog Devices, Inc., Morwood, Massachusetts, 1982, pp. 14-3 to 14-29 and pp. 16-3 to 16-12.

QUESTIONS

11-1. Define the terms *modulation* and *demodulation*.

11-2. Describe the AM waveform referring to its expression given by Equation (11.2-3).

11-3. Sketch waveforms to illustrate PAM, PDM, PPM, and PCM, and briefly explain these types of pulse modulation.

11-4. Describe a typical balanced modulator.

11-5. Discuss how distortion can be avoided when a practical AM diode detector is designed.

11-6. Explain the operation of the precision-rectifier AM demodulator.

11-7. What is the major difference between frequency modulation and phase modulation?

11-8. What are the advantages of FM over AM?

11-9. Briefly explain how the phase detector (Figure 11.5-1) works and how its output voltage is related to phase difference.

11-10. Briefly explains the operation of the basic PLL FM detector.

11-11. Draw a transistor circuit showing a PAM system based on a one-shot multivibrator. Briefly explain the operation of the circuit.

11-12. Sketch a PDM system and explain its operation.

11-13. Sketch a PDM demodulating system and explain its operation.

11-14. Sketch a PPM modulating system and explain its operation.

11-15. Sketch a PPM demodulating system and explain its operation.

11-16. Describe the process for multiplexing and demultiplexing M information channels onto a single transmission channel.

11-17. Explain the term *commutation*.

11-18. Explain intersymbol crosstalk.

11-19. Sketch the TDM waveforms of three PDM channels.

11-20. What is a ring counter?

11-21. Sketch a block diagram and the waveforms for a four-channel TDM and PDM system. Explain the system operation.

11-22. Sketch the block diagram and waveforms of a decoding system for four-channel time-multiplexed PDM signals. Explain the system operation.

11-23. Explain the companding process used for PCM transmission.

11-24. Explain the PCM encoding process.

11-25. Draw the block diagram of a PCM transmission system and describe the function of each block.

11-26. Describe the operating principle of the digital modulator circuit shown in Figure 11.10-1.

11-27. Explain the term *laser*. What is the difference between the GaAlAs laser and the GaAlAs LED?

11-28. Draw the block diagram of an optical fiber communication system and describe the purpose of each functional element.

11-29. What results will be obtained if an analog signal is first modulated onto a subcarrier in an optical fiber system?

11-30. What are the signal-to-noise ratio requirements for telephone and TV signals and for a music channel?

11-31. Explain the method used to reduce the S/N ratio requirement for an analog signal.

11-32. Explain the S/N ratio required for a digital signal in an optical fiber system.

11-33. What is the quantum efficiency? What is the quantum limit?

11-34. Describe the construction and features of a PIN photodiode.

11-35. Describe the features of an avalanche photodiode (APD).

11-36. (a) Sketch a sixth-channel PDM biotelemetry system with a three-data-channel multiplexer; (b) sketch its encoder waveforms; (c) explain the operation of the system.

11-37. Sketch the functional diagram of a possible multichannel PDM decoder with an FM radio receiver at the input. Explain the operation of the decoder.

PROBLEMS

11-1. Calculate the instantaneous frequency of the signal $\phi(t) = A \cos(20\pi t + \pi t^2)$.

Answer.   10 Hz.

11-2. Calculate the instantaneous frequency of the following signal at $t = 0$: $\phi(t) = 5 \cos(15t + \sin 5t)$.

Answer.   20 rad/s.

11-3. Prove that the circuit of Figure 11.1-1b does not introduce amplitude modulation.

11-4. A signal containing frequency components ranging from 50 Hz to 5 kHz is to be transmitted by means of an AM system with a 800-kHz carrier frequency. What frequency range is required to transmit the AM RF signal?

11-5. A carrier wave of 10-V peak and 1-MHz frequency is amplitude modulated by a sine wave of 5-V peak and 3-kHz frequency. Determine the modulation index for the wave, the sideband frequencies, and the sideband amplitude.

11-6. A 100%-modulation AM transmitter can produce 10 W of output power. If the modulating signal is sinusoidal, what is the maximum power carried by one sideband for (a) an AM signal, (b) a carrier-suppressed signal, and (c) a single sideband signal.

Answer.   (a) $P_{LSB} = P_{USB} = 1.667$ W; (b) $P_{USB} = P_{LSB} = 5$ W; (c) $P_{SB} = 10$ W.

11-7. A constant-amplitude 3-kHz sine wave is used to phase modulate a carrier for a constant modulation index m_p. If the frequency of the modulating signal is increased to 6 kHz at some later time, what will happen to the output signal from a receiver in order for it to receive the modulated wave? Explain.

Answer.   The amplitude of the output signal will increase by a factor of 2.

11-8. Derive Equation (11.5-8) and explain its meaning.

11-9. In the 565 PLL FM detector circuit (Figure 11.6-2), $R_1 = 10$ k, $C_1 = 220$ pF, $C_2 = 330$ pF, and $V^+ = 6$ V. Calculate the VCO free-running frequency (f_o), lock range (f_L), and capture range ($f_{c'}$).

Answer. $f_o = 136.36$ kHz; $f_L = 181.8$ kHz; $f_{c'} = 156.1$ kHz.

11-10. Derive Equation (11.7-1) and explain its meaning.

11-11. Add a bias circuit to cause voltage v_1 to have the proper negative bias for the positive output waveform from the pulse amplitude modulator shown in Figure 11.7-2.

11-12. In a PCM system, the signal-to-noise (quantization noise) ratio is to be held to a minimum of 40 dB (a power ratio of 10,000): (a) find M, the number of levels (steps) needed; (b) if the number I in the expression $M = 2^I$ is not an integer, determine the actual value of M and the actual signal-to-noise ratio.

Answer. (a) $M = 81.7$; (b) actual value of $M = 128$; actual signal-to-noise ratio $= 49.9$ dB.

11-13. The sampling gate shown in Figure p11-13 uses a $\frac{1}{2}747$ operational amplifier (see Table 5.7-1) and a 2N4856 FET. The switching of the FET is controlled by a pulse train. When the FET is off, the output voltage is

$$V_o = \frac{R_2}{R_1} \times V_s.$$

When it is on, the output becomes

$$V_{o(\simeq 0)} = \frac{R_2 // R_{D(on)}}{R_1} \times V_o \simeq 0,$$

where $R_2 // R_{D(on)} \ll R_1$. The voltage gain is $A_v = 12$. The maximum signal voltage is $V_s = 0.5$ V. Determine: (a) R_1, R_2, R_3; (b) control voltage V_{CG}; and (c) zero error—that is, the output error due to $R_{D(on)}$.

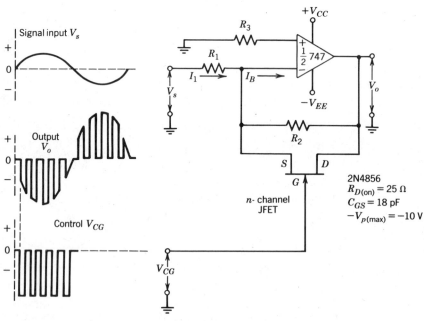

FIGURE P11-13

Hint. Let $I_1 = 100 \times I_{B(max)}$; $R_3 = R_1 /\!/ R_2$; $V_{CG} > V_{p(max)}$;

$$\text{zero error} = \frac{V_{o(\approx 0)}}{V_o} \times 100\%.$$

Answer. (a) $R_1 = 10K$, $R_2 = 120K$, $R_3 \simeq 10K$; (b) $V_{CG} > 10 \text{ V}$; (c) 0.021%.

11-14. The circuit of Figure 11.12-4 uses two 741 operational amplifiers (see Table 5.7-1) and an *n*-channel FET whose $R_{D(on)} \simeq 30\,\Omega$, $C_{GS} \simeq 10 \text{ pF}$, and $V_{p(max)} = 10 \text{ V}$. The signal amplitude to be sampled is $V_s = \pm 1 \text{ V}$. The holding time is 450 μs. Find (a) the hold-capacitor value, (b) the minimum sampling time, and (c) the % error due to C_{GS}.

Answer. (a) $0.225 \ \mu F$, (b) $47.25 \ \mu s$, (c) 0.0533%.

11-15. PDM can be generated by applying trigger pulses (at the sampling rate) to control the starting time of pulses from an emitter-coupled monostable multivibrator ($Q_1 - Q_2$, npn) and feeding the signal to be sampled to control the duration of these pulses. The CR network between Q_1 and Q_2 is connected with capacitor C from the Q_1 collector (C_1) to the Q_2 base (B_2) and resistor R from B_2 to the $+V_{CC}$ supply. The stable state is with Q_1 off and Q_2 on. The applied trigger pulse switches Q_1 on while Q_2 is switched off by regeneration action. As soon as this happens, C begins to charge up to $+V_{CC}$ via R. After a time determined by the V_{CC} supply and the RC time constant, B_2 becomes sufficiently positive to switch Q_2 on. Q_1 is simultaneously switched off by regeneration action. The Q_1 base bias is governed by the instantaneous changes in the applied signal voltage. Thus the applied modulation voltage controls the voltage to which B_2 should rise to switch Q_2 on. Since this voltage rise is linear, the modulation voltage is seen to control the period of time during which Q_2 is off, that is, the pulse duration $T(\ll 1/f_{sig,max})$. Assume that the two series combinations R_B–C_B and R_A–R_C are connected between the Q_1 base and ground and between $+V_{CC}$ and ground, respectively. The secondary of a signal input transformer is connected between both junctions of the two combinations. (a) Draw the circuit of the monostable multivibrator generating PDM; (b) explain how the changes of the Q_1 base bias and the emitter voltage affect the operation of the PDM generator; and (c) sketch the PDM waveform corresponding to a given sinusoidal signal. Assume a trigger pulse applied at the end of each half sine-wave.

12

ANALYSIS OF TELEVISION WAVE FORMING AND WAVE PROCESSING CIRCUITS

12.1 INTRODUCTION TO TELEVISION SYSTEMS

Scanning Pattern and Flying-Spot Scanner

In a TV transmitting system, the camera tube breaks an image down into picture elements so that the elements can be transmitted sequentially. The scanning action of the camera tube is similar to that of the picture tube (or cathode-ray tube, CRT); it can be explained referring to the flying-spot scanner shown in Figure 12-1. The electron beam in the CRT sweeps from left to right and from up to down if sawtooth deflection voltages are sequentially applied to both horizontal and vertical deflecting plates. *Horizontal sawtooth frequency* is the line frequency equal to 15750 ($= 60 \times 262\frac{1}{2}$) Hz, and *vertical frequency* is the field frequency equal to 60 Hz. Hence the electron beam sweeps more rapidly from left to right than from up to down. The retrace (or flyback) interval is very short compared with the forward-trace interval. When the beam reaches the lower right-hand corner of the screen, it quickly returns to the top of the screen. Then the scanning pattern is repeated. This pattern is termed a *raster*; it has $262\frac{1}{2}$ glowing lines.

In Figure 12.1-1, the raster on the screen of the CRT is the light source. Since a short persistence type of screen is employed, the phototube effectively responds to a moving spot of light instead of a complete raster. The optical lenses provide sharp focusing of the image.

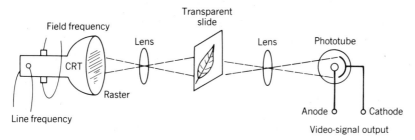

FIGURE 12.1-1 The flying-spot scanner.

Picture Elements, Frames, and Fields

Federal Communication Commission (FCC) standards stipulate that the basic or ideal TV image shall include 211,000 picture elements. A *picture element* is the smallest area in an image that can be reproduced by the video signal. The FCC standard entails the use of 525 scanning lines, a maximum video frequency of 4 MHz, and a well-focused electron beam in the picture tube. To minimize flicker, the basic frame of 525 lines is broken down into two fields, each of which contains $262\frac{1}{2}$ lines. This is known as the *interlaced scanning method*, which is specified by the FCC. The frame frequency is 30 Hz, or half the field frequency. The time from the beginning of one scanning line to the next is 1/15750 or 63.5 µs. The time from the beginning of one field to the next is 1/60 or 16,667 µs. Actually, only about 483–495 lines are active, and so each actual field contains about $241\frac{1}{2}$ to $247\frac{1}{2}$ lines. The actual scanning time along each line is 53.5 µs. The actual scanning time over each field is about 15,500 µs. Each scanning line is blanked for 10 µs ($= 63.5 - 53.5$). There are 1167 µs ($= 16,667 - 15,500$) of blanking time between successive fields.

Synchronizing Pulses

In order to reproduce the picture in the TV receiver, the scanning spot in the picture tube must be kept exactly in step with the scanning beam in the TV camera. Thus synchronizing pulses must be used to control correct timing. A sync pulse is transmitted at the start of each forward scan. Sync waveform specified by the FCC is shown in Figure 12.1-2 and described in the next paragraph.

The equalizing pulses serve to stabilize the operation of the vertical scanning generator (or vertical oscillator) in a TV receiver. These pulses are followed by a wide vertical sync pulse, which triggers the vertical oscillator. Serrations (or slots) are placed in vertical sync pulse to stabilize the operations of the horizontal scanning generator (or horizontal oscillator) during vertical retrace. Each horizontal sync pulse follows its predecessor by 63.5 µs. The duration of a horizontal sync pulse is 5.1 µs. At the end of both even and odd fields, the last equalizing pulse is at changed spacing from the vertical sync pulse. The sync

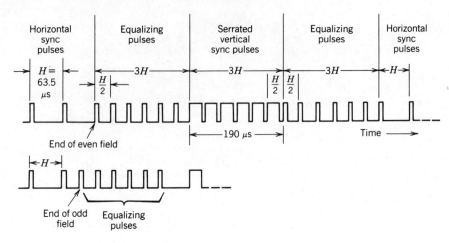

FIGURE 12.1-2 Synchronizing waveform specified by FCC.

pulses extend upward (above the video signal) into the blacker-than-black region. They extend above the blanking pedestals. Each blanking pedestal extends from white signal level to black signal level. A sync pulse has 25% of the total height of the waveform. Each horizontal blanking pedestal is 10 μs. The vertical sync waveform extends from a very broad blanking pedestal, often termed the *vertical blanking pulse*. The vertical sync pulse has six component pulses with a total width equal to 190 μs. These component pulses are separated by five serration pulses, each of which has the same width as an equalizing pulse, or approximately 2.5 μs.

Complete TV Signal

The TV composite waveform shown in Figure 12.1-3 characterizes negative transmission, as specified by the FCC. The composite video signal is a blanked picture (or camera) signal combined with the sync signal. The complete TV signal consists of the composite video signal and the sound signal. The video signal is amplitude modulated on the picture carrier 4.5 MHz below the sound carrier. The audio signal is frequency modulated on a sound carrier 4.5 MHz above the picture carrier. The composite video signal is encoded into the VHF signal and transmitted from the TV transmitter of Figure 12.1-4. At the TV receiver of Figure 12.1-5, the intercepted signal is amplified through a super-heterodyne arrangement.

Function of Sound Section in a TV Receiver

An FM sound section is associated with the AM picture section in a TV receiver. The maximum frequency deviation of the TV FM sound signal is ± 25 kHz. (The maximum deviation of broadcast FM sound signals is ± 75 kHz.) After amplification through an RF amplifier, the antenna input sound signal is

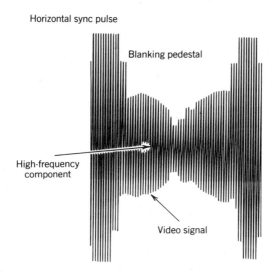

FIGURE 12.1-3 TV signal waveform through the RF tuner.

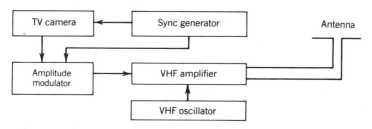

FIGURE 12.1-4 Block diagram of a composite video signal transmitting system.

converted to a typical intermediate frequency (IF) of 41.25 MHz. The 41.25-MHz signal is stepped up through the IF amplifier and fed to the video detector. This sound signal undergoes a second conversion in the video detector, where it is heterodyned with the picture IF to produce the 4.5-MHz intercarrier sound signal.

The picture and sound signals branch from the video detector into separate channels. A 4.5-MHz tuned amplifier circuit "picks off" the intercarrier sound signal and feeds it to the FM detector. The intercarrier sound signal is demodulated by the FM detector and applied to the audio amplifier. Since the IF amplifier can provide only limited gain for the sound-IF signal, a 4.5-MHz intercarrier sound amplifier is used before the FM detector.

Frequency Relations in a TV Receiver

The TV VHF range is from 54 to 216 MHz for channels 2 to 13; each TV channel is 6 MHz in width. An example illustrating the frequency relation is described

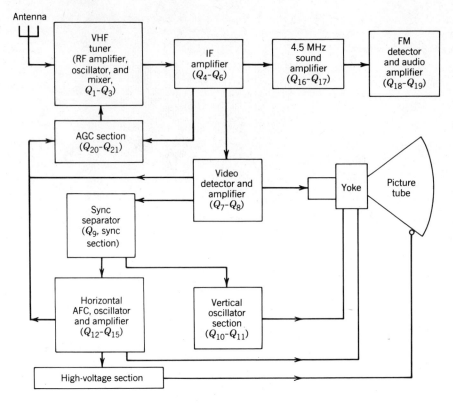

FIGURE 12.1-5　Block diagram of a TV receiver.

as follows:

IF picture carrier frequency = PIF = 45.75 MHz.

IF sound carrier frequency = SIF = 41.25 MHz.

Local oscillator frequency = f_o.

Picture carrier frequency = PCF.

Sound carrier frequency = SCF.

f_o = PIF + PCF = SIF + SCF.

SCF − PCF = PIF − SIF = 4.5 MHz.

Channel 7 = 174–180 MHz.

PCF = 175.25 MHz.

SCF = 179.75 MHz.

f_0 = 221 MHz.

Channel 8 = 180–186 MHz.

PCF = 181.25 MHz.

SCF = 185.75 MHz.

$f_o = 227$ MHz.
Channel 9 = 186–192 MHz.
PCF = 187.25 MHz.
SCF = 191.75 MHz.
$f_o = 233$ MHz.
Adjacent picture carrier = $227 - 187.25 = 39.75$ MHz.
Adjacent sound carrier = $227 - 179.75 = 47.25$ MHz.

Functions of RF Tuner

An *RF tuner* is an RF section called the front end of the TV receiver. This section consists of an RF amplifier, a local oscillator, and a mixer. The oscillator merely provides a source of high-frequency voltage to beat against the incoming TV signal in the mixer stage. The output from the mixer IF transformer contains the 45.75-MHz PCF and the 41.25-MHz SCF signals. An RF tuner has full amplification at both carrier frequencies and provides selectivity so that interference is rejected. Full amplification means a gain of 20 dB or more in the RF amplifier and a gain of about 16 dB in the mixer stage.

Most of the RF passband accommodates the picture signal, as shown in Figure 12.1-6. The picture-signal sidebands extend to the left and right of the picture carrier. The picture signal has a bandwidth of approximately 4 MHz, whereas the sound signal occupies only a small "slice" in the RF passband; the sound signal has a bandwidth of approximately 50 kHz. The picture signal includes comparatively few lower sideband frequencies and all of the upper sideband frequencies because most of the lower sidebands are suppressed with vestigial sideband transmission employed by the TV transmitter. The difference and the sum of a picture carrier frequency and a video frequency in an amplitude-modulated wave are called the *lower sideband* and the *upper sideband*, respectively.

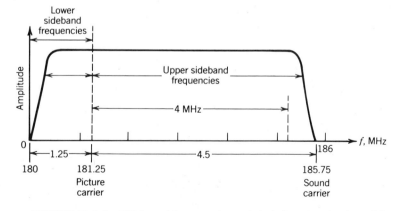

FIGURE 12.1-6   TV channel frequency intervals and allocations for channel 8.

IF Amplifier Principles

When tuned to a weak TV signal, the IF input voltage is approximately 0.25 mV and the IF output voltage is about 1 V. Hence the maximum available gain of an IF amplifier is on the order of 4000 times. Amplification takes place at a center frequency of approximately 43 MHz. The three IF amplifier stages (Q_4–Q_6) operate in cascade. The gains of an IF stage and the RF tuner can be changed automatically over a wide range when the input signal level changes. This is the function of the AGC (automatic gain control) section (Q_{20}–Q_{21}).

As depicted in Figure 12.1-6, the lower sideband of the video signal extends 1.25 MHz below the picture carrier frequency. Alignment of the IF amplifier is made to compensate for the incomplete lower sideband, insofar as practical considerations permit. Since the local oscillator in the RF tuner generally operates on the high side of the incoming signal, the frequency progression shown in Figure 12.1-6 becomes reversed through the mixer, and a typical IF response curve exhibits the picture carrier at a higher frequency than the sound carrier, as shown in Figure 11.1-7. The slope of the picture IF response curve is such that it falls to about zero over a 0.75-MHz interval and rises to maximum over a 0.75-MHz interval, on either side of the picture carrier. The sound carrier is passed at 10% of maximum response on the IF curve, at 41.25 MHz. Therefore both the picture and sound signals are processed through the IF amplifier.

Video Detector and Amplifier

Semiconductor diodes are commonly employed as video detectors. The input circuit of a video detector is part of the IF system, whereas the output circuit

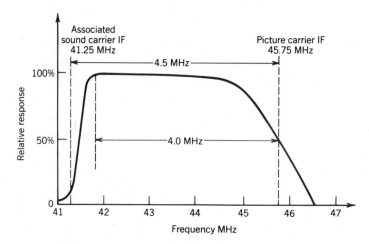

FIGURE 12.1-7 A typical picture IF response curve, showing the bandwidth, the sound carrier IF, and the picture carrier IF.

of the detector is part of the video-amplifier system. A video-amplifier section is commonly composed of two stages. It amplifies the video-input signal from the video detector from a level of about 1.5-V peak to peak to an output level at the picture tube of approximately 50-V peak to peak. The amplitude of the FM intercarrier sound signal is restricted to approximately 10% of the composite video-signal amplitude. The 4.5-MHz sound signal is formed by heterodyning the PIF carrier with the SIF signal in the video-detector diode. The video amplifier frequency response extends from almost zero frequency (dc) to 4.5 MHz. The sound takeoff circuit operates as a trap and produces the minimum overall response at 4.5 mHz. The bandwidth ($\simeq 3.5$ MHz) is measured between the 50%-of-maximum points on a television response curve. The collector of the video output stage (Q_8) is coupled to the cathode of the picture tube. Vertical and horizontal blanking pulses are injected into the video amplifier system. Blanking pulses function to cut off the picture tube during retrace, so that retrace lines are made invisible. The picture tube characteristic is nonlinear, as shown in Figure 12.1-8.

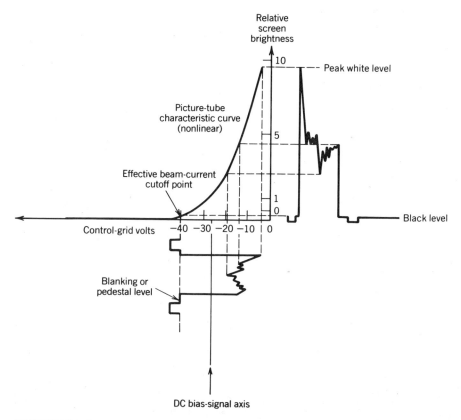

FIGURE 12.1-8 The curve for screen brightness versus control-grid voltage, showing the non-linear characteristic of a picture tube.

Functions of Sync Sections

The sync separator (Q_9) effectively slices the sync tips from the composite video signal. Most TV receivers employ sync takeoff in the collector branch of a video-amplifier stage and use RC filters to separate horizontal and vertical sync pulses from the stripped sync waveform. As shown in Figure 2.9-1, horizontal sync pulses are passed by an RC differentiator, and vertical sync pulses are passed by an RC integrator. A horizontal automatic frequency control (AFC) circuit is employed between the differentiator output and the horizontal sweep oscillator input.

Functions of Vertical and Horizontal Sweep Sections

Electromagnetic deflection is almost always used in a picture tube. Vertical motion of the scanning beam is determined by the vertical sweep section, which is a vertical oscillator section synchronized by a vertical sync section. This sweep section produces a sawtooth current flowing through the vertical deflection coils (or yoke).

The horizontal motion of the scanning beam in a picture tube is basically determined by the horizontal AFC, oscillator, and amplifier sections. The output section produces a sawtooth current flowing through the horizontal deflection coils (or yoke). The horizontal deflection coils are shunted by a damper diode. This diode conducts only during the flyback interval (or retrace time). A typical large screen receiver employs a horizontal deflection system that operates at a peak power level of more than 1000 VA. Silicon transistors are generally used in the output circuit and can withstand a peak reverse voltage of about 100 V. Hence the peak-to-peak current value is typically over 10 A. The accelerating voltage (HV) for the picture tube is generally obtained from a subsection of the horizontal deflection system. This high-voltage arrangement is energized by the horizontal flyback pulse. In addition, a booster subsection is provided for the generation of dc potentials of 150–300 V.

12.2 RF TUNER CIRCUIT

Antenna Input and Trap Configuration

A VHF tuner circuit is shown in Figure 12.2-1. The 300-Ω antenna input section consists of a *balun* (balanced unit), or VHF transformer arrangement (L_1), which is constructed of bifilar windings. The antenna-input signal passes through the trap configuration before it is applied to the RF amplifier Q_1. L_2 and C_1 operate as a parallel trap to prevent feedthrough interference in the IF range (41 to 47 MHz) into the IF amplifier (Figure 12.3-1). L_3 and C_2 function as a high-pass filter. C_3 and L_4 function as a series trap. Interference signals in the 88- to 108-MHz range (FM broadcast band) are greatly attenuated. The input signal is applied to the base of Q_1 through a series-resonant circuit.

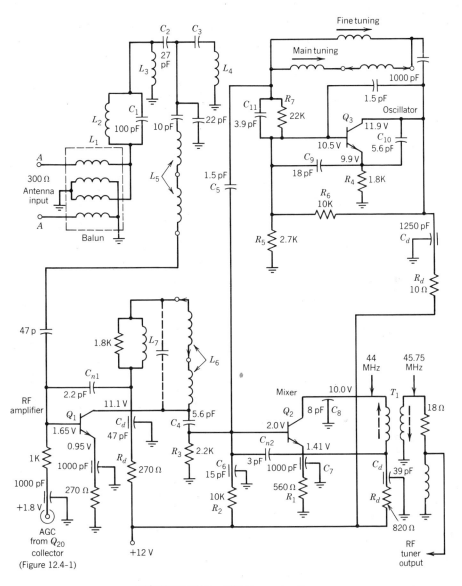

FIGURE 12.2-1   VHF tuner circuit.

RF Amplifier and Mixer

Transistor Q_1 provides gradual control with forward AGC (automatic gain control). Forward AGC operation consists of increasing the forward bias on a transistor, with simultaneous reduction in collector voltage and shift of operation into the collector saturation region. Forward AGC requires substantial series resistance in the collector circuit. Therefore, as the collector current is increased, more of the collector supply voltage appears as an IR voltage drop across this series resistance. The collector voltage may approach zero at the high value of forward bias. The dc current gain beta of a transistor reduces when either collector current or collector voltage is greatly decreased. The forward bias V_{BE} of Q_1 is derived from the AGC amplifier output (Q_{20}, Figure 12.4-1). The reverse-bias V_{CB} of Q_1 is derived from the +12-V dc source. Biases V_{BE} and V_{CB} of Q_2 and Q_3 are also derived from the same +12-V dc source through different voltage dividers of two resistors in series. Q_1 and Q_2 are biased in series-feed, rather than shunt-feed mode, as for oscillator Q_3. When trace an ac signal path, all the dc bias-source lines can be considered as at ac ground potential.

Since the frequencies are very high, all by-pass capacitors are connected with short leads. In the amplifier and mixer stages (Q_1 and Q_2), neutralization capacitance C_{n1} and C_{n2} are used to balance out the positive feedback via the interelectrode capacitance between collector and base (C_{CB}), so that parasitic oscillations are prevented. The equivalent neutralization circuit is a capacitance bridge. In order to avoid parasitic oscillations due to positive feedback via the internal resistance of the dc voltage source, the supply voltage V_{CC} must be connected to an R_d-C_d decoupling network before it is applied to the collector circuit of each stage. In the Q_1 collector circuit, coil L_7 shunted by the 1.8-kΩ resistor serves to suppress regenerative feedback.

The amplified signal in the collector circuit of Q_1 is coupled to the mixer stage (Q_2) via capacitor C_4. Oscillator (Q_3) voltage is injected into the Q_2 base circuit by capacitive coupling. C_5 and C_6 form a capacitive voltage divider, with the oscillator voltage across C_6 for the mixer. In the Q_2 emitter circuit, R_1 and C_7 produce emitter bias, which is reverse voltage for the emitter-base junction of the *npn* transistor. However the R_2-R_3 voltage divider provides positive forward bias for the base, resulting in an effective value of forward bias V_{BE}. Transistor Q_2 operates in a nonlinear region. Its base-emitter junction serves as a diode rectifier to heterodyne the RF TV and oscillator signals. In the mixer collector output circuit, the IF transformer T_1 with C_8 is tuned to the IF signal frequencies. Tuned circuits in the RF tuner must resonate at suitable frequencies to provide the standard double-humped frequency response curve, which is obtained by stagger tuning the RF amplifier and mixer resonant frequencies in an alignment procedure.

Local Oscillator

The *npn* transistor Q_3 is operated as a local oscillator. The V_{CC} source is supplied in shunt-feed mode. The emitter resistor R_4 is not bypassed to provide ac feedback. Forward bias for the base and reverse bias for the collector are supplied by R_5 and R_6, respectively. In the local oscillator, the feedback network is an *LC* tuned circuit between the base and collector with the emitter as a reference zero potential. Between the base and collector is the C_9-C_{10} voltage divider, with its midpoint connected to the emitter. This capacitive divider and the tuned circuit complete the positive (regenerative) feedback loop, so that oscillations can be produced in the regenerative *CE* configuration.

The output of the oscillator is at the resonant frequency (f_r) of its tuned circuit. Using

$$f_r = \frac{1}{2\pi\sqrt{LC}}$$

or

$$f_{\text{MHz}} \simeq \frac{159}{\sqrt{L_{\mu H}C_{\text{pF}}}}, \tag{12.2-1}$$

we can calculate resonant frequency ($\simeq$ oscillator frequency). Assume the effective inductance $L = 0.8\ \mu H$ and the effective capacitance $C = 2.4\ \text{pF}$. Then the resonant frequency is

$$f_r = f_{\text{MHz}} \simeq \frac{159}{\sqrt{0.8 \times 2.4}} = 114.75\ \text{MHz}.$$

12.3   OSCILLATOR WITH AUTOMATIC FINE TUNING (AFT) SYSTEM

An oscillator with AFT is shown in Figure 12.3-1. The varactor diode is employed for automatic frequency tuning. The dc reference voltage of 3 V is furnished from the AFT circuit. When the AFT correction voltage changes, the varactor capacitance varies to change the oscillator frequency. Capacitors marked OTC and −TC have a zero temperature coefficient and a negative temperature coefficient (150 PPM/°C), respectively. The feedback network is the effective LC tuned circuit between the base and collector of the CE configuration (Q_3' stage).

As shown in Figure 12.3-1b, an AFT system contains an AFT discriminator with an input CE stage (Q_4'). The 45.75-MHz IF signal is coupled from the third IF stage to the discriminator input coil L. The capacitive divider C_1–C_2 serves to couple the picture IF carrier signal at a suitable level to the Q_4' base. Q_4' provides a voltage gain of about 10 and a power gain of approximately 20 dB. The collector connects to the primary of transformer T_1 at a tap point for a suitable Q value. The primary is coupled to the secondary both inductively and capacitively so that in-phase and out-phase discriminator action is developed. Diodes D_1 and D_2 rectify the output signal; D_1 is reverse biased, and D_2 is forward biased. Under no signal condition, about 3 V will appear at the test point. With a normal signal present, the AFT control voltage varies between 1 and 8 V. The forward and reverse bias voltages applied to D_1 and D_2 modify the conventional discriminator S curve. The modified response curve is shown in Figure 12.3-1c.

Let us follow the conventional circuit shown in Figure 12.3-2 to explain

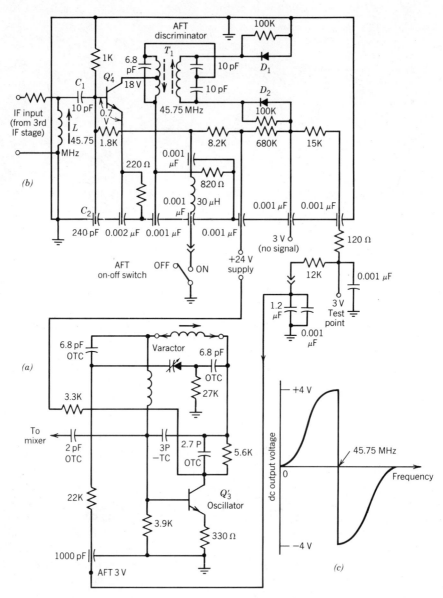

FIGURE 12.3-1 (a) Local oscillator; (b) AFT (or automatic frequency control, AFC) circuit; (c) frequency response curve of an AFT discriminator.

basic discriminator action. Signal voltage V_p is applied to the primary of the discriminator transformer. Current I_p lags V_p by 90°. Signal voltage V_i is induced in the secondary. In the secondary tuned circuit, current I_s is in phase with V_i. The voltage V_s across the secondary leads current I_s by 90°. $V_{s1} = V_{s2} = V_s/2$. The phase relationship among these voltages and currents is shown in Figure

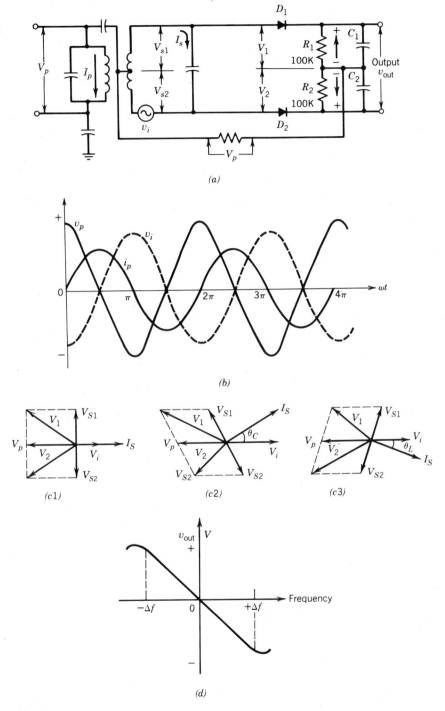

FIGURE 12.3-2 Conventional discriminator: (*a*) circuit; (*b*) phase relationship among voltage v_p, v_i, and current i_p; (*c*) phasor diagrams: (*c1*) at IF, $|V_1| = |V_2|$, $V_{out} = 0$; (*c2*) $f < $ IF, $|V_1| > |V_2|$, $V_{out} = $ positive value; (*c3*) $f > $ IF, $|V_1| < |V_2|$, $V_{out} = $ negative value; (*d*) discriminator frequency response curve.

12.2-3b. The phasor diagrams of the voltages and currents are shown in Figure 12.3-2c. Their quantitative relationships are as follows:

$$v_p = V_m \cos \omega t;$$

$$i_p = I_m \sin \omega t;$$

$$v_i = -M \frac{di_p}{dt} = -M\omega I_m \cos \omega t = -V_m \cos \omega t;$$

$$V_{s1} = V_{s2} = \frac{V_s}{2}.$$

The AFC (automatic frequency control) action can be seen from the frequency response curves shown in Figures 12.3-1c and 12.3-2d. When the oscillator frequency is decreased or when the input frequency f_{in} is less than IF, the discriminator dc output voltage V_{out} is positively increased. The higher V_{out} value applied to the cathode of the varactor diode in the oscillator will reduce the effective capacitance of the tuned circuit, thus increasing the oscillator frequency f_o. On the other hand, when f_o is increased or when f_{in} is higher than IF, V_{out} is decreased in a negative direction. The lower V_{out} value applied to the varactor cathode will increase the effective capacitance of the oscillator tuned circuit, thereby reducing the oscillator frequency.

12.4   PICTURE IF AMPLIFIER CIRCUIT

A picture IF amplifier circuit is shown in Figure 12.4-1. A trap is provided to reduce interference from the adjacent-channel sound carrier 47.25 MHz. Another trap may be added to reduce interference from the accompanying sound IF carrier 41.25 MHz. Coil L_{22} in series with a 12-pF capacitor is tuned to the IF signal frequency. L_{23}, L_{24}, L_{25}, and the mixer tank (T_1, Figure 12.2-1) provide IF tuned circuits. The necessary wide-band response can be obtained with single-tuned stages by staggering their resonant frequencies. The primary and secondary inductance of transformer L_{25} tune the shunt capacitances to the desired frequencies in the IF range. The damping resistors across the windings of L_{25} provide the required wide band. The neutralization capacitor C_{n3} connects between the base and collector circuit of transistor Q_6 to cancel out positive feedback via interelectrode capacitance C_{CB}. The forward voltage across the base-emitter junction of transistor Q_4 is derived from the Q_5 emitter bias. Diode D_{21} is connected between the base of the second PIF stage and the collector of the AGC amplifier in order to provide temperature compensation.

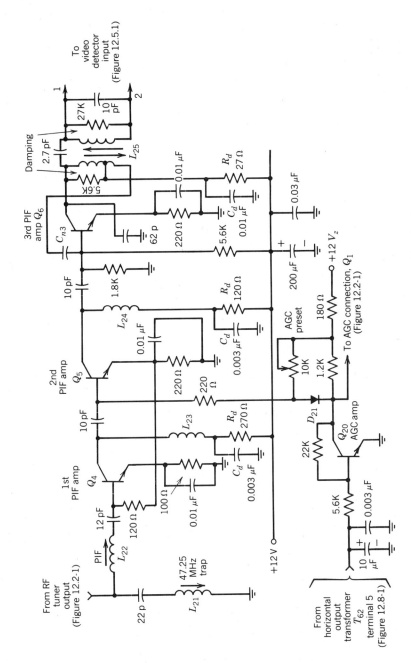

FIGURE 12.4-1 Picture IF (PIF) amplifier circuit.

407

12.5   VIDEO DETECTOR AND AMPLIFIER CIRCUIT

The video detector and amplifier are shown in Figure 12.5-1. The input circuit of the detector is the output part of the PIF amplifier. The video detector is driven by the PIF amplifier, and the detector drives the video amplifier. Diode detection is preferred because of its good detection linearity. The series diode detector is basically a peak detection configuration. The detector tends to charge the filter capacitors to the negative peak of the applied PIF signal. Between negative peaks, these capacitors discharge via the resistance-inductance network contained in the base circuit of Q_7 stage.

The video detector also serves as a frequency converter to produce the 4.5-MHz second sound IF signal (which is called the *intercarrier sound signal*). In terms of intercarrier sound, the 45.75-MHz picture carrier in the detector stage is equivalent to a local oscillator. The heterodyning action of the 45.75-MHz picture carrier beating with the 41.25-MHz center frequency of the sound signal results in the lower center frequency of 4.5 MHz. This FM signal is then coupled through the emitter follower Q_7 to the 4.5-MHz sound IF section for amplification and detection to recover the audio modulation. Since a load resistor is connected to the collector of Q_7, this transistor also operates as a CE configuration to provide positive sync pulses applied to the AGC keyer Q_{21}. Therefore Q_{21} produces amplified negative flyback pulses through the collector diode D_{34}. These negative pulses combine in series with positive flyback pulses from the horizontal output circuit, producing a resultant voltage applied to the AGC amplifier Q_{20}. Thus AGC action is accomplished. The AGC keyer output voltage is free from noise pulses that can occur between the horizontal sync pulses. The AGC bias varies only with the strength of the RF carrier signal on the antenna. In the forward AGC method, the transistor (Q_1) gain is decreased by increasing bias at the base, toward saturation.

The input circuit of the video amplifier (Q_7-Q_8) has a low-pass filter comprising C_{31}, L_{31}, C_{32}, L_{32}, L_{33}, and the base input capacitance of Q_7. This network prevents the passage of the feedthrough-IF signal into Q_7, thus eliminating the possibility of overload and distortion from this source. The series peaking coils L_{31} and L_{32} and shunt peaking coil L_{33} serve to maintain full response at midband and high-video frequencies. Class A operation is utilized for Q_7 to avoid video signal distortion; its operating point is set by potentiometer R_{32}. Since the primary of T_{71} (Figure 12.9-1) provides 4.5-MHz trap action in the emitter circuit of Q_7, most of the intercarrier sound signal branches into the sound takeoff channel. If excessive intercarrier sound signal passes through to the picture tube, "sound grain" will appear in the image.

The collector circuit of video output transistor Q_8 drives the picture tube input circuit and also feeds the sync takeoff branch circuit. The contrast control R_{33} is supplemented by the high-frequency compensating capacitor C_{33}. Brightness control is provided by R_{34}. Vertical blanking pulses are injected into the picture tube grid circuit via diode D_{32}. Uniform response at higher frequencies is maintained by the inductance–capacitance–resistance network between the tap of R_{33} and the picture tube cathode.

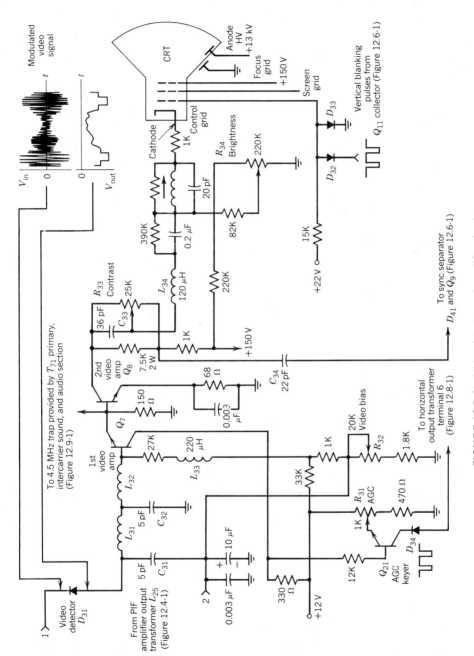

FIGURE 12.5-1 Video detector and amplifier circuit.

12.6 SYNCHRONOUS SEPARATOR AND VERTICAL OSCILLATOR CIRCUIT

A circuit consisting of a sync separator and a vertical oscillator is shown in Figure 12.6-1. The composite video signal applied to the sync separator (which is also called a *sync clipper*) is obtained from the collector branch of the video-amplifier output stage (Q_8, Figure 12.5-1). D_{41} serves as a diode clipper. The RC time constant of the anode input circuit is (220 kΩ) × (22 pF) or 4.84 μs, which approximates the duration (5 μs) of a horizontal sync pulse. The cathode output circuit is a voltage divider connected to the base of transistor Q_9. The dc biases for Q_9 are derived from the +12-Vz supply through the voltage divider R_{41}-D_{41}-R_{42}-R_{43}. The base forward bias is small so that Q_9 does not conduct when no signal is applied. The vertical sync pulses from the collector circuit of Q_9 are passed by the R_{46}-C_{42} integrator (see Sections 12.1 and 2.9). The low-pass integrator has an RC time constant of (6.8 kΩ) × (0.05 μF) or

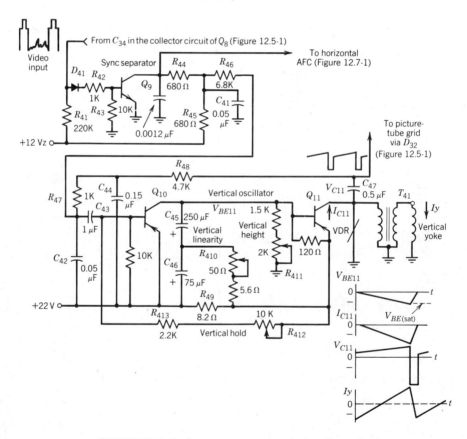

FIGURE 12.6-1 Sync separator and vertical oscillator circuit.

340 μs, whereas the vertical sync pulse with a width of 190 μs has five serrations, each of which is 2.5 μs wide. A horizontal sync pulse only 5 μs wide will produce a negligible charge on the integrating capacitor (C_{42}). The integrator output signal is used to synchronize the vertical oscillator (Q_{10}-Q_{11}).

The two-pnp transistor oscillator (Q_{10}-Q_{11}) is a modified astable multi-vibrator. The Q_{10} collector is direct coupled to the Q_{11} base, while the Q_{11} emitter is coupled to the Q_{10} base via R_{412} and R_{413}. Also the collector of Q_{11} is coupled back to the Q_{10} emitter through C_{47}, R_{48}, and C_{44}. When the +22-V source is switched on, C_{46} and C_{45} are charged. Charging linearity control is provided by R_{410}. As the negative-going voltage for the base of Q_{11} reaches $V_{BE(\text{sat})}$, Q_{11} and Q_{10} sequentially saturate; thus the capacitors rapidly discharge, and then oscillation repeats. The approximate oscillating waveforms for Q_{11} are shown in Figure 12.6-1. The discharge time of capacitors C_{45} and C_{46} is the brief flyback interval, during which the magnetic field in the primary of transformer T_{41} collapses rapidly. This collapse generates a counter emf pulse ($-L\,di/dt$) across the winding. The sawtooth component of Q_{11} collector voltage V_{C11} can be regarded as deriving from the V_{BE11} sawtooth. The negative flyback pulse from the Q_{11} collector is applied to the picture tube control grid for vertical blanking. The vertical hold and height controls are provided by R_{412} and R_{411}, respectively. VDR (e.g., a carbide varistor) is a voltage-dependent resistor that serves to protect transformer T_{41}.

12.7   HORIZONTAL AFC AND OSCILLATOR CIRCUIT

The circuit of a horizontal AFC and an oscillator is shown in Figure 12.7-1. The negative horizontal sync pulses are passed by the differentiator comprising capacitor C_{51} and the effective resistance associated with the AFC circuit. This horizontal AFC is an example of a single-end sync discriminator. Its operating principle was discussed in Section 3.6.

The horizontal oscillator is an emitter-coupled astable multivibrator comprising the two npn transistors Q_{12} and Q_{13}. The Q_{12} collector is coupled to the Q_{13} base through C_{53}, and the Q_{13} emitter is coupled to the Q_{12} base through R_{59} and R_{510}; thus a regenerative (positive) feedback loop is completed. When the +12-Vz source is switched on, the horizontal hold R_{54} is adjusted to let the collector current of Q_{12} (I_{C12}) increase. Then, due to regenerative feedback, the collector current of Q_{13} (I_{C13}) rapidly goes down while I_{C12} quickly rises up. Eventually Q_{12} is saturated while Q_{13} is off. This condition will be maintained until capacitor C_{53} has discharged or the reverse bias for Q_{13} base has been removed. When the Q_{13} base junction is forward biased to cutin voltage ($V_{BE,\text{cutin}}$), Q_{13} starts conducting, causing I_{C12} to reduce due to positive feedback. In turn, the decrease of I_{C12} forces I_{C13} to increase, and this increase causes I_{C12} to further decrease, until Q_{13} is saturated and Q_{12} is at cutoff. Following the sequence just described, oscillations repeat from one state to another.

The multivibrator operation is explained referring to the approximate

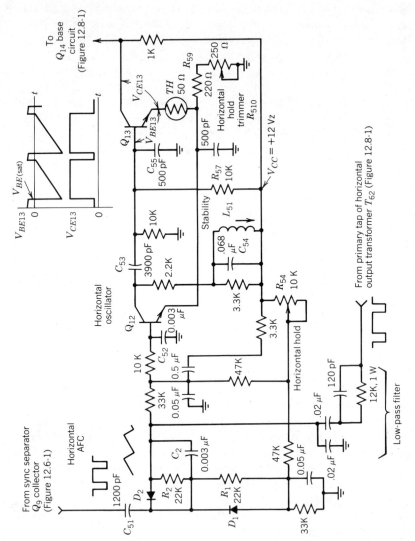

FIGURE 12.7-1 Horizontal AFC and oscillator circuit.

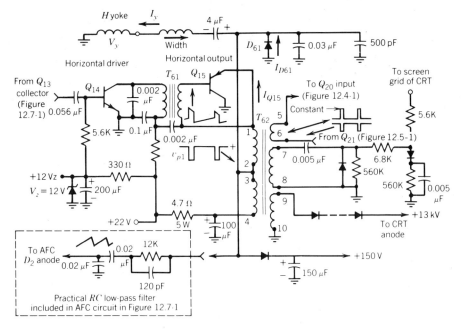

FIGURE 12.8-1   Horizontal output stage and high voltage circuit.

waveforms shown in Figure 12.7-1. When Q_{12} is in saturation and Q_{13} is at cutoff, the negative charge of C_{53} holding reverse bias for Q_{13} base junction discharges through R_{57} toward $+V_{CC}$. When the base-emitter voltage of Q_{13} reaches $V_{BE(sat)}$, Q_{13} saturates while Q_{12} is cut off due to positive feedback. Then coupling capacitor C_{53} rapidly charges via Q_{13} until Q_{13} is cut off. Q_{13} conducts for only a short time because of the small time constant for the charge of C_{53}. When Q_{13} is cut off, C_{53} slowly discharges through R_{57}. Q_{13} saturates for a brief interval and stays off for a long duration, resulting in a negative flyback pulse from its collector, as shown in Figure 12.7-1. The horizontal hold and stability controls are provided by R_{54} and L_{51}, respectively. The oscillator output signal is coupled from the collector of Q_{13} to the base of horizontal driver Q_{14} through a 0.056-μF capacitor (Figure 12.8-1).

12.8   HORIZONTAL OUTPUT AND HIGH VOLTAGE CIRCUIT

Figure 12.8-1 shows a commonly used horizontal output and high-voltage arrangement for the typical TV receiver. Transistor Q_{15} operates as an emitter follower in the horizontal output stage. The high-voltage transformer T_{62}, deflection coil (or yoke), and damper D_{61} are the load circuit for Q_{15}. The approximate waveforms for the horizontal deflection circuit are shown in Figure 12.8-2.

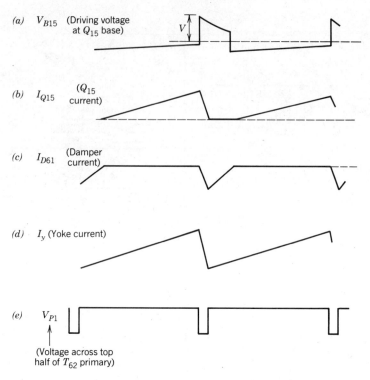

(a) V_{B15} (Driving voltage at Q_{15} base)

(b) I_{Q15} (Q_{15} current)

(c) I_{D61} (Damper current)

(d) I_y (Yoke current)

(e) V_{P1}

(Voltage across top half of T_{62} primary)

FIGURE 12.8-2 Approximate waveforms for horizontal deflection circuit in Figure 12.8-1. The yoke current sawtooth waveform results from the combined effect of Q_{15} current and damper current through the top half of T_{62} primary. (Use Lenz's law to explain the transient when the inductor circuit is abruptly switched off.)

Referring to Figure 2.12-5a, we can see that the low-frequency equivalent circuit of the driver transformer T_{61} in Figure 12.8-1 is a high-pass RL network. Hence the flat top of the pulse through T_{61} is exponentially decayed. The driving voltage at Q_{15} base shown in Figure 12.8-2a can be expressed as

$$V_{B15} \simeq V e^{-R_p t/L_M}, \tag{12.8-1}$$

where L_M is the magnetizing inductance and R_p is the ohmic value of the primary winding.

The output *pnp* transistor is an emitter follower. It is on during the trace time. The sawtooth current continues to rise linearly until the instant that Q_{15} is cut off by the driving pulse from T_{61}. This positive pulse switches damper diode D_{61} into conduction. D_{61} not only eliminates undesired oscillations but also fills the missing component for trace, as shown in Figure 11.8-2c. The current of Q_{15} falls to zero with a delay due to the minority charge storage time. The voltage across the top half of the T_{62} primary is the pulse v_{p1}. When this is applied to the horizontal deflection-coil circuit, a current of sawtooth waveform flows in the yoke. This sawtooth waveform results from the combined effect of Q_{15} current and damper current through the top half of the T_{62} primary. The sawtooth current in the yoke provides the complete trace from left to right.

When the pulse v_{p1} is applied to the effective low-pass RC filter, a sawtooth voltage is produced across the output capacitor; thus the horizontal oscillator frequency can be compared with the sync pulse repetition rate in the AFC circuit. The pulse from the top secondary of T_{62} is opposite that from the AGC keyer (Q_{21}, Figure 12.5-1); these two opposite signals combine in series, with the resultant voltage applied to the base circuit of the AGC amplifier Q_{20} (Figure 12.4-1).

12.9 INTERCARRIER SOUND AND AUDIO CIRCUITRY IN A TV RECEIVER

Introduction

The 4.5-MHz intercarrier sound signal is actually available first in the mixer (Q_2) output. The 41.25-MHz conversion product is stepped up through the IF amplifier (Q_4-Q_6), and the 4.5-MHz intercarrier sound signal becomes available next as a conversion product from the video detector. In the Q_4-Q_6 stages, the sound IF level is restricted to 10% or less of the picture-IF level in order to avoid objectionable amplitude modulation of the FM sound signal by the AM picture signal. Since the bandwidth of the intercarrier sound signal is approximately 50 kHz, the sound takeoff circuit has fairly high Q value, typically about 80.

An intercarrier sound and audio circuit is shown in Figure 12.9-1. Intercarrier sound takeoff is located at the video emitter follower (Q_7) output, from which the FM signal is coupled to the 4.5-MHz sound-IF amplifier through a 4.5-MHz trap in the primary of transformer T_{71}. The 4.5-MHz sound carrier goes through the trap rather than the video output stage to the picture tube. The secondary sound-IF (4.5-MHz) amplifier comprises two stages (Q_{16}-Q_{17}), with the Q_{16} collector direct coupled to the Q_{17} base. Two tuned circuits are connected to the Q_{16} base and Q_{17} collector, respectively. The forward bias for the base-emitter junction of Q_{16} is supplied from part of the Q_{17} emitter voltage. The output 4.5-MHz signal is inductively coupled to the ratio detector. Demodulation of the FM detector produces the audio signal sent to the direct-coupled, two-stage audio amplifier (Q_{18}-Q_{19}).

Ratio Detector

A basic ratio detector circuit is shown in Figure 12.9-2. The transformer includes a tertiary winding L_T connected to the secondary center tap. Two diodes D_1 and D_2 at both ends of the secondary are connected in opposite polarity and in

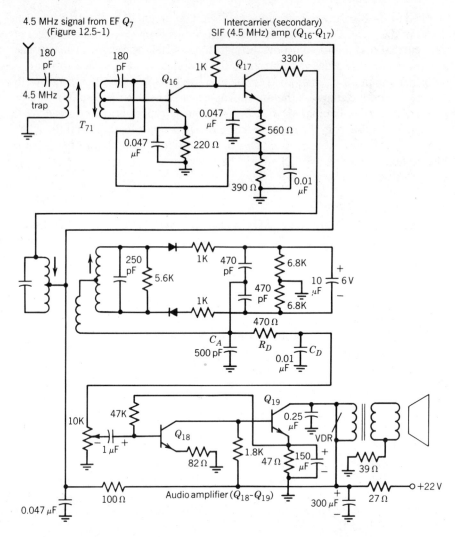

FIGURE 12.9-1 A circuit of intercarrier sound and audio section.

series with load resistors R_1 and R_2 for half-wave rectification. Capacitors C_1 and C_2 complete the circuit for 4.5-MHz voltage developed across the tertiary winding so that the voltage at the low end of the tertiary winding is applied to the C_1-C_2 sides of diodes D_1 and D_2. A large capacitor C_s is connected across the R_1-R_2 load. C_s is used for stabilizing. It is largely responsible for the AM-limiting properties of the ratio detector. The output is taken from the tertiary winding. An audio signal is developed across capacitor C_A. R_D and C_D form a deemphasis network to reduce the ratio detector response at high frequencies. Both the primary and secondary circuits are tuned to the secondary sound-IF

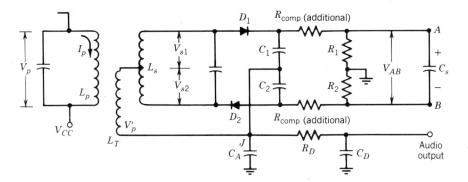

FIGURE 12.9-2 Basic ratio detector circuit. When the circuit action is explained, the additional compensation resistors (R_{comp}) can be neglected.

center frequency (f_c) of 4.5 MHz. The primary voltage is applied to the secondary center tap by L_T, the tertiary winding, which consists of a few turns of wire coupled tightly to the low end of L_p. Since L_T is tightly coupled to L_p, no phase shift occurs from L_p to L_T; the tertiary voltage is in phase with the primary voltage. Thus a voltage applied to the secondary center is in phase with the primary voltage. Let the tertiary voltage be V'_p, which is in phase with the primary voltage V_p. Let the primary current, secondary input voltage, and current be I_p, V_i, and I_s, respectively. The waveforms of i_p, v'_p, and v_i are shown in Figure 12.9-3, where

$$i_p = I_m \sin \omega t,$$

$$v_i = M \frac{di_p}{dt} = \omega M I_m \cos \omega t = V_m \cos \omega t,$$

and

$$v'_p = V'_{pm} \cos \omega t.$$

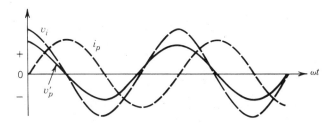

FIGURE 12.9-3 Waveforms of i_p, v'_p, and v_i.

At resonance or the center frequency f_c, the primary line current is in phase with V_p (or V'_p) while the circular current I_p lags V_p (or V'_p) by 90°. Let V_{s1} and V_{s2} be the two equal halves of the secondary induced voltage due to I_s, and let V_1 and V_2 be the forward voltages applied to D_1 and D_2, respectively. V_1 is the phasor sum of V'_p and V_{s1}. V_2 is the phasor sum of V'_p and V_{s2}.

Phasor diagrams for the detector circuit are shown in Figure 12.9-4. When input frequency f_{in} equals f_c, equal voltages ($|V_1| = |V_2|$) at 4.5 MHz are applied to the diodes, as shown Figure 12.9-4a. When $f_{in} > f_c$, I_s lags V_i by θ_L, and so $|V_1| > |V_2|$, as shown in Figure 12.9-4b. When $f_{in} < f_c$, I_s leads V_i by θ_C, and so $|V_1| < |V_2|$, as shown in Figure 12.9-4c.

When $f_{in} = f_c$, $|V_1| = |V_2|$, diodes D_1 and D_2 conduct equally. Current flow to the right through D_1 is equal to the flow to the left through D_2. Therefore no current flows in the tertiary winding, and C_A remains uncharged. Current only flows around the loop L_s-D_1-R_1-R_2-D_2, with equal voltage drops $|V_1|$ and $|V_2|$ across R_1 and R_2, respectively. The voltage between points A and B is $V_{AB} = V_1 + V_2$. Thus the stabilizing capacitor C_s (a few μF or more) charges to voltage V_{AB}. Since C_s charges fast and discharges slowly, it can hold V_{AB} constant.

When $f_{in} > f_c$, $|V_1| > |V_2|$, or $r_{d1} < r_{d2}$; hence D_1 conducts more than D_2. This indicates a positive output taken from point J. A current equal to the difference in D_1 and D_2 currents flows via the tertiary-winding terminal J. Capacitor C_A is charged positively, and the voltage developed on C_A represents the output of the detector.

When $f_{in} < f_c$, $|V_1| < |V_2|$, or $r_{d1} > r_{d2}$; hence D_2 conducts more than D_1. This indicates a negative output taken from point J or C_A charged negatively.

When an input FM signal is applied, the diode rectification causes voltage V_1 and V_2 to develop across R_1 and R_2, respectively. The polarity of V_1 and V_2 is such that they add to give V_{AB}. The average amplitude of the input signal determines the value of V_{AB}.

Now the sum of V_1 and V_2 equal to V_{AB} is held constant by the large capacitor C_s, so that as f_{in} changes to increase V_1, V_2 must necessarily decrease. When f_{in} varies, the ratio of V_1 to V_2 varies accordingly. The long time constant of the R_1-R_2-C_s circuit makes the ratio detector unresponsive to amplitude modulation. This is its inherent limiting action. If the conventional discriminator shown

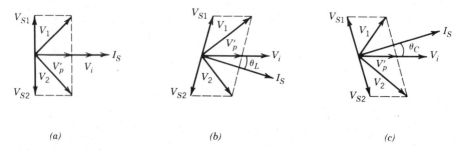

(a)   (b)   (c)

FIGURE 12.9-4 Phasor diagrams for a ratio detector: (a) $f_{in} = f_c$; (b) $f_{in} > f_c$; (c) $f_{in} < f_c$.

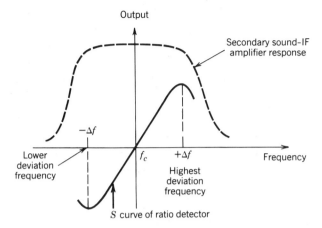

FIGURE 12.9-5 A typical plot of ratio detector output as a function of frequency. The dotted-line curve indicates secondary sound-IF amplifier response.

in Figure 12.3-2 is used as an FM demodulator, the response to amplitude modulation must be avoided by employing a separate limiter.

Ratio detector circuits can sometimes become "overstabilized" so that when an increase in amplitude suddenly occurs, the output of the ratio detector actually drops momentarily. This can be avoided by connecting compensation resistances (R_{comp}) as shown in Figure 12.9-2. These limit maximum diode current and reduce AM sensitivity at high signal levels.

Figure 12.9-5 shows a ratio-detector response curve or S curve. The central linear region is the active or useful portion of the curve. The output is zero when the input frequency equals the center frequency (f_c) or 4.5 MHz. When the input frequency increases, the output becomes positive and increases with the frequency to the positive peak of the S curve. This peak occurs at the upper limit of the IF amplifier bandwidth. If the frequency becomes higher, the input amplitude falls off quite sharply. This produces the downward turn of the response, forming the upper portion of the S curve. A similar response is obtained as the input frequency falls below the center frequency. The output becomes increasingly negative when the frequency becomes lower. The negative peak occurs at the lower limit of the secondary sound-IF amplifier bandwidth.

12.10 INTRODUCTION TO COLOR TELEVISION SYSTEMS

General Description

A commonly used color television receiver can be considered as a conventional black-and-white receiver plus a chroma section and a color picture tube, as shown in Figure 12.10-1.

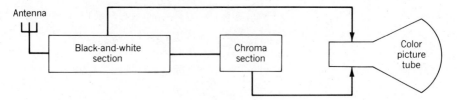

FIGURE 12.10-1   Basic arrangement of a commonly used color television receiver.

The term *chroma*, or *chrominance*, is used to indicate both *hue* and *saturation* of a color. The color itself is its hue or tint. The saturation indicates how little the color is diluted by white. White light can be regarded as a mixture or blend of the red, green, and blue primary colors in the proper proportions. Saturated colors are vivid or deep. Weak colors have little saturation. For instance, a vivid red is fully saturated or 100% saturated. As the red is diluted by white, the result is a desaturated red called pink. A fully saturated color has only its own hue, without white or any other components that could be added by the red, green, and blue of white. Each color has a certain wavelength. Different hues result from the visual sensations produced by different wavelengths of light. The color of any object is distinguished primarily by its hue. In color television, the chroma signal or C signal is the 3.58-MHz modulated subcarrier. This C signal includes the hue and saturation for all colors. Before modulation and after demodulation, the color information is in red, green, and blue color video signals. The range of these modulation frequencies of the color baseband is from practically 0 to 0.5 MHz. The amount of light intensity indicated by luminance is perceived by the eye as brightness. Some colors appear brighter than others. For example, a saturated green appears brighter than a saturated red. The total brightness of a color blend equals the sum of brightness of all the hues making up the particular blend. In color television, the luminance information is in the luminance or Y signal and the 3.58-MHz C signal is multiplexed with the Y signal, since both signals modulate the main picture carrier wave. The Y signal is sometimes called the *monochrome signal*. Color television is compatible with black-and-white television. Color TV receivers can use a monochrome signal to reproduce the picture in black and white. Black and white receivers can automatically process the complete color signal as if no chroma component were present.

In a color television receiver, the color subcarrier is placed approximately 3.58 MHz above the picture carrier, as shown in Figure 12.10-2. The C signal has two components—the I *signal* and the Q *signal*. I stands for "in phase," and Q stands for "quadrature phase." Both the I signal and the Q signal are modulated on the color subcarrier. The I signal has a vestigial sideband form, while the Q signal has a double sideband form. Both the I signal and the black and white signal are transmitted in vestigial sideband form. The I and Q signals have bandwidths of 0–1.5 MHz and 0–0.5 MHz, respectively. The I signal has the extra bandwidth between 0.5 and 1.5 MHz. This extra bandwidth is seldom

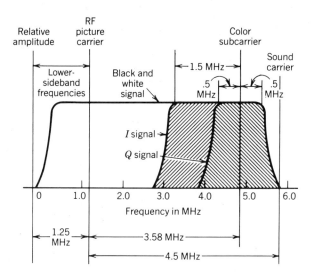

FIGURE 12.10-2 Bandwidths for the black-and-white signal and chroma (I and Q) signals in a color television receiver.

used in color receivers, since the color circuits are much simpler when all the color video signals have the same bandwidth of 0.5 MHz, the practical baseband for color.

Functions of *I/Q* Color Television Transmitter

A simplified block diagram for the I/Q color television transmitter is shown in Figure 12.10-3. The color camera receives red (R), green (G), and blue (B) light corresponding to the color information in the scene to produce the primary color video signals. The matrix unit combines the R, G, and B voltages in specific proportions to form three video signals—the Y, I, and Q signals:

$$Y = 0.30R + 0.59G + 0.11B, \qquad (12.10\text{-}1)$$

$$I = 0.60R - 0.28G - 0.32B, \qquad (12.10\text{-}2)$$

and

$$Q = 0.21R - 0.52G + 0.31B. \qquad (12.10\text{-}3)$$

Positive I signal is orange; negative I signal is cyan. Positive Q signal is purple; negative Q signal is yellow green. The Q signal modulates the color subcarrier 90° out of phase with the I signal. The negative signs for subtracting R, G, and B signals mean adding the video voltages in negative polarity. The bandwidth of 1.5 MHz for the I signal and 0.5 MHz for the Q signal are specified by the FCC.

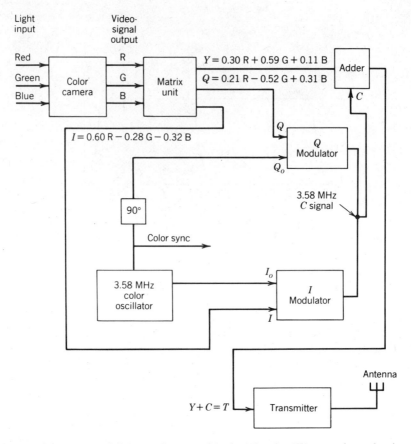

FIGURE 12.10-3 Simplified block diagram of basic I/Q color TV transmitter, showing its functions in encoding the 3.58-MHz C signal.

Color TV receivers may use R-Y and B-Y demodulators. The R-Y and B-Y signals are also in quadrature with each other, but they have a different phase angle than the I and Q signals. The hue of R-Y is a purplish red. Subtract Y from R to obtain the formula for the R-Y signal voltage:

$$R - Y = 1.00R - 0.30R - 0.59G - 0.11B,$$

or

$$R - Y = 0.70R - 0.59G - 0.11B. \qquad (12.10\text{-}4)$$

Similarly, obtain the formula for the B-Y and G-Y signal voltages:

$$B - Y = 1.00B - (0.30R + 0.59G + 0.11B),$$

or

$$B - Y = -0.30R - 0.59G + 0.89B, \tag{12.10-5}$$

$$G - Y = 1.00G - (0.30R + 0.59G + 0.11B),$$

or

$$G - Y = -0.30R + 41G - 0.11B. \tag{12.10-6}$$

Note that the R, B, and G signals are the video-signal outputs from the red, blue, and green camera tubes, respectively. If the red, blue and green chroma signals contain no brightness information, they are the R-Y signal, B-Y signal, and G-Y signal, respectively.

In the I/Q color TV transmitter, the I and Q signals are transmitted to the receiver as the modulation sidebands of a 3.58-MHz subcarrier, which in turn modulates the main carrier wave. The value of 3.58 MHz is selected as a high video frequency to separate the chrominance signal from low video frequencies in the luminance signal. The output from the 3.58-MHz color subcarrier oscillator is coupled to the I and Q modulators, which also have I and Q video signal inputs from the matrix unit. Each modulator produces amplitude modulation of the 3.58-MHz subcarrier, with a combined output containing the 3.58-MHz modulated C signal. The 3.58-MHz oscillator output Q_o coupled to the Q modulator is shifted by 90° with respect to I_o, thus keeping the I and Q signals separate from each other. To reduce interference at 3.58 MHz, the subcarrier is suppressed, and only the modulation sidebands are transmitted. Hence, for the detection of the C signal, the color TV receiver must have a 3.58-MHz subcarrier oscillator. A sample of the 3.58-MHz subcarrier is transmitted with the C signal as a phase reference for the color oscillator at the receiver. In color television, phase angle is hue. The frequency and phase of the 3.58-MHz oscillator at the receiver is controlled by the transmitted color sync burst of 8 to 11 cycles of the subcarrier. The C signal with the color information and the Y luminance signal are both coupled to the adder or color-multiplexer section. This section combines the Y signal with the 3.58-MHz C signal to form the total colorplexed video signal marked T. This signal is transmitted to the receiver by amplitude modulation of the picture carrier.

12.11   PRINCIPLE OF COLOR TELEVISION RECEIVERS

The color TV receiver shown in Figure 12.11-1 reproduces a color image when it is tuned to a color TV broadcast signal, and it reproduces a black and white

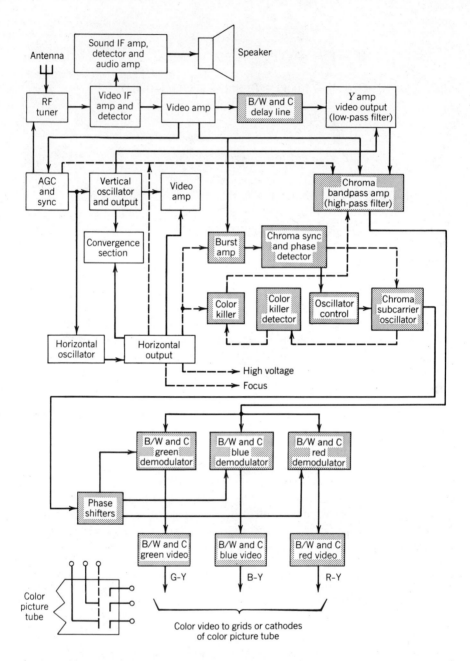

FIGURE 12.11-1 Block diagram for a color TV receiver. Shaded blocks indicate stages used only in a color receiver; - - - indicates the control signal or voltage path, and ——— indicates the signal path.

424

(B/W) image when tuned to a B/W TV broadcast signal. The shaded blocks indicate stages utilized only in color receivers. The color picture tube generally includes three electron guns. The three electron beams simultaneously scan the trios of three phosphors with which the screen is formed. The screen grows white when all three electron guns are operating. It grows yellow if the blue gun is cut off, and it grows green if both the blue and red guns are cut off. When the receiver is tuned to a color broadcast signal, the composite color signal is demodulated through the video detector to develop the envelope information. The 3.58-MHz component of this composite color signal is encoded into the Y component. The first step in signal processing consists in passing the composite color signal via a low-pass filter (a Y amplifier) and a high-pass filter (a chroma bandpass amplifier) to separate the Y information from the chroma information. The chroma demodulators respond to a band of video frequencies from 3.0 to 4.2 MHz. Since the chroma section has a considerably narrower bandwidth than the Y section, the chroma signal is comparatively delayed in its arrival at the color picture tube. Hence a delay line of typically 1 μs is inserted in the Y channel. In addition to the delay line, a color-subcarrier trap in Y channel is used to attenuate the response at 3.58 MHz.

The chroma bandpass amplifier has approximate signal amplitudes of 1.5 V and 4 V at its input and output, respectively. The bandpass amplifier operates in association with the color-killer section, which functions as an electronic switch. When the receiver is tuned to a color signal, the bandpass amplifier channel is switched on. When the receiver is tuned to a B/W signal, the bandpass amplifier is switched off by the color killer. This electronic switch is utilized to reject possible reference during B/W reception. The color killer section also serves as an automatic chroma control (ACC) arrangement during color reception. Thus the output chroma signal is held at about the same level whether the input signal is weak or strong.

The 3.58-MHz signal is coupled from the chroma bandpass amplifier to the chroma demodulators. Through the demodulator circuits, the chroma signals are decoded to recover the original color information. Since the subcarrier is suppressed at the color television transmitter, a local subcarrier oscillator should be employed at the receiver in order to reconstitute the complete chroma signal. The subcarrier is passed through phase-shifting circuits in order to recover the red, blue, and green chroma signals. The Y signal must combine with the complete chroma signal to form the complete color signal.

The color sync circuit contains the oscillator control and chroma-subcarrier oscillator sections. The color TV receiver can lose color sync without losing black and white sync action. On the other hand, if black and white sync is lost, color sync action is also lost.

As in the case of a B/W picture tube, a deflection yoke is mounted on the neck of the color picture tube. In addition, magnet assemblies are mounted behind the yoke. These magnets provide correct convergence adjustment for the three electron beams.

12.12 INTEGRATED CIRCUITS USED IN TELEVISION

The conventional symbol for an integrated circuit (IC) is shown in Figure 12.12-1. This symbol identifies only the terminals. Various integrated circuits used in television are briefly described in this section.

The μA3064 (SN76565) is a monolithic TV automatic fine-tuning circuit. It combines all the automatic fine-tuning circuitry, except transformers, in one integrated circuit. Systems with low-level IF amplifiers can now achieve tuning accuracies of ± 25 kHz due to the 3064's high sensitivity.

The TBA970 is a monolithic video amplifier for TV receivers. The circuitry includes a video preamplifier, dc contrast control utilizing a linear potentiometer (which can be ganged to the chroma gain control), and beam current limiting via contrast. Beam current limiting can be obtained with either positive or negative control voltage. Black level control is achieved by a clamped feedback circuit combined with the brightness control. Emitter follower output can be used to directly drive the video output stage.

The TDA1170 and TDA1270 are monolithic integrated circuits designed for use in TV vertical deflection systems. The TDA1170 is designed primarily for large and small screen B/W TV receivers and industrial TV monitors. The TDA1270 is designed primarily for driving complementary vertical deflection output stages in color TV receivers and industrial monitors.

The TBA920's horizontal oscillators are monolithic integrated circuits designed for TV receiver applications. They accept the composite video signal, separate sync pulses, and provide a sync output for the vertical integrator. Also incorporated is the horizontal oscillator along with two phase comparators, one to compare flyback pulses to the oscillator and the other for sync pulse comparison. The devices will interface with both SCR and transistor deflection systems.

The TBA510 chroma processor is a monolithic integrated circuit designed to perform the chroma amplifier function for TV receivers. A dc chroma gain control, which can be ganged to the receiver contrast control, is provided. Also incorporated is a variable gain automatic color control (ACC) stage, chroma blanking, burst gating, and burst output stage. Two single-output transistors provide burst and chroma output.

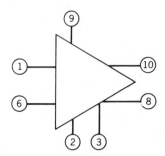

FIGURE 12.12-1 Conventional IC symbol.

The TBA560C luminance and chroma control combination is a monolithic integrated circuit used in the decoding system of color TV receivers. The circuit consists of a luminance and a chroma amplifier. The luminance amplifier input is matched to the delay line. Dc contrast, brightness, black level clamping, blanking, and beam current limiting functions are provided by the luminance amplifier portion of the circuit. The chroma amplifier performs functions such as gain controlled amplifying, chroma gain control tracking with contrast control, separate saturation control, PAL delay line driving burst getting, and color killing.

The TAA630S is a synchronous demodulator for direct drive of color video output stages. It is designed for use in color TV receivers operating on the phase alternate line (PAL) system. The circuit consists of two synchronous demodulators, a decoding matrix, and a PAL switch with an internal multivibrator and a color killer switch.

The TBA990 is an integrated color demodulator circuit for color TV receivers. It consists of two active synchronous demodulators for the R-Y and B-Y chroma signals, a matrix (producing the G-Y color difference signal), a PAL phase switch, and a flip-flop. It is suitable for dc coupled drive to the picture tube. When associated with the matrix integrated circuit (TBA530), it provides RGB output signals.

The TBA530 RGB matrix preamplifier is an integrated circuit for color TV receivers incorporating a matrix preamplifier for the RGB cathode or grid drive of the picture tube without clamping circuits. The chip layout has been designed to insure tight thermal coupling between all the transistors in each channel to minimize and equalize thermal drifts between channels.

The μA780 (SN76242) is a monolithic phase locked loop designed for use as a color TV subcarrier regenerator. This integrated circuit, which uses an automatic phase control (APC) loop, accepts the composite NTSC (National Television Systems Committee) color video signal, extracts the color subcarrier reference, and generates a CW (continuous wave) signal suitable for use as a chroma demodulation reference. Other features include control of the CW phase (tint) by a dc voltage, blanking of the CW output during burst time, and synchronous generation of an automatic color control (ACC) voltage.

The μA781 (SN76243) is a monolithic gain controlled chroma amplifier for color TV. The first section is a gain controlled chroma signal amplifier whose output is used to drive a subcarrier regenerator circuit. The gain of the second section is controlled by means of an external dc voltage to set the chroma level. In addition, the second stage may be gated off to provide "color killing" action in the absence of a color signal with the trip point of the gate adjusted externally.

The μA746 (CA3072) IC chroma demodulator demodulates the chroma subcarrier information contained in a color television video signal and provides color-difference signals at the outputs. The low voltage drift of the dc output insures excellent performance in direct-coupled chroma output circuitry.

The TDA1190 and TDA1190Z are silicon monolithic integrated circuits in 12-pin, plastic power packages. They supply all the components needed for TV sound systems, including the IF limiter amplifier, FM detector, AF preamplifier, and power output stage. The TDA1190 is specified for 5.5-MHz (PAL) sound systems, and the TDA1190Z is specified for 4.5-MHz (NTSC) sound systems.

12.13 BASIC TV TERMINALS

The TV terminal (TVT) is primarily used as the terminal of a computer or microprocessor (MPU) for displaying the output information on the TV screen. There are two methods for TV terminal interfacing: one is that using an RF modulator, and the other is of direct video technique. A typical TVT block diagram is shown in Figure 12.13-1.

The memory stores the external input information, which, after a suitable conversion, will be displayed when applied to the TV set. One of the commonly used memory is 2102 MOS RAM. In order that our information can be received by the usual TV set, the TVT must have a system timing circuit for converting and transmitting the signals. Since the timing frequency generally exceeds the extent capably handled by the microprocessor, it is necessary to add an auxiliary circuit containing 7 or 8 TTL or CMOS devices for the horizontal synchronization between line and line.

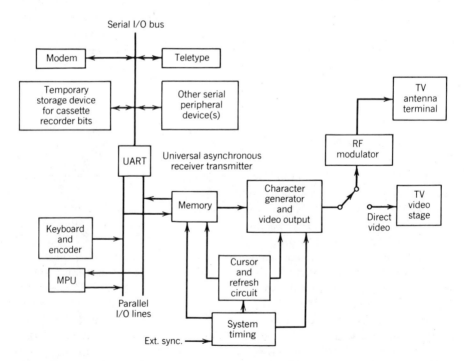

FIGURE 12.13-1 Block diagram of TVT using either an RF modulator or the direct video method.

The information stored in the memory is primarily of characters. The character generator and the video output stage are used to divide these characters into a group of dots, which appear on the screen as a view of complete characters. The character generator output, passing through a shift register, becomes a number of serial data. These data mix with the sync signal, resulting in the output video signal. This video signal can be applied to the video amplifier stage of a TV set, or modulated with an RF carrier and then sent to the antenna terminal.

The cursor and refresh circuits are used to send the new information into the memory and the corresponding display section. In order that the user knows the location of the next character appearing on the TV screen, the cursor circuit should display a flashing square at a suitable position on the screen. Both the cursor and refresh circuits have two different constructions: one is the frame-rate system in which only one character per sweep (1/60 s) is inserted, the other is DMA (direct memory access) system in which the information of a frame is transferred within 1/60 s. The DMA system is much faster and so more complicated than the frame-rate type.

In designing TVT we usually prefer parallel operation, with which 8 bits are conveyed each time. The parallel operation is not only faster and cheaper than the serial operation, but also directly compatible with the MPU data bus, keyboard, etc. When use a serial input, the input bits must pass through a serial interface so that the operation is converted to parallel type. An IC designed for this purpose is called the universal asynchronous receiver transmitter (UART).

ICs Used in TVTs

The following ICs are used in the basic TV terminal.

1. *Baud Rate Generator.*   The number of bits per second being transmitted is commonly referred to as the baud rate. Typical rates are 100, 150, 300, 600, and 1200 bits/sec or baud. The baud rate generator is usually crystal-controlled to maintain an exact reference frequency. This frequency is used to control the serial-parallel interface of the UART, telephone modem, teletype (TTY), cassette recorder, etc. Standard Microsystem 5061: 18 pins; two outputs; two reference frequencies. Motorola 14411: 24 pins; 16 different sync signals; frequency selection: X1, X8, X16, X64.

2. *Character Generator.*   The memory stores the characters of ASCII (American Standard Code for Information Interchange) and a few of timing instructions. They are all converted to the dot-character shapes by the character generator for the information transmitted to the TV. Signetics or GI 2513 converts 64 characters to 5×7 dot-characters; $V_{CC} = +5$ V. (GI = General Instrument.) Standard Microsystem 5004

contains a section for temporary output storage and uses 7×9 dot-matrix; $V_{CC} = +5$ V. Monolithic Memory 6072 converts 128 characters to 7×9 dot-characters.

3. *Keyboard Encoder.* It is used to convert the open and close keying signal into the ASCII code. GI and Standard Microsystem 2376; 3600: encoder for 88 keys.

4. *Output Drive Stage and Receiver.* AMD 26S10 (quad), equivalent to Motorola 3443, TI T5138. (AMD = Advanced Micro Devices; TI = Texas Instrument.) Motorola (TI, AMD) 1488, 1489 used in RS232-C standard.

5. PROM. Intersil 5600; NS5330; TI74188: BJT; permanent memory; capacity $= 32 \times 8$ bits; access delay $= 50$ ns. EPROM (Erasable PROM) Intersil (AMD) 1702; NS5202: memory capacity = 2048 bits ($= 256 \times 8$); data are erased when irradiated with ultraviolet light.

6. RAM with 1024 bits, the main memory. Intel (AMD) 2101; TI4039: 22 pins; 256 words; 4 bits/word. Intel (AMD, Fairchild, Synertek) 2102; Intersil 7552; TI4033: 16 pins, 1 bit/word. Intel 5101, CMOS, for long-term storage with battery source.

7. UART.
American Microsystem S1883
General Instruments AY-5-1012
Signetics 2536
Standard Microsystems COM 2502
Texas Instruments TMS 6012
Western Digital TR 1602

12.14 TVT OPERATION PRINCIPLE

Interlaced Scanning for TV Raster

The objective of a TVT is to convert the characters or bits into a video signal which can be received by a TV set. In a TV receiver a television picture is formed by scanning an electron beam across the face of the picture tube. A scanning line is one sweep of the electron beam from the left of the picture tube to the right and is initiated by the horizontal synchronization, as shown in Figure 12.14-1. The horizontal sync pulse causes termination of a line, horizontal retrace (flyback) of the electron beam back to the left side of the screen, and the start of the new line. During the retrace time the beam is blanked so that the retrace will not be seen. The time allotted for each complete line (including retrace) is 63.5 μs. Of this about 16% is taken by retrace, leaving 53.5 μs of usable line. Video information in the form of a voltage fed to the picture tube controls the brightness of the beam as it is swept across the screen.

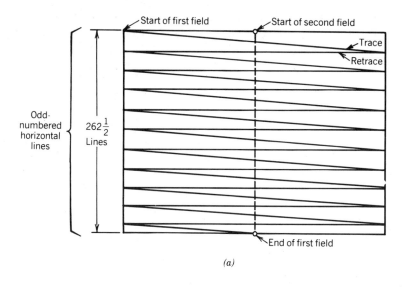

(a)

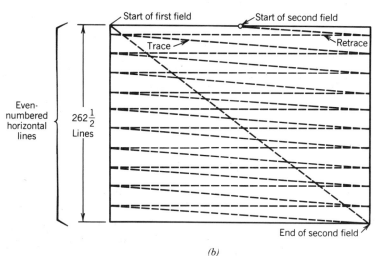

(b)

FIGURE 12.14-1 Interlaced scanning process for the basic raster: (a) structure of the first field, (b) structure of the second field. A complete frame contains two consecutive fields.

To minimize flicker, a frame (complete picture) is broken down into two fields. To trace out of a field, the electron beam is slowly deflected from the top of the screen to the bottom as it rapidly sweeps horizontal lines. This vertical sweep is allotted 16.67 ms (60 Hz), thus there are $262\frac{1}{2}$ lines in one field. In a manner similar to the horizontal sync, the vertical sync causes the beam to be returned from the bottom of the screen to the top to start a new field. The beam is blanked during vertical retrace which takes about 1250 μs. This leaves 242 usable lines in each field. The two consecutive fields are interlaced with each other, so that the horizontal lines of one field fit in between the horizontal lines of the other field. This is called the interlaced scanning method. The result is 30 frames every second of about 484 usable (525 total) lines each. The horizontal-line frequency is 15750 Hz for B/W TV and 15735 Hz for color TV.

The video composite signal applied to the input of the video amplifier contains the information needed to generate the vertical and horizontal sync, blanking, and video. The TVT may simulate this signal by supplying a composite waveform containing the same information normally present except sound.

Interlace   Modes

When the same information is stored in both fields of a scan frame, the display is in the interlace mode. This mode is effective in improving the quality of the picture. If the even lines of a character are displayed in the even fields and the odd lines displayed in the odd fields, the mode is an interlace sync and video mode. This effectively doubles the character time density on the screen since no information is duplicated. These are some restrictions in interlace mode.

Noninterlace Mode

Noninterlace mode commonly uses 262 or 264 scanning lines per frame. Since 60 frames per second are maintained precisely, the horizontal line frequency is 15720 or 15840 Hz. The horizontal and vertical sync signals must be locked each other.

Dot-Matrix Pattern

The scanning nature of the TV raster requires that the video (or brightness) information be sent in serial form to control the electron beam as it sweeps lines across the screen. Suppose, for example, that the character "K" is to be displayed as shown in Figure 12.14-2. The first line can be represented as 10001, ones signifying light spots (dots) and zeros signifying dark spots. The remaining six lines are similarly to be represented as a series of dots and dark spaces. When the seven lines are displayed one above the other, the 7×5 character "K" is seen. This method is called the dot-matrix. The tedious job of deciding where to put the dots (ones versus zeros) to generate a given character is done by an IC called the character generator (containing ROM and some control circuit).

Color Technique

A horizontal line frequency of nearly 15735 Hz may result in a good color display in a TVT system. Basically, it needs adding a color subcarrier (about 3.58 MHz) to the video output. In order to produce various colors the subcarrier should alter its phase with respect to itself after the last sync pulse has elapsed. Figure 12.14-3 shows the difference between the frequency responses of the color TV

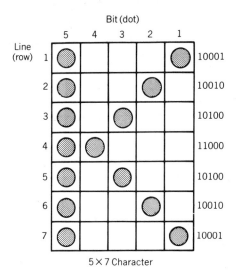

	Bit (dot)					
	5	4	3	2	1	

5 × 7 Character

FIGURE 12.14-2 Example of a dot-matrix pattern generated by a TV display.

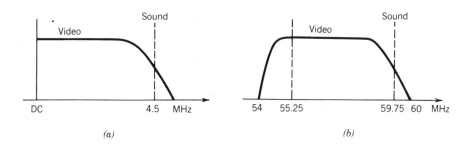

(a)

(b)

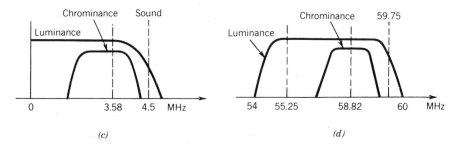

(c)

(d)

FIGURE 12.14-3 Difference between the frequency responses of the black and white TV and color TV. (*a*) B/W TV—base band. (*b*) B/W TV—second channel. (*c*) Color TV—Base band. (*d*) Color TV—second channel.

and the B/W TV. The B/W TV has a 4-MHz bandwidth of video baseband and a 4.5-MHz narrow sound-subcarrier. The video signal is amplitude-modulated, while the sound is frequency-modulated. The total RF channel is 6 MHz, as shown in Figure 12.14-3*b*. A color subcarrier is added at 3.57945 MHz, as shown in Figure 12.14-3*c*. The video signal to which no color subcarrier is added, is referred to as the luminance (brightness). The chroma information is modulated on the subcarrier. The color subcarrier with its modulation is called the chrominance, which determines the hue and saturation of a color.

Hue corresponds to the resultant subcarrier phase with respect to a reference point. Saturation corresponds to the resultant subcarrier amplitude. Brightness corresponds to the amplitude of the monochrome (black and white) video signal (Y).

After each horizontal sync pulse, there is a color burst of 8 cycles continuous at least, as shown in Fig. 12.14-4. This color burst is a sync signal, utilized as timing reference in the chroma circuits of a color TV set. The magnitude of the color burst should be 25% of the maximum amplitude, corresponding to the peak-to-peak value of the horizontal sync pulse.

When a color TV set is tuned to a color-broadcast signal, the waveform applied to the video detector consists of a composite color signal modulated on the picture-IF carrier. This composite color signal is demodulated through the video detector to develop the envelope information. By means of a bandpass

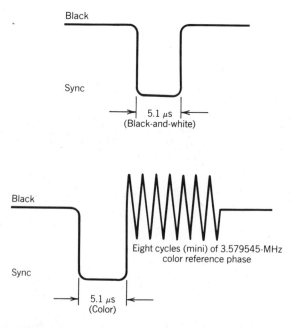

FIGURE 12.14-4 Color burst after each horizontal sync pulse.

amplifier (whose response peaked at the high frequency end of the video channel), the chroma signal is taken from the demodulated video signal. Then the chroma signal is synchronously demodulated by the chroma demodulators, which respond to a band of video frequencies from 3.0 to 4.2 MHz, with the 3.58 MHz as reference. The ratio of the currents through the red, blue and green electron guns in the picture tube is determined by the phase and amplitude of the chroma signal, which correspond to the hue and saturation of the color, respectively.

Because the chroma section has a considerably narrower bandwidth than the B/W section, the chroma signal is comparatively delayed in its arrival at the color picture tube. Hence a delay line is inserted in the B/W channel. The delay time provided is typically 1 μs. This delay provides optimum "color fit" in the black and white image. In addition to the delay line, a sound-subcarrier trap and a color-subcarrier trap are used to attenuate the responses at 4.5 and 3.58 MHz, respectively. The output luminance signals simultaneously modulate the cathodes of three electron guns to determine the brightness of the picture tube.

Basically, a color demodulator is a phase detector and an amplitude detector. After each horizontal sync pulse, the 3.58-MHz reference signal in the TV set immediately finds the color burst and compares with the latter at the phase detector, so that the reference signal and the color burst interlock. Colors corresponding to the phases and delays of color sync signals are listed in Table 12.14-1. Since the required colors are not many, the phase-shift of the reference signal can be accomplished by using appropriate digital gates rather than analog circuits.

TABLE 12.14-1
Colors Corresponding to Phases and Delays of Color Sync Signals

Color	Approximate Phase	Approximate Delay
Burst	0°	0
Yellow	15°	12 ns
Red	75°	58 ns
Magenta	135°	105 ns
Blue	195°	151 ns
Cyan	255°	198 ns
Green	315°	244 ns

REFERENCES

1. Liff, A. A., *Color and Black & White Television Theory and Servicing*, Prentice-Hall, Englewood Cliffs, New Jersey, 1979

2. Roddy, D. and J. Coolen, *Electronic Communications*, 2nd ed., Reston Publishing Company, Reston, Virginia, 1981, Chap. 18.

3. Grob, B., *Basic Television Principles and Servicing*, 4th ed., McGraw-Hill, New York, 1975.

4. Herrick, C. N., *Television Theory and Servicing*, 2nd ed., Reston Publishing Company, Reston, Virginia, 1978.

5. Bierman, H. and M. Bierman, *Color Television Principles and Servicing*, Hayden Book Company, Rochelle Park, New Jersey, 1973.

6. Chiang, H. H., *Electrical and Electronic Instrumentation*, Wiley, New York, 1984, Chap. 15.

7. Scott, B. and M. Bergan, Linear One-Chip Modulator Eases TV Circuit Design, *Electronics*, pp. 76–78, Dec. 29, 1981.

8. *Linear Circuit Data Book*, RCA Semiconductor Corp., Somerville, New Jersey, 1978.

9. Linear Integrated Circuit Data Book, National Semiconductor, Santa Clara, California, 1980, Sec. 10.

10. *Linear Integrated Circuit Data Book*, Motorola Inc., Phoenix, Arizona, 1979, Sec. 5.

QUESTIONS

12-1. Electromagnetic waves travel approximately ——— cm/s; the wavelength of a 200-MHz signal is ———.

12-2. The impedance of a standard television twin lead is ——— Ω.

12-3. The TV VHF range is from ——— to ——— MHz for channels 2 to 13; each TV channel is ——— MHz in width.

12-4. IF picture carrier frequency is ——— MHz; IF sound carrier frequency is ——— MHz; intercarrier sound frequency is ——— MHz.

12-5. IF picture carrier frequency is a difference of ——— and ———; IF sound carrier frequency is a difference of ——— and ———.

12-6. TV pictures are transmitted at a rate of ——— frames/s.

12-7. The TV picture is made up of ——— lines/frame.

12-8. Each visible raster includes ——— lines.

12-9. Each field includes ——— lines.

12-10. The line frequency of the TV raster is ——— Hz.

12-11. The scanning rate for the electron beam on the face of the picture tube is ——— lines/s.

12-12. The complete TV signal consists of ———.

12-13. A TV station employs an ——— signal for the picture and an ——— signal for the sound.

12-14. The picture signal has a bandwidth of about ——— MHz.

12-15. The bandwidth of the intercarrier sound signal is about ——— kHz.

12-16. The purpose of the AGC voltage in Figure 12.2-1 is to control ———— .

12-17. The antenna signal passes through ———— before it is applied to the RF amplifier.

12-18. Draw the equivalent neutralization circuits of the Q_2 and Q_3 stages in Figure 12.2-1.

12-19. The purpose of the AFT varactor-diode shown in Figure 12.3-1 is to ———— .

12-20. Describe the operation of the AFT circuit shown in Figure 12.3-1.

12-21. Explain the operation of the conventional discriminator shown in Figure 12.3-2.

12-22. What is the purpose of diode D_{21} used in the picture IF amplifier of Figure 12.4-1?

12-23. Wide band is obtained in the picture IF amplifier by ———— tuning.

12-24. Explain the operation of the AGC keyer (Q_{21}) and AGC amplifier (Q_{20}) system. (See Figures 12.5-1, 12.4-1, and 12.8-1.)

12-25. How does the first PIF amplifier stage (Q_4, Figure 12.4-1) obtain its V_{BE} forward-bias voltage?

12-26. What are the purposes of the series peaking coils L_{31} and L_{32} and the shunt peaking coil L_{33} used in the video detector circuit of Figure 12.5-1?

12-27. How does most of the intercarrier sound signal branch into the sound takeoff channel?

12-28. Explain the functions of the contrast and brightness controls in the video amplifier circuit.

12-29. How are the vertical and horizontal sync pulses processed after they have left the sync separator (Q_9 in Figure 12.6-1)?

12-30. How does the vertical oscillator of Figure 12.6-1 work?

12-31. Explain the horizontal AFC action in the circuit of Figure 12.7-1.

12-32. How does the horizontal oscillator of Figure 12.7-1 work?

12-33. The comparison sawtooth voltage at the horizontal AFC circuit is originally a flyback pulse from the primary of the horizontal output transformer in Figure 12.8-1. How does it become such a waveform?

12-34. What is the purpose of the damper diode D_{61} used in the horizontal output section of Figure 12.8-1?

12-35. Explain the operation of a ratio detector.

12-36. What is a monochrome signal?

12-37. The bandwidth required for color transmission is about —— that required for black and white transmission.

12-38. What is the color-burst signal?

12-39. What is the purpose of the delay line in the block diagram of Figure 12.11-1?

12-40. What is the function of the color killer section in a TV receiver?

12-41. What is the basic function of a color demodulator?

PROBLEMS

12-1. Calculate: (a) frequencies corresponding to the periods of 33333 μs, 16667 μs, 63.5 μs, and 53.5 μs; (b) difference frequencies between 83.25 and 87.75 MHz, between 45.75 and 41.25 MHz, between 83.25 and 86.83 MHz, and between 45.75 and 42.17 MHz.

12-2. In a television receiver, the local oscillator frequency for channel 10 (from 192 to 198 MHz) is 239 MHz. Find the picture carrier frequency and sound carrier frequency.

12-3. The regenerative feedback loop in Figure p12-3 is opened to help one analyze the oscillator circuit. Write (a) the barkhausen criteria for oscillation, and (b) an expression for the loop gain (v_1/v_t).

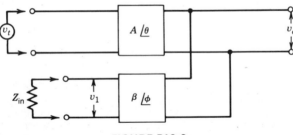

FIGURE P12-3

12-4. An oscillator circuit is shown in Figure p12-4. Verify that (a) the oscillator frequency is given by

$$\omega_o = \frac{1}{\sqrt{C_1 C_2 L/(C_1 + C_2)}},$$

and (b) $h_{fe} \geq C_2/C_1$ if oscillation is to occur.

> *Hint.* (1) The approximate equivalent circuit is shown in Figure p12-4b.
> (2) Reasonable assumptions:

$$R_L \gg 1/\omega C_2; \quad R_1//R_2 \gg h_{ie}; \quad h_{re} \simeq 0; \quad 1/h_{oe} \gg R_L.$$

Then the circuit in Figure p12-4b can be simplified.
(3) Loop gain $|A\beta| = v_1/v_t \geqslant 1$ for oscillation. Node analysis is recommended.

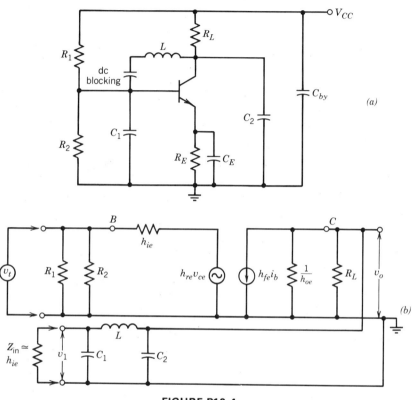

FIGURE P12-4

12-5. A 221-MHz oscillator has a 0.6-μH coil in the tuned circuit. Find the capacitance needed.

12-6. Show that a voltage of 3 V under no signal condition will appear at the test point in Figure 12.3-1b.

12-7. Draw the equivalent neutralization bridge for the neutralization arrangement of the final PIF stage (Q_6) in Figure 12.4-1.

12-8. The circuit in Figure p12-8 is used to pass the PIF ($= 45.75$ MHz) signal to the video circuits and prevent the sound IF ($= 41.25$ MHz) signal from passing. Calculate the reactances and explain the circuit action. Does the circuit achieve its purpose?

Answer. At 45.75 MHz, $X_{C2p} = X_{Lp} = 1.739$ kΩ; at 41.25 MHz, $X_{C1s} = X_{Ls} = 1.568$ kΩ.

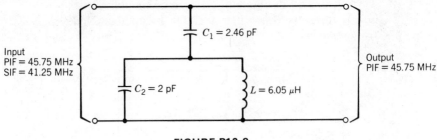

FIGURE P12-8

12-9. Figure p12-9a shows a video amplifier. It consists of two transistors (Q_1 and Q_2) coupled by an LC series circuit. Q_1 is regarded as a 15-V generator with a 200-Ω internal resistance, as shown in the equivalent circuit of Figure p12-9b: (a) find the voltage V_{BE} at 2.5 MHz; (b) find V_{BE} at 4.5 MHz; (c) compare both the V_{BE} values and explain the circuit function when used in a TV receiver.

Answer. (a) At 2.5 MHz, V_{BE} = 1.259 V $\underline{/65.19°}$; (b) at 4.5 MHz, V_{BE} = 3 V $\underline{/0°}$; (c) the circuit is used to separate sound and picture signals.

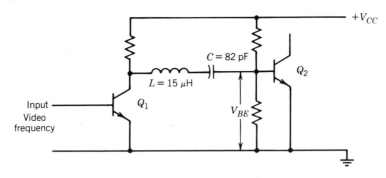

FIGURE P12-9A

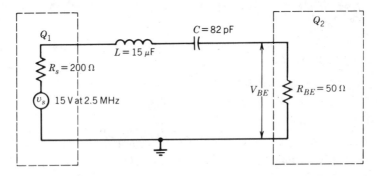

FIGURE 12-9B

12-10. Design a blocking oscillator as a horizontal oscillator.

12-11.   Design a limiter stage to clip amplitude-modulation variations for a conventional discriminator used as an FM detector.

12-12.   There is a practical *RC* low-pass filter in Figure 12.8-1. Change it to a simplified equivalent *RC* low-pass circuit.

12-13.   (a)   Draw the whole circuit of the TV receiver including the sections shown in Figures 12.2-1, 12.4-1, 12.5-1, 12.6-1, 12.7-1, 12.8-1, and 12.9-1.

(b)   Trace the dc bias circuits and input and output signal paths contained in these sections, and then explain the operation of the circuitries that you suppose most important.

13

PULSE-HEIGHT ANALYZERS

13.1 SINGLE-CHANNEL PULSE-HEIGHT ANALYSIS

Introduction

The height of the output pulse from a linear amplifier connected to a radiation detector is proportional to the detector pulse amplitude, and hence to the energy dissipated by the nuclear radiation within the detector. The energy-distribution curves (the nuclear spectra) may then be obtained by measuring the amplitudes of the output pulses. The linear or analog pulses now require selection by pulse height. Once the linear pulses have been selected according to pulse height, their amplitude is of no further interest and they are fed into a trigger circuit; for each pulse, this circuit gives an output "logic" pulse of standardized height and duration. The logic pulses carry time information about the original nuclear events, and so they are still randomly distributed in time. The preferred practice standards recommended by the Nuclear Instrument Modules (NIM) Committee for logic signals to ensure compatibility of interconnected circuit modules are given in this section.

For medium-speed positive logic signals, module outputs must deliver -2 to $+1$ V (4 to 12 V) and module inputs must respond to -2 to $+1.5$ V (3 to 12 V) for logical O (1). For fast negative logic signals, the NIM standards relate to the case of an output working into a 50-Ω impedance and an input having an input impedance of 50 Ω to provide matching for an interconnecting cable of this characteristic impedance; in such a case, module outputs must deliver $+1$ to -1 mA (-14 to -18 mA) and module inputs must respond to $+20$ to -4 mA (-12 to -36 mA) for logical 0 (1).

The measuring instrument for obtaining pulse-height-distribution curves is known as a *pulse-height analyzer*. This apparatus is used to determine the total number (or sometimes the rate of arrival) of pulses whose heights fall in selected intervals or channels throughout the range of pulse heights of interest. If this analysis is made by sampling one channel at a time, the apparatus is called a *single-channel pulse-height analyzer* (SCA). Thus the SCA is capable of recording only those pulses falling within a single channel or section of an energy spectrum; all other pulses are rejected.

SCA System

A block diagram of an SCA is shown in Figure 13.1-1. This arrangement includes two discriminators. They are the *lower level discriminator* (LLD), with the channel level E, and the *upper level discriminator* (ULD), with the level $E + \Delta E$. Here E is the threshold setting and ΔE is the channel window or window width; this is a small, usually adjustable difference in the triggering levels of ULD and LLD. These discriminators are followed by the anticoincidence circuit (inhibitor), which is arranged to pass a signal only when a pulse

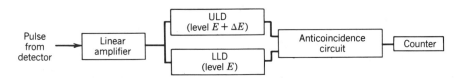

FIGURE 13.1-1   Block diagram of a single-channel analyzer (differential mode).

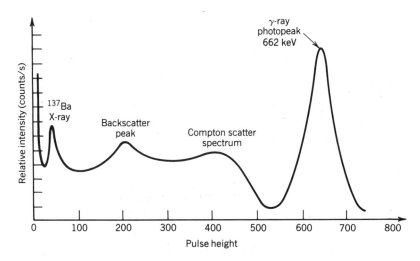

FIGURE 13.1-2 Typical differential spectrum of cesium-137 radiation source from a NaI(Tl) scintillation detector. The gamma-ray photopeak, broad Compton scatter spectrum, and ^{137}Ba X-ray are shown.

passes the LLD but not the ULD and to reject all signals that simultaneously pass both the LLD and the ULD. Therefore the SCA generates an output pulse if and only if an input pulse is received that has an amplitude within a selected voltage window ΔE. Thus the counter (scaler) registers only those pulses whose height lies in the interval or window ΔE.

In obtaining a differential spectrum, a series of 20 observations can be made in which each measurement of intensity is taken in 0.5-V increments for a 0.5-V window width over a 10-V range. Plotting the intensity (or count rate) versus the voltage or pulse height (channel level or threshold setting) yields a differential spectrum. These increments are referred to as *channels*, and 20 channels are measured. Single-channel analyzers are used to make the measurements a single channel at a time, while multichannel analyzers allow pulses in all channels to be recorded simultaneously.

The analyzer of Figure 13.1-1 is operated in a differential mode, and so is a differential pulse-height analyzer. Figure 13.1-2 shows the schematic representation of a spectrum from a NaI(Tl) scintillation detector obtained with a cesium-137 (^{137}Cs) radiation source.

13.2 TRANSISTOR CIRCUIT OF TYPICAL SINGLE-CHANNEL PULSE-HEIGHT ANALYZER

Figure 13.2-1 shows the basic transistor circuit of a typical single-channel pulse-height analyzer (SCA). The lower-level discriminator circuit consists of a Schmitt trigger (Q_1 and Q_2), a gate (an OR gate containing Q_4 and Q_5 with an emitter follower Q_3), and a feedback amplifier (Q_6 and Q_7). The upper-level discriminator circuit consists of a Schmitt trigger (Q_8 and Q_9), a stretcher (Q_{10} and Q_{11}), and a reset stage (Q_{12}). Both discriminator circuits are connected to an anticoincidence circuit (Q_{13} and Q_{14}). The analyzer generates an output signal when the lower-level trigger recovers. The lower gate and the feedback amplifier are combined into a feedback system to force the trigger to recover only at the zero input level, which is the condition for zero time shift. The waveforms in the analyzer are shown in Figure 13.2-2. The stretcher is a one-shot multivibrator that generates a rectangular output pulse when the input pulse from the amplifier is over the upper level $E + \Delta E$. The reset stage terminates the rectangular pulse produced by the stretcher when the lower-level trigger recovers.

The source of the window supplies a positive voltage to the base of Q_9 in the upper discriminator. The source of the threshold supplies a negative voltage to the bases of the *npn* transistors Q_8 and Q_1 in the upper and lower discriminators, respectively. Then both Q_8 and Q_1 in the Schmitt triggers are normally cut off, and both Q_9 and Q_2 are normally conducting. The anticoincidence circuit consists of Q_{13} and Q_{14} connected in series, and the emitter junction of Q_{13} is normally conducting while Q_{14} is cut off. When a positive pulse that has an amplitude within a selected voltage window is applied to the bases of Q_1 and Q_8, it is not sufficient to trigger the upper Schmitt circuit, and thus no output

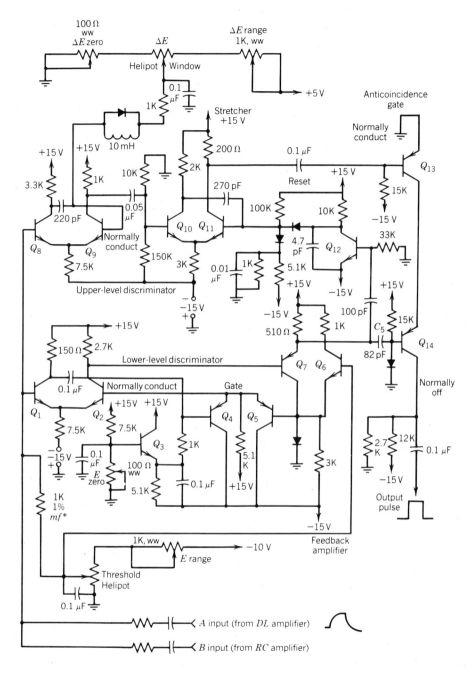

FIGURE 13.2-1 Basic circuit of a typical single-channel analyzer: Q_1, Q_2, Q_8, and Q_9 = GA568; Q_3 and Q_6 = 2N1306; Q_4 and Q_5 = 2N1307; Q_7 = 2N1143; Q_{10} and Q_{11} = 2N1308; Q_{12} = 2N835; Q_{13} and Q_{14} = 2N779. *mf = metal film. (Refer to H. H. Chiang, *Basic Nuclear Electronics*, Wiley, New York, 1969, pp. 206–207.)

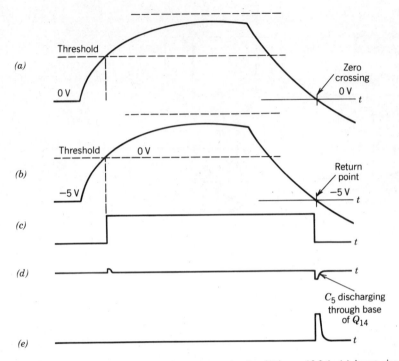

FIGURE 13.2-2 Waveforms in the basic analyzer circuit of Figure 13.2-1: (*a*) input signal at point *A* or *B*; (*b*) signal at base of Q_1; (*c*) signal at emitter of Q_7; (*d*) signal at base of Q_{14}; (*e*) output pulse from collector of Q_{14}. Note that the base line is assumed to be at 5 V (threshold potentiometer at 500 divisions). The maximum dynamic range of the analyzer is 0.1 to 10 V.

pulse is produced by the upper discriminator. However the lower discriminator is triggered; thus an output positive pulse is produced in the driver stage (Q_{14}) by the trailing edge of the discriminator output (emitter of Q_7).

When the input pulse has an amplitude above the upper level, the upper-level discriminator triggers, which in turn triggers the stretcher. The stretcher cuts off Q_{13}, preventing the output from being generated.

The analyzer circuit of Figure 13.2-1 has a maximum dynamic range of 0.1–10 V, a temperature stability of 0.5 mV/°C, and an output-pulse time stability of less than 10-ns shift for a 1- to 10-V double delay-line input pulse with a 0.25-μs or less rise time with the 3-V output.

13.3 SINGLE-CHANNEL PULSE-HEIGHT ANALYZERS BUILT ON LINEAR INTEGRATED DIFFERENTIAL COMPARATORS

Dual IC Comparator Used as Discriminator

The single-channel pulse-height analyzers built on IC differential comparators can have a large dynamic range of the threshold level with an allowable input pulse width of 50 ns or less. This results primarily from the low input offset voltage, high temperature stability, large differential input voltage, and short response time of the comparators. In many applications, both very high accuracy and speed, with which the comparator can distinguish between an input signal and the reference signal, are required; more important, the comparator should deliver only one pulse at its output in response to each input pulse. It is very often impossible to fulfill these demands in manufactured integrated comparators.

However these difficulties can be overcome by the introduction of two positive-feedback paths in a dual IC comparator. Both paths are from the common output of the comparator; one is a resistor connected to the noninverting input of the first comparator, and the second is a capacitor connected to the noninverting input of the second comparator. The first feedback path improves the voltage resolution of the comparator; although it makes the circuit very sensitive to any noise appearing on the signal to be discriminated, the second feedback path and some hysteresis in the transfer characteristic will cause the comparator to have excellent voltage resolution for short pulses and less sensitivity to noise.

Analyzer with Threshold of Each Discriminator Controlled Independently

In the analyzer circuit of Figure 13.3-1, one dual IC comparator is used as the lower level discriminator (LLD) and the second is used as the upper level discriminator (ULD). The level of each discriminator can be controlled independently for operation in differential (window) mode. The output pulse of the analyzer should appear only when the input pulse amplitude exceeds the lower level threshold and not the upper one. The anticoincidence circuit may use an IC dual one-shot with clear (such as SN74123) and other gates. The analyzer may accept either unipolar or bipolar input pulses within the threshold range from 50 mV to 10 V. The output pulse is delivered at the end of the analyzed pulse. Thermal stability of the LLD and ULD thresholds can be better than 0.1 mV/°C.

The analyzer circuit in Figure 13.3-1 can be modified as follows. The two separating networks are shorted, and the input pulse is directly applied to both noninverting inputs of the discriminators via the resistor R. An LLD threshold-setting network is connected referring to that shown in Figure 13.2-1. The LLD reference voltage V_{ref} is applied to the inverting terminal via a voltage follower, whose output connects a resistor R_1 to the inverting input of a summer. Similarly a voltage follower with a window control input ($-\Delta E'$) also has an output resistor R_1 connected to the inverting terminal of the summer. A third resistor R_1 is connected between the output and the inverting input of the summer. Then the summer output applied to the ULD inverting input is $V'_{ref} = \Delta E = \Delta E' - V_{ref}$. The window ΔE is a positive voltage, usually less than 1 V.

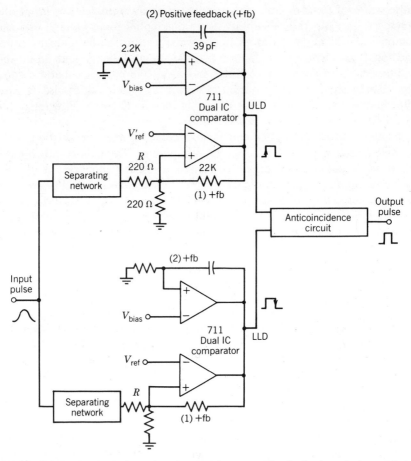

FIGURE 13.3-1 The single-channel analyzer with two positive feedback paths in an IC dual comparator used as a discriminator.

13.4 MULTICHANNEL PULSE-HEIGHT ANALYSIS

Introduction

Nuclear pulse spectrometry may be carried out using either a single-channel or a multichannel pulse-height analyzer. The single-channel analyzer (SCA) is a very useful tool for routine counting and for scanning samples to determine the energy spectra of radioactive sources. About 15 to 30 min are required per scan. This is because the SCA is capable of recording only those pulses falling within a single channel while all other pulses are rejected. Thus to cover 100 channels, one must count point by point 100 individual channels. This takes time. Therefore when energy spectra of low intensity are studied, inordinately long counting times are required to obtain a certain accuracy. This time element makes the use of an SCA impractical for low-activity samples, as well as for specimens with short half-lives.

Multichannel analyzers (MCAs) help to avoid this limitation. In an MCA, determinations of the counting rates can be made simultaneously in all channels; that is, MCAs are capable of recording virtually every pulse from an energy spectrum as the pulses occur. The only pulses not recorded are those that occur while the analyzer is busy handling a previously acquired pulse. Since such analyzers can be constructed to require analysis time on the order of 10 μs/ pulse, few pulses are ignored when reasonable source intensities are employed (up to about 30,000 pulses/s).

The analyzers usually have more than 100 channels for single-parameter devices, and thus time can be saved when all channels are permitted to accumulate information. The result of a single-parameter multichannel analysis is a spectrum, usually in the form of a histogram; the number of events is given as a function of one or more of such parameters as the height of the pulse, its shape, and its time of occurrence, or of the height, shape, and time of occurrence of coincident or otherwise time-related events. Multiparameter multichannel analysis has for some time been used by the larger laboratories. In determining a complex nuclear decay scheme, one can examine the source with two detectors whose outputs serve as the two inputs to a two-parameter analyzer. When the events involving coincident pulses in the two counters are stored, a three-dimensional record from which a great deal of information regarding the inter-correlation of the events in the two detectors can be extracted results.

All single-parameter multichannel analyzers consist of some or all of the following elements: (1) an analog-to-digital converter that associates the height of each input pulse with a specific channel; (2) a memory- or data-storage device that preserves the number of pulses that fall in their associated channels and, thus, contains information on the number of pulses in each channel until this information is recorded otherwise; (3) a data-display device that gives the experimenter information about the data stored in memory without removing it from memory; and (4) auxiliary data readout devices such as curve (X, Y) plotters, paper-tape and card punches, magnetic tape, typewriters, and so on.

Basic Principles of Operation

Figure 13.4-1 shows the chief components of a modern one-parameter MCA. The most widely used system employs the pulse-height-to-time converter as the analog-to-digital converter (ADC). This converter was developed by Wilkinson, Hutchinson, and Scarrott. Figure 13.4-2 illustrates the process of the Wilkinson type of analog-to-digital conversion. First, a small capacitor is charged up to the peak voltage of the pulse; it is then discharged at constant current. While the discharge is in progress, clock pulses from a stable oscillator are counted by a scaler; the number of clock pulses counted are proportional to the time the capacitor took to discharge and hence to the original height of the pulse.

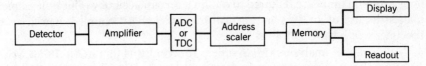

FIGURE 13.4-1 Chief components of a one-parameter multichannel analyzer. ADC and TDC stand for analog-to-digital converter and time-to-digital converter, respectively.

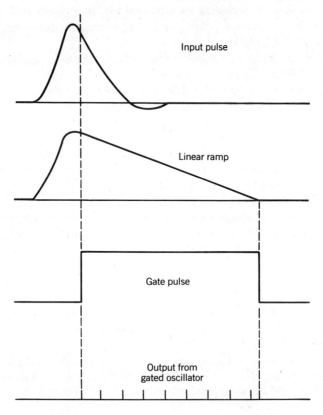

FIGURE 13.4-2 Illustration of the discharge type of analog-to-digital converter.

Another way of looking at this is that a capacitor is charged linearly until it reaches the height of the stretched input pulse. By starting the ramp in synchronism with a clock pulse, we can avoid the one-channel uncertainty in the start time. The number of clock pulses that occurred before the oscillator gate closed is the number of the channel that the count is to be stored in. This number is called the *address*.

One-parameter analyzers conventionally use memories consisting of three-dimensional arrays of small ferromagnetic cores. Typically the memory cores of one address, as selected by the address scaler (register), are read out during the read current pulse when the information from the cores is transferred to the memory register. Thus a count is stored by reading out the contents of the appropriate channel, adding one, and restoring the result. This interrogation and the restorage process takes perhaps 10 to 30 μs. But a much greater contribution to pulse-height-analyzer dead time is the result of the conversion process. The faster clock oscillator circuits in general use give pulses at a 4- or 5-MHz rate, making the conversion time for a pulse in channel 400, for instance, equal to 80 or 100 μs. Thus it is necessary to prevent new pulses from entering the converter while it is converting, so that storage rates are limited to less than 10^4/s.

In time-of-flight work, the type of time-to-digital converter (TDC) used depends to some extent on the magnitude of the flight times involved. For flight times running, say, 20 μs or longer, it is feasible to start a clock oscillator at time zero (when the neutron starts its flight) and to stop it when the neutron is detected. The count is then stored as described previously. For shorter flight times, ranging down into the nanosecond region, digital circuits are not fast enough. However one can start a linear ramp at time zero and stop it when the neutron arrives. The result is a pulse whose amplitude is proportional to the time of flight and which can be analyzed by conventional pulse-height analysis methods.

To convert the digital information of the address and memory registers from digital to analog form, two digital-to-analog converters (DACs) are usually provided. These are the horizontal and vertical DACs, which are referred to as the *address decoder* and *data decoder*, respectively. The operation principle of DACs was explained in Section 9.10.

Semiconductor (Integrated-Circuit) Memories Versus Magnetic Core Memories

The chief disadvantages of magnetic core memories are that the cores are susceptible to influence from electrical noise and difficult to string on their wires. As more advanced types of semiconductor memory become available, particularly nonvolatile devices, the use of cores will gradually diminish.

The two main types of semiconductor memory are the read/write store usually termed RAM (random access memory) and the read only memory, ROM. There are bipolar memories, both RAM and ROM, various types of MOS memories, both RAM and ROM, and many varieties of programmable read only memory (PROM). Semiconductor RAM circuits are available in two basic forms; they are classified as static or dynamic. In static memory circuits, power is applied constantly to all flip-flop elements to retain stored data. Dynamic memory circuits rely on semiconductor capacitors, not on flip-flop arrays, for bit storage. Microscopic capacitors present data by being charged for logic 1 and discharged for logic 0. With such a system, power need not be constantly applied to a capacitor for it to retain its charge (and thus its data).

Because charge eventually dissipates in any capacitor, a dynamic memory requires periodic refreshing to counteract the slow discharge. The chief advantages of dynamic memories over static ones are smaller basic storage cell size, lower power requirement per bit, lower heat dissipation per bit, and lower cost per bit. Offsetting these advantages is the requirement for additional circuitry to enable refreshing.

In general, bipolar memories are faster than static MOS memories, but they have smaller capacities per chip (currently 256 bits arranged as 256 × 1, or 64 bit arrays of 16 × 4 bits, with access times in the region of 120 ns for low power units). So-called dynamic MOS memories, however, are larger in capacity than bipolar units and have faster access times. Static RAMs are designed to overcome the refreshing requirement for dynamic ones, but they accomplish this only at the expense of speed, as might be anticipated. In general, unless it is used in conjunction with a microprocessor, when it will have a special function, RAM will not be used in large quantities in interfacing projects.

The commonly used semiconductor RAMs consist of an array of flip-flop memory cells. They are volatile memories: if their power sources are removed, any stored data will be lost. Read-only memory (ROM) may be viewed as a form of random access memory that can be encoded permanently with data and is unchangeable after manufacture. While ROM is nonvolatile, nevertheless it cannot be written with new data, and therefore its applications are more limited than those for read/write RAM. Usually ROM is used for permanent storage of instructions. A popular PROM technique utilizes a special mask applied to a microcircuit transistor matrix at the time of manufacture. The mask represents a data pattern that is to be recorded.

Erasable PROM (EPROM) are very sophisticated devices in which the recorded data is nonvolatile but erasable. A commonly used technique involves the recording of data in insulated gate MOS transistors, which are the floating gate avalanched injection MOS devices. The basic storage process involves charging up the gate of the transistor, thereby causing the device to conduct; since the gate is insulated, the charge will not leak and the conducting path through the transistor will remain until the charge is removed, whether or not power is applied to the system. Exposure of the circuit to sufficiently high intensity ultraviolet (UV) radiation, or to similar short wavelengths such as "soft" X-rays, will cause a photocurrent to flow from the insulated gate to the silicon substrate, thus restoring its original uncharged condition. Programming (charging the relevant gates) is accomplished by applying a high voltage pulse (about 50 V in magnitude) between the substrate and the drain of the particular transistor that is to hold a conducting state. This induces an avalanche of electrons through the p^+ channel region onto the insulated gate, which remains charged after the pulse subsides. Thus, if a transistor has a charged gate, it will appear on; otherwise it will be off. Read-out of the state of the memory cell (transistor) is achieved by applying a negative voltage, well below the programming threshold, to the drain (of p-channel type) to sense the conductance of the device. Since no insulator is absolutely perfect, the charge on the floating gate will eventually leak away, resulting in the loss of the programmed state. Tests have indicated that of the amount of charge stored on a typical programmed gate (about 10^{-7} coulomb/cm^2) approximately 30% will decay over 10 years at temperatures constantly in excess of 100°C, well within the normal lifetime requirements of such devices. Memories of this type can readily be made with capacities of 2048 bits, and capacities of 8192 (1 K by 8 bits) are available. The 1702A (256 × 8-bit ROM matrix) is an example of an EPROM that can be erased by exposure to high-intensity ultraviolet light for approximately 10 to 20 min. When the 1702A is erased, its storage cells contain all 0s (which are lows). The 1702A is available in a dual in-line package integrated-circuit (DIP) with a quartz lid over its storage cells to allow erasure.

13.5   MULTICHANNEL PULSE-HEIGHT ANALYZERS

Single-Parameter Multichannel Analyzers

In one-parameter analyzers the number of channels ranges from about 100 to about 1000, with capacities of 10^5 or 10^6 counts per channel. Most instruments offer selective storage—the ability to divide the memory into halves, quarters, or eighths, and to "route" the input to any desired segment. Thus it is possible to store up to eight spectra in the same memory. Background subtraction can be accomplished by first storing background in the negative mode and then counting the sample for the same length of time in the positive mode. Automatic programming may also be carried out when counting a series of samples using an automatic sample changer. Some systems permit *spectrum stripping*—that is, the process of analyzing a spectrum by subtracting from it, one by one, the spectra of known components. A computer in the analyzer permits any desired multiple of the standard spectrum to be subtracted from the unknown.

All the analyzers can display the memory contents on a cathode-ray tube (CRT) while no input is being analyzed. In addition, most of them have a live display, which shows each pulse as a brief flash on the cathode-ray screen as it is being stored. If the counting rate is fairly high, one can watch the spectrum growing.

While the analyzer is busy converting and storing a pulse, it is rendered insensitive to new input pulses. For accurate quantitative work it is often insufficient to know the clock-time duration of the count: one needs to know the *live time*, which is the clock time minus the dead time—the actual length of time the analyzer was ready to accept input pulses. This can be measured rather simply by means of an accurate, usually crystal-controlled, free-running oscillator, whose pulses are counted only if the input is "live" when the pulse occurs. Thus, within statistics, the number of live-timer pulses accepted is proportional to the live time, provided that the pulses to be analyzed arrive at random times. If the pulses arrive in bursts, as, for instance, from a pulsed accelerator, the live time will give a misleading indication.

The option of multichannel scaling is offered by most of the instruments. With this feature in operation, the channels become time channels instead of pulse-height channels. At the start of counting, all input pulses above the lower discriminator setting are registered as counts in channel 1. Channel-advance signals, internally or externally generated, switch the counting to channel 2, and so on, at a predetermined rate. The result is a record of the counting rate as a function of time such as one would want for determining the half-life of a radioactive source.

Multiparameter Multichannel Analyzers

The advent of multiparameter analyzers has greatly speeded data taking in certain types of experiments, and it has, in fact, made possible a new range of experiments that could probably not be done otherwise. For the most part, multiparameter analyzers use the principles of operation developed for single-parameter instruments; the major difference is in the required memory size. The magnitude of the difference can be appreciated by noting that for 100-channel coverage in each of n dimensions, one needs 100^n channels, making 10^4 channels for two dimensions and 10^6 for three. Frequently 100-channel resolution is not good enough.

Memories for multichannel analyzers can be of two types—accumulating and open ended. As examples of the former we have the commonly used magnetic core memory and the less widely used rotating magnetic drum. In such a memory, each channel has its own set of locations, and whenever a count occurs, the contents of the appropriate channel are increased by one. A typical open-ended system records all the information about each event digitally on magnetic tape (or occasionally on punched paper tape, moving film, etc.). An accumulating memory has the distinct advantage that its contents can be interrogated at will, to see how the experiment is going, whereas the open-ended tape must be processed later in auxiliary equipment.

Figure 13.5-1 shows an experimental arrangement for two-parameter analysis.

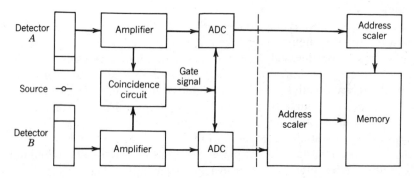

FIGURE 13.5-1   Typical experimental setup with a two-parameter analyzer.

13.6   A 100-CHANNEL PULSE-HEIGHT ANALYZER

Introduction

This section discusses the 100-channel pulse-height analyzer used for displaying particle-size distributions from the output of a model ZBI Coulter counter.[1] The Coulter counter[2] is widely used in medical and biological laboratories for counting and sizing cells and microorganisms. However much more useful information about the distribution of particle sizes present can be obtained by feeding the pulse train[3] produced by the counted particles into a multichannel pulse-height analyzer.

The block diagram of the 100-channel pulse-height analyzer is shown in Figure 13.6-1. The analyzer can accumulate up to 10^6 counts in each channel; the channels are individually addressable via a thumbwheel switch, and their contents are read out on a six-digit-channel LED (light-emitting diode) display. The pulse height distribution can be displayed on a CRO (cathode-ray oscilloscope) and applied to a YT or XY plotter. The analyzer uses standard CMOS logic almost exclusively. The memory (implemented with three 128×8-bit RAMs) contains 100 six-digit (24-bit) BCD (binary-coded decimal) numbers corresponding to the accumulated counts in each channel.

Principle of Operation

During operation the analyzer (Figure 13.6-1) is either in memory update cycles for updating channel counts or in display cycles for reading channel counts from memory.

Display Cycles.   Most of the analyzer's time is spent implementing CRO display cycles, which occur at the rate of 10^4 channels/s; thus every channel is accessed 100 times/s. The memory channel-count locations are addressed sequentially by the channel display counter, and the corresponding channel counts are loaded from memory into the six-digit BCD counter. The channel-count DAC produces an analog signal proportional to the channel count in the six-digit loadable BCD counter and is used to drive the CRO Y input. Since maximum channel counts in distributions obtained by the analyzer may vary greatly, the analog output is scaled by choosing two adjacent BCD digits from the channel-count data bus and inputting them to the eight-bit channel-count DAC. The scale factor, determined by the pair of digits chosen, is switched manually. The channel number DAC provides the X-input signal for the CRO.

---

[1] Manufactured by Coulter Electronics Ltd., Coldharbour Lane, Harpenden, Herts, England.

[2] See J. G. Webster, *Medical Instrumentation Application and Design*, Houghton Mifflin, Boston, 1978, pp. 550–553.

[3] See Problem 13-4.

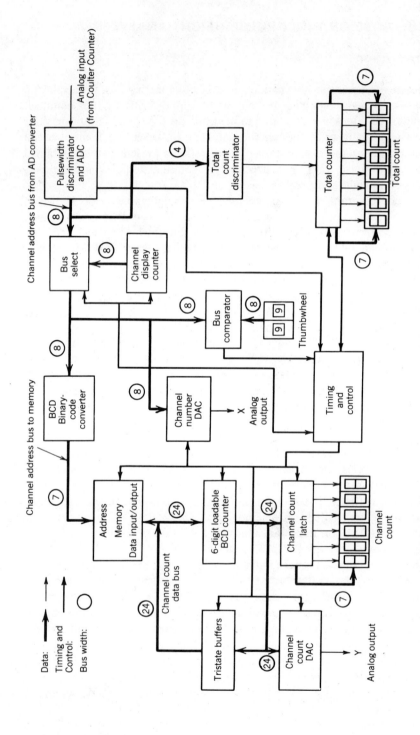

FIGURE 13.6-1 Block diagram of the 100-channel pulse-height analyzer.(Figures 13.6-1 to 13.6-5 are from S. J. Rackham, and **R. A. Sherlock**, *IEEE Transactions on Biomedical Engineering*, Vol. BME-26, No. 7, July 1979, pp. 436–440.)

When the address on the channel address bus corresponds to the channel number selected by the thumbwheel switch, the bus comparator causes the channel count to be latched into the channel-count latch and displayed on the channel-count LED display. The bus comparator also drives a cursor on the CRO display; this is achieved by strobing the channel-count DAC into saturation at the selected channel, thus enabling easy identification of channel positions. The plotter display is implemented by simply slowing down the channel display counter and gating it so that it makes one complete cycle only through all memory locations.

Memory Update Cycles.   The timing and control section is alerted when an acceptable pulse from the Coulter counter has been detected and its height converted to a two-digit BCD number by pulsewidth discriminator and ADC. In order to ensure that all channels are of equal width, an ADC with high differential linearity is required. The current display cycle is interrupted, and the digitized height of the newly acquired pulse (i.e., the address of the channel in which the pulse belongs) is gated to the memory address input. The corresponding channel count is loaded from memory into the six-digit loadable BCD counter, which is then incremented by one. The updated channel count is then written back into the same memory location via the tristate buffers, and the displaced display cycle is restored, thereby completing the memory update cycle. A tristate buffer has three possible conditions: high (1), low (0), and off.

The eight-digit total counter accumulates the total number of counts in all channels with index numbers greater than or equal to the value set on the total current discriminator. It can be preset by a switch to channel numbers 0, 10, 20, or 40; its purpose is to get out lower channel noise counts that are characteristics of Coulter counter pulse-height distributions. The Coulter counter in its basic form counts the number of particles in a preselected size range in a sample volume of fluid suspension. Figure 13.6-2 shows the chart recorder output of pulse-height distribution obtained from a suspension of 2.0-μm diameter latex spheres. The counts in the low channels are due to Coulter transducer noise.

Pulsewidth Discriminator and ADC

Random current fluctuations at the transducer output are accentuated due to the differentiator action of the ZBI Coulter counter. This effectively results in a relatively large number of small amplitude ($\leq 5\%$ of max) pulses in the ZBI output. These noise pulses give rise to large numbers of counts in the low channels (see Figure 13.6-2), and it is necessary to prevent the analyzer's processing time being swamped by them. Happily, a large proportion of the noise pulses are of relatively short duration (≤ 5 μs) compared with those arising from typical biological particles, and, thus, they can be removed by a pulsewidth discriminator. The discriminator permits only pulses whose widths lie within a preset range to be accepted by the analyzer.

Figure 13.6-3 shows the analyzer analog input circuit, including a detailed

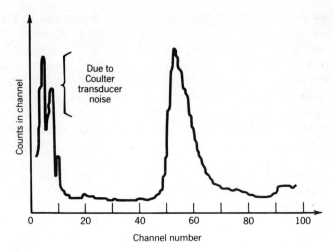

FIGURE 13.6-2 Chart recorder output of pulse-height distribution obtained from a suspension of 2.0-μm diameter latex spheres.

schematic of the pulsewidth discriminator. The sequencer is an eight-state counter whose main task is the control of the ADC. When an acceptable pulse is recognized by the discriminator, a cycle of the sequencer is initiated. When the analog-to-digital conversion is complete, the timing and control section is alerted and the sequencer is reset. Figure 13.6-4 shows the waveforms of the peak detector and hold circuit. Figure 13.6-5 shows waveforms of the pulse-width discriminator.

The seven control signals shown in Figure 13.6-3 perform the following functions:

1. "Peak detect" switches the peak detector from the tracking to the peak detect mode (Figure 13.6-4).
2. "Hold" causes the sample and hold to be switched to the hold mode (Figure 13.6-4).
3. "Sequence reset" initiates one cycle of the sequencer by going to logic 0 when an acceptable pulse has been detected by the discriminator.
4. "Sequencer last state" signals the discriminator that the sequencer has completed its cycle. The discriminator then resumes control and resets the sequencer by returning sequencer reset to logic 1.
5. "Start conversion" initiates an analog-to-digital conversion.
6. "ADC ready" informs pulse-height analyzer timing control that a pulse height has been converted to a channel number.
7. "EOC" signals the sequencer that the analog-to-digital conversion is complete.

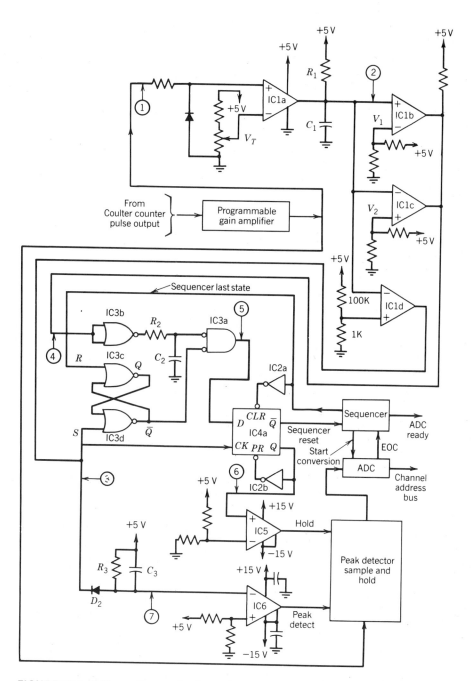

FIGURE 13.6-3 The analyzer analog input circuit, including a detailed schematic of the input pulsewidth dicriminator: IC1 = LM339, IC2 = MM74CO4, IC3 = MM74CO2, IC4 = MM74C74, IC5 = LM311, and IC6 = LM311. (From S. J. Rackham, and R. A. Sherlock, *IEEE Transactions on Biomedical Engineering*, Vol. BME-26, No. 7, July 1979, p. 438.)

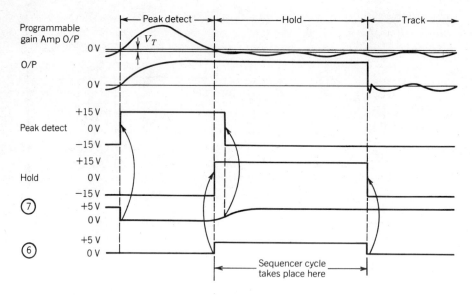

FIGURE 13.6-4 Peak-detector and hold-circuit waveforms.

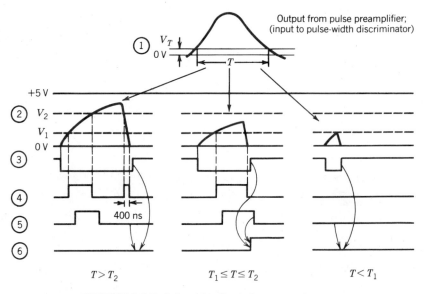

FIGURE 13.6-5 Pulsewidth-discriminator waveforms.

When the programmable gain amplifier output rises above the small input threshold voltage V_T, ICla ($\frac{1}{4}$LM339 open collector output comparator) "floats" its output and C_1 charges through R_1 until the incoming pulse falls below V_T, at which point C_1 is rapidly discharged (see Figure 13.6-5). The maximum voltage transferred to C_1 provides a measure of the pulsewidth, and the front panel adjustable thresholds V_1 and V_2 on the voltage window comparator (IClb and IClc) set the pulsewidth discriminator upper and lower thresholds (T_1 and T_2). The comparator ICld generates a rising edge as C_1 is reset, and it clocks the output of IC3a onto IC4a. If the output of the window comparator was at logic 1 ($+5$ V) just before C_1 was reset (i.e., $T_1 \leqslant T \leqslant T_2$, "acceptable pulse"), then a logic 1 would be clocked onto Q (IC4a) and sequencer reset would initialize a sequencer cycle; otherwise no action would occur.

IC3a, $R_2 C_2$, and IC3b serve two purposes:

1. They delay the window comparator output such that the value it held at the time C_1 was discharged is presented to the D input of IC4a as it is clocked.

2. They smooth the 400 ns "glitch" generated as C_1 is reset for the case $T > T_2$ (see Figure 13.6-5).

As soon as IC4a is set, it is "locked" into this state by IC2b and can only be "unlocked" by the sequencer last state signal (via IC2a).

The RS flip-flop (IC3c and IC3d) prevents the pulsewidth discriminator from reinstating the sequence until no pulses are present at the pulsewidth discriminator input. This prevents a second sequencer cycle from immediately following the first and not allowing time for the sample and hold to acquire the new pulse height.

The operation of hold and peak detect signals is illustrated in Figure 13.6-4. The necessary overlap between these two signals is generated by the pulse stretcher D_2, C_3, and R_3.

REFERENCES

1. Bayer, R., Simple, Precise, Single-Channel Pulse-Height Analyzer with High Dynamic Range, *Nuclear Instruments and Methods*, Vol. 146, 1977, pp. 469–471.

2. Cole, H. A., A Differential Pulse-Height Discriminator with Full Channel Logic Using Two Integrated Circuits, *Nuclear Instruments and Methods*, Vol. 138, 1976, pp. 551–556.

3. Rackham, S. J. and R. A. Sherlock, A Pulse Height Analyzer for Displaying Coulter Counter Particle Size Distributions, *IEEE Transactions* on *Biomedical Engineering*, Vol. BME-26, No. 7, July 1979, pp. 436–440.

4. Kieser, W. E., R. P. Beukens, and T. E. Droke, A Multichannel Analyzer to Computer Interface, *Nuclear Instruments and Methods*, Vol. 164, 1979, pp. 587–590.

5. Toyoshima, K., Adaptation of a Conventional Multichannel Pulse Height Analyzer for 4096 × 4096 Two-Parameter Measurement, *Nuclear Instruments and Methods*, Vol. 138, 1976, pp. 557–560.

6. Greenfield, J. D., *Practical Digital Design Using IC's*, John Wiley & Sons, New York, 1977, Chap. 8.

QUESTIONS

13-1. What is the relationship between the height of the output pulse from a linear amplifier connected to a radiation detector and the energy deposited by the nuclear radiation within the detector?

13-2. What is a pulse-height analyzer?

13-3. What is a single-channel pulse-height analyzer?

13-4. Sketch a block diagram of a basic single-channel pulse-height analyzer and briefly describe it.

13-5. Sketch a block diagram to indicate the basic analyzer circuit shown in Figure 13.2-1 and explain the circuit action referring to the related waveforms.

13-6. What is the advantage of a single-channel analyzer using a dual IC comparator with two positive feedback paths for one of the discriminators?

13-7. What is a multichannel pulse-height analyzer?

13-8. Explain the operation of a time-to-digital converter.

13-9. What is the channel number or the address?

13-10. Explain the chief difference between the single-parameter and multi-parameter multichannel analyzer?

13-11. What is the purpose of the channel count DAC used in the analyzer of Figure 13.6-1?

13-12. In the analyzer of Figure 13.6-1, the analog output is scaled. Why? How can it be scaled?

13-13. What is the function of the bus comparator in the analyzer of Figure 13.6-1?

13-14. Why is an ADC with high differential linearity required in the analyzer (Figure 13.6-1)?

13-15. How does the eight-digit total counter (in Figure 13.6-1) function? Why is it preset by a switch to channel numbers 0, 10, 20, or 40?

13-16. What is the function of the pulsewidth discriminator shown in Figure 13.6-3? Why can it remove the noise pulse?

13-17. What is the sequencer in Figure 13.6-3?

13-18. Describe the functions performed by the seven control signals shown in Figure 13.6-3?

13-19. Describe the operation of the voltage window comparator (IC1b and IC1c in Figure 13.6-3) and the directly related circuit.

13-20.   What are the purposes of IC3a, R_2C_2, and IC3b used in the discriminator (Figure 13.6-3)?

13-21.   Describe the function of the RS flip-flop (IC3c and IC3d in Figure 13.6-3) and its effect on the sequencer cycle.

PROBLEMS

13-1.   Change the analyzer circuit of Figure 13.3-1 into a practical circuit with the LLD threshold and the window controlled independently.

13-2.   Design an anticoincidence circuit using a dual one-shot SN74123 (Figure p13-2), NAND gates, and inverters for the analyzer of Figure 13.3-1.

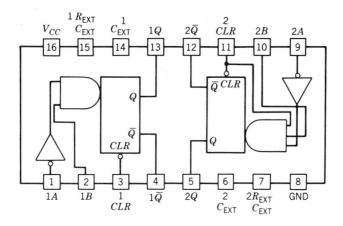

(a)

Function Table

Inputs			Outputs	
Clear	A	B	Q	$\bar{Q}$
L	X	X	L	H
X	H	X	L	H
X	X	L	L	H
H	L	$\uparrow$	$\sqcap$	$\sqcup$
H	$\downarrow$	H	$\sqcap$	$\sqcup$
$\uparrow$	L	H	$\sqcap$	$\sqcup$

(b)

FIGURE P13-2 The 74123 dual retriggerable one-shot multivibrator: (a) circuit pin layout; (b) function table.

13-3. For $C_{EXT} > 1000 \, pF$, the pulse width of a one-shot $\frac{1}{2}74123$ (Figure p13-2) is

$$T_W = 0.28 R_T C_{EXT} \left(1 + \frac{0.7}{R_T}\right),$$

where T_W is the pulse width in ns, R_T is the timing resistor in kΩ, and C_{EXT} is the timing capacitor in pF. Find the pulse width: (a) if $R_T = R_{EXT} = 30 \, k\Omega$ and $C_{EXT} = 3 \, \mu F$; (b) if $R_T = R_{EXT} = 5.6 \, k\Omega$ and $C_{EXT} = 1200 \, pF$.

13-4. Figure p13-4 shows the electrode placement for one type of the Coulter counter. A vacuum pump draws a carefully controlled volume of fluid from the WBC (white blood cell)-counting bath through the aperture. A constant current passes from the electrode E_1 in the WBC-counting bath through the aperture to the second electrode E_2 in the aperture tube. As each WBC passes through the aperture, it displaces a volume of the solution equal to its volume. The resistance of the WBC is much greater than that of the fluid ($R_C \gg R_F$), so that, in the circuit connecting the two electrodes, a voltage pulse V is created whose magnitude is related to the volume of the WBC. Assume that the circuit between the two electrodes is a dc source V_S in series with a load $R_L = R_F$ and that the ratio of R_C to R_F is n:1. Find the expression for the magnitude of the pulse V.

Answer. $\quad V = R_F \Delta I = V_S \left(\dfrac{1}{2} - \dfrac{1}{n+1}\right).$

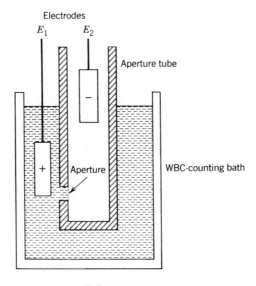

Electrodes

E_1 E_2

Aperture tube

Aperture

WBC-counting bath

FIGURE P13-4

14

TIME-OF-FLIGHT (TOF) MASS SPECTROMETERS— THE PULSE-TIMING SYSTEMS

14.1 INTRODUCTION

General

The mass spectrometer is one of the most important tools in modern chemistry. It is used in both basic research and industry. By means of this instrument, the researcher can gain insight into molecular structures, ionization phenomena, and chemical kinetics. Many industrial processes require a constant chemical analysis of the product to check quality and monitor chemical reactions. In many cases, chemical techniques are time consuming and laborious. Chemical analysis breaks down completely in identifying isotopes of the same element since all isotopes of a single element have identical chemical properties. The mass spectrometer provides fast chemical analysis and identification of isotopes by separating the positive ions of test gases and vaporized chemicals according to mass.

This instrument has also been used in other fields, such as materials research, nuclear engineering, space science, environmental science, and biomedical engineering. Mass spectrometry has provided much of the important data relating to the mass scale, the positive isotopic identification of nuclides that result from the fission of heavy elements, the measurement of nuclear cross section, and other functions fundamental to nuclear engineering. Figure 14.1-1 is the mass spectrum of a 0.3-μg lead sample, indicating its mass number.

Mass spectrometers can be divided into eight classes: single magnetic analyzers, double-focusing spectrometers, multiple magnet systems, cycloidal

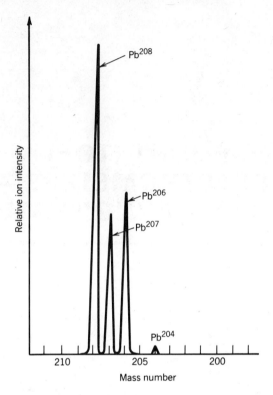

FIGURE 14.1-1 Mass spectrum of a 0.3-μg lead sample. The spectrum was obtained using a surface-ionization source in the mass spectrometer.

mass spectrometers, cyclotron resonance types, time-of-flight (TOF) spectrometers, quadrupole and rf mass filters, and special types. Among these, the TOF spectrometer is one of the newer ones and the most common type.

The TOF instrument provides a simple, rapid measurement of the abundance of various isotopes or elements comprising a sample. It operates in the following manner. A quantity of positive ions are introduced into an accelerating field and shot into a long field-free vacuum chamber. By ingenious techniques, all the ions are given equal energy, and thus they have substantially the same initial velocities, starting out from the same point in space. In the field-free chamber the ions separate according to mass: the lighter ions receive greatest acceleration and arrive at the end of the chamber first. The heavier ions are not sped up so much in the accelerating field and arrive at a later time.

Relationship between Time-of-Flight and Mass-to-Charge Ratio of Ions

The kinetic energy acquired by the ions in their drop through an accelerating potential of V (in erg/esu) is

$$W_k = m \frac{v^2}{2} = eV,$$

where v is the velocity (cm/s) of the ions of mass m (in g) and charge e (in esu).
The time of flight t_f (in s) is given by

$$t_f = \frac{d}{v} = \frac{d}{\sqrt{2eV/m}} = d\sqrt{m/2eV}$$

or

$$t_f = d(\sqrt{1/2V}\sqrt{m/e}). \qquad (14.1-1)$$

If d and V are maintained constant, then

$$t_f = k\sqrt{m/e}.$$

Thus the time of flight of the various ions is simply proportional to the square
root of the mass-to-charge ratio of the ions. The time taken for each bundle of
ions to make the trip is measured by a pulse-timing technique.

14.2   BASIC PRINCIPLE OF TOF MASS SPECTROMETERS

The Basic TOF instrument is shown in Figure 14.2-1. It consists of a vacuum
chamber (tube), including an electron gun, an ion gun, and a detector. The
sample to be analyzed is a gas or vapor. A small volume of the vapor is admitted
into the chamber through the gas inlet. The positive ions may be produced by
passing a beam of electrons through the gas at a pressure of about 10^{-5} to
10^{-6} mm Hg. The energy of the electron beam is determined by the difference
(e.g., 100 V) between the dc potential on the filament and the ground potential
on the second grid of the electron gun. If this energy is greater than the ionization potential of the gas, ionization may occur by collision with these electrons
in a narrow column perpendicular to the direction in which the ions will
eventually be accelerated.

Many models of TOF machines operate in the repetition frequency range of
10,000–100,000 Hz. Thus they will produce 10,000–100,000 complete mass
spectra each second, and flight times are measured with an entire single cycle
of operation lasting 10 to 100 μs. The first event in each cycle is the creation of
the ionizing electron beam. The electrons are generated from a tungsten wire
(e.g., diameter $= 0.3$ mm) heated by an appropriate current (e.g., 4.5 A). At the
beginning of each cycle, a short positive pulse (0.1–1.0 μs) is applied to the
control grid, allowing the electron beam to pass through the ionization region.
After the electron gun is pulsed, a waiting time (delay) is provided. During the
electron-beam time and the waiting period, no ion-accelerating pulse is applied

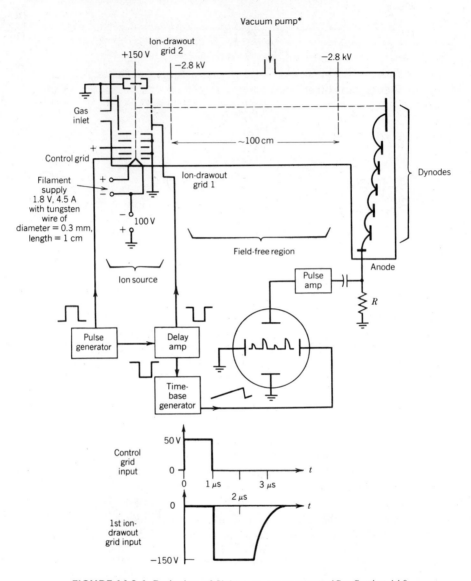

FIGURE 14.2-1 Basic time-of-flight mass spectrometer. *See Section 14.8.

to the ionized region. This permits the faster ions to slow down due to collision with residual gas molecules so that at the end of the waiting period the ions that have not recombined with electrons have about equal velocities. After a time delay of 0–3 μs, the first grid in the ion gun is pulsed to about −150 V, and this pulse remains long enough (about 1.5 μs) for all ions of interest to pass through this grid into the accelerating region of the ion source. Here they are accelerated to an energy of about 2800 V. The final grid in the ion gun is about 1 cm away from the electron beam, and the ion detector is about 100 cm farther down the tube.

Mass separation results only from the mass dependent velocities. The ratio of the two fields in the ion source is adjusted by the height of the ion accelerating pulse so that two ions of the same mass, initially at different positions in the ion source, will reach the detector simultaneously. If all ions were devoid of initial velocities, the depth of a single mass bunch at the detector would be very narrow.

In the field-free region the ions separate according to their mass-charge ratio (m/e) and arrive at the right end of the tube in discrete bundles—light ions first, followed by the heavier ions.

The electron multiplier is used as an ion-signal detector. The ions strike a target liberating secondary electrons, which are multiplied by a group of dynodes. The multiplier is similar to that of a photomultiplier except that the electron path between dynodes may be controlled by a magnetic field. The electrons collected at the multiplier anode develop voltage pulses across the load resistor R that are amplified and applied to the vertical deflection plates of a cathode ray tube (CRT). The beam of the CRT is deflected horizontally by a linear time base initiated at the time the ions are ejected into the accelerating zone. The horizontal axis of the trace is calibrated in mass units since the time of flight is proportional to mass.

Example 14.2-1. A TOF mass spectrometer has a flight length of 100 cm. The accelerating potential is 2800 V. How long will it take a PH_3^+ ion to traverse the spectrometer from ion source to detector?

Solution. $d = 100$ cm;

300 practical volts $= 1$ erg/esu;

$V = 2800/300 = 9.33$ erg/esu;

$m = 30.97 + 1.008 \times 3 = 34$;

Avogadro's number $=$ number of molecules in one mole or gram molecular weight of a substance

$= 6.024 \times 10^{23}$ per gram mole.

Electronic charge $= 4.8 \times 10^{-10}$ esu;

$e = 4.8 \times 10^{-10} \times 6.024 \times 10^{23} = 28.9152 \times 10^{13}$ esu.

From Equation (14.1-1),

$$t_f = d \left(\frac{1}{2V}\right)^{1/2} \left(\frac{m}{e}\right)^{1/2}$$

$$= 100(1/2 \times 9.33)^{1/2}(34/28.915 \times 10^{13})$$

$$= 100 \left(\frac{1}{0.1866}\right)^{1/2} \left(\frac{34}{2.8915}\right)^{1/2} \times 10^{-8}$$

$$= 100(5.36 \times 11.76)^{1/2} \times 10^{-8} = 100 \times 2.31 \times 3.43 \times 10^{-8}$$

$$= 7.92 \times 10^{-6} \text{ s} = 7.92 \ \mu s$$

Example 14.2-2. What is the difference in the times of arrival of the PH_2^+ and PH_3^+ ions, using the same data as given in Example 14.2-1?

Solution. $t_{f34} = 7.92 \ \mu s;$

$$t_{f33} = 7.92 \ (33/34)^{1/2}$$
$$= 7.92 \ (0.9706)^{1/2} = 7.92 \times 0.98 = 7.76 \ \mu s$$
$$\Delta t = t_{f34} - t_{f33} = 7.92 - 7.76 = 0.16 \ \mu s$$

14.3 ELECTRON MULTIPLIERS

Electrostatic Electron Multipliers

The electrostatic electron multiplier can be successfully operated in a pulsed counting mode for quantitative mass spectral work provided that the following obtains:

1. The first dynode has a suitable ion-secondary electron yield.
2. A sufficient number of multiplying stages are used with reasonable gain per stage.
3. The multiplier is shielded from magnetic fields.
4. The arrival rate of ions is restricted to values commensurate with the resolving times of both multiplier and associated circuitry.

The dynodes are often constructed of commercial-grade 2% beryllium copper and magnesium silver. They may be sensitized by heating to near red heat in vacuum, inert atmospheres, or in hydrogen. Multiplication factors (current gains) of from 10^5 to about 10^6 are usually obtained with 10 or 12 dynodes contained in the electron multipliers, which are shielded from magnetic fields.

Figure 14.3-1 shows the electrode system of a 13-stage electron multiplier designed by James S. Allen. The electrodes were formed from BeCu alloy sheets. The BeCu surfaces were polished and given the usual heat treatment in vacuum. Electrode 0 is connected within the tube to electrode 1 and prevents distortion of the field inside the electrode structure. The metal shield (not shown) covering electrodes 0 through 4, likewise is connected to electrode 1 and serves to shield the field inside the electrode structure. An additional function of this shield is to prevent electrons or positive ions from drifting into the electrode system. The overall multiplication of this type tube is about $10.^7$

The operation of the electron multiplier usually involves a rather large negative potential (e.g., -2000 V) on the cathode, with the anode held at or near ground potential. This is convenient in the mass spectrometer operation with positive ions, for it adds to the energy of the ion beam and gives somewhat greater electron emission from the cathode.

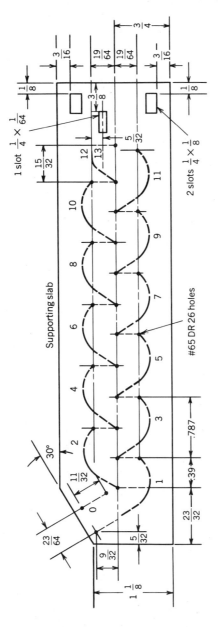

FIGURE 14.3-1 Electrode system of a 13-stage electron-multiplier tube. 1–12: dynodes. (From James S. Allen, *The Review of Scientific Instruments*, Vol. 18, No. 10, 1947, pp. 739–749.)

Wiley Magnetic Electron Multiplier

The Wiley magnetic electron multiplier is one of the crossed field type that makes use of both electrostatic and magnetic fields to determine the secondary trajectories. Such multipliers are especially attractive if the detection of ions must be made in a magnetic field, as purely electrostatic multipliers are rather severely defocused in a magnetic field of only a few gauss. The original Wiley magnetic electron multiplier had a thin metallic oxide film coated on glass backing plates approximately 4 in in length. When a high voltage is applied across the terminals of this high-resistance strip, a high voltage gradient results. The addition of a superposed magnetic field, with field lines perpendicular to the voltage gradient, causes electrons to multiply in cascade along the length of electrode. This device was subsequently developed in commercial form as the basic detector for Bendix time-of-flight mass spectrometers. Figure 14.3-2 shows such a multiplier detector.

As the individual groups of ions arrive at the end of the field-free flight region in the time-of-flight instrument, they collide with the plane ion cathode of the magnetic electron multiplier, which eliminates ion transit time variations

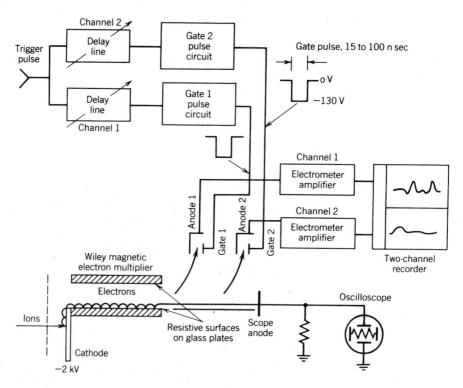

FIGURE 14.3-2 Wiley magnetic electron multiplier used as the detector for the time-of-flight mass spectrometer.

encountered with a curved ion cathode of the conventional electrostatic multiplier. Each collision produces a group of secondary electrons, and because of the crossed magnetic and electric fields present, the electrons follow cycloidal paths down the dynode strips of the multiplier. In this manner, a current gain of the order of 10^6 is obtained before the group of electrons reaches the gating section of the multiplier.

Because each mass signal is less than 50 ns wide at its base, a bandwidth significantly in excess of 20 MHz ($= 1000/50$) is needed in order to avoid distorting the shape of the pulses. This requirement will be satisfied if the Wiley multiplier is employed, since it can provide a bandwidth of direct current to about 500 MHz. Such a wide bandwidth permits many stages of multiplication to be used without loss of resolution; thus a lower gain per stage is allowable. The Wiley multiplier with field trips provides a "continuous dynode," which is used instead of the large number of dynodes of the conventional multiplier. Each of the field trips consists of a high resistance coating that has been fired onto a glass insulating support. The resistance coating is thin but very tenacious, both chemically and physically, thereby permitting chemical cleaning as well as nongritty abrasive cleaning. Exposure to humid air has been found not to affect the amplifying power of this multiplier.

The gate anodes in the Wiley multiplier operate in the following manner. A gate pulse applied to the gate anode will cause the group of electrons passing the assembly at that moment to be gated onto the corresponding anode. The electrical capacity of these gate-anode assemblies is sufficiently low so that very short pulses (pulse width = 10 to 100 ns, approximately) can be applied to the gate, thus allowing the anode to collect the electron current due to a particular mass in the sample. Complete mass spectra can be obtained by applying the gating pulse in the magnetic electron multiplier a little later in each operating cycle of the time-of-flight instrument.

The electron current at the gate anode can be measured by precision electrometers. The electrometer output is commonly 0–10 V dc, which is sufficient to drive most recorders. The output from the scope anode is displayed on an oscilloscope synchronized and triggered by the master pulsing system of the mass spectrometer. Any portion of the mass spectrum can be observed in detail by adjusting the sweep controls of the oscilloscope and the control of the continuously variable delay of the mass spectrometer between the ion accelerating pulse and the oscilloscope trigger.

14.4 FUNCTIONS IN TOF MASS SPECTROMETRY

General

In order to accomplish time-of-flight mass spectrometry in a quantitative manner over a large range of peak heights, it is necessary to integrate the current from individual peaks over a large number of mass spectra and to present the

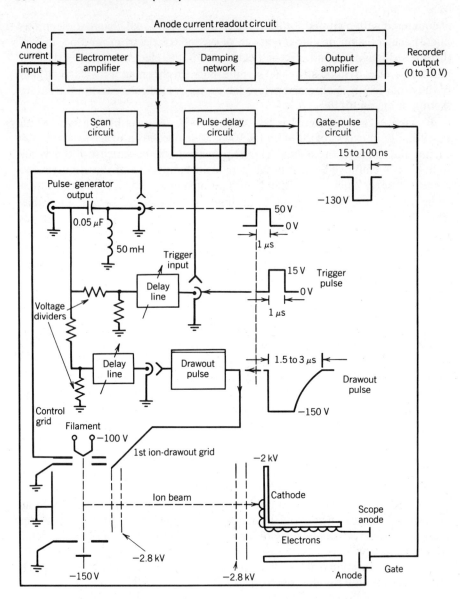

FIGURE 14.4-1 Basic functions of the arrangements of the time-of-flight mass spectrometer. The leading edge of the drawout pulse is coincident with the trigger pulse input.

integrated signal in analog form. Using a gated electron multiplier, it is possible to selectively gate a particular mass peak from each spectrum and cause it to be collected by one of the multiplier gate anodes. The current at the anode can then be measured by an electrometer circuit.

Scanning is achieved by varying the relative time at which the gate is pulsed on during each cycle of operation. A trigger pulse into the scanner circuit is coincident with the leading edge of the ion drawout pulse, which is a negative pulse applied to the first drawout grid of the ion source. A pulse delay circuit delays the gating pulse, allowing all or any part of the spectrum to be scanned at a variable rate. The gate literally "jumps" along, but the increments are so small that a continuous scan is the apparent result. With an appropriate strip chart recorder, permanent recordings of the mass spectra can be accomplished.

Typical Output Scanner

Figure 14.4-1 shows the basic functions of the arrangements of the TOF mass spectrometer with the output scanner Bendix type MA-001. The output scanner includes the electrometer and output circuits and the scanner circuit.

The electrometer amplifier detects and amplifies the current pulses from one of the multiplier gate anodes. The output circuit amplifies the signal to provide a maximum output of 10 V for recorder operation.

The scanner circuit provides the modes of operation for manual, recurrent, or triggered scanning of all, or any part, of the spectrum. When functioning in either the "recurrent" or "triggered" mode, the spectrum can be scanned at a variable rate.

Specifications of the output scanner are listed below.

Linear range: 1×10^{-5} to 3×10^{-3} A full scale.

Log range: 10^{-7} to 10^{-12} A.

Output: 0 to 10 V into a 1000-Ω minimum-load impedance.

Gate pulse width: Variable from approximately 10 to 100 ns.

Spectral scan rates: 10 scan rates, either recyclic or triggered, 2 ns/s to 5 μs/s; 2–200 amu in 1.5 s to 1 hr. The unit is capable of monitoring mass peaks to 500 amu. 1 amu = 1.66×10^{-24} g.

Quantitative reproducibility: $\pm 1\%$ in the range 100 to 1%; detection to 1 ppm below 1%.

14.5   TYPICAL ELECTROMETER AND OUTPUT CIRCUIT

Figure 14.5-1 shows the anode current readout circuit of the output scanner MA-001. This circuit consists of an electrometer amplifier (with associated feedback components) and a line driving output amplifier with an input damping network. Integrated current from the gated multiplier anode flows through the feedback components, around the electrometer amplifier, and drives the amplifier input slightly negative. The input draws practically no current and remains at virtual ground at all times.

The electrometer amplifier has very high gain and, with a low impedance output, is capable of driving the damping network and output amplifier, the feedback network, and the overload circuit.

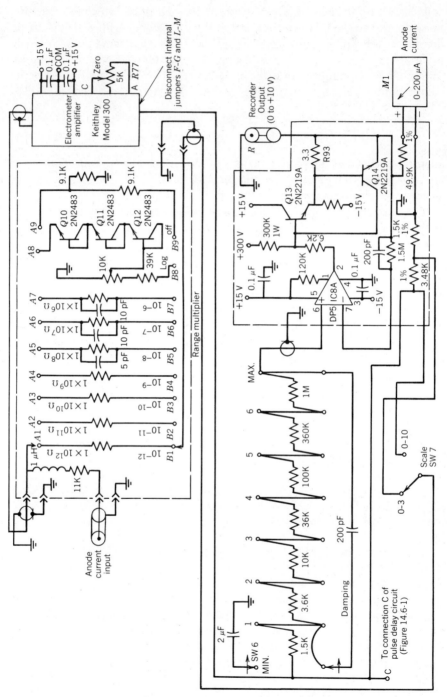

FIGURE 14.5-1 Anode current readout circuit of the output scanner Bendix MA-001. (Courtesy CVC products, Inc.)

The feedback circuit consists of resistors selected to give full scale readings of from 10^{-13} to 10^{-6} A. Diode-connected transistors in the log position provide a logarithmic range usable from 10^{-12} to 10^{-7} A. A 10% feedback resistance range is also available for zero-adjusting the electrometer amplifier (with zero anode current input). The range multiplier switch is used to control the gain of the electrometer amplifier, change the feedback characteristic, and select the logarithmic range.

The output amplifier provides a current gain to drive the line and recorder loads. The voltage gain is unity. The output of the electrometer amplifier is passed through a variable damping network to filter out statistical (or other undesirable) signal variations. It is then fed into the noninverting input of a high input impedance operational amplifier (DP-5). The output of DP-5 drives current amplifier Q_{13}. Both the recorder output and the feedback are taken from Q_{13}. In addition, 100% feedback from the output is applied to the input of DP-5, resulting in a voltage gain of one for the output amplifier.

Overload protection is provided by transistor Q_{14}, which turns on if the output current exceeds a level determined by resistor R_{93}. Conduction of Q_{14} limits the current of Q_{13} to the selected level, thus preventing damage. The electrometer amplifier output is also applied to the pulse delay circuit as an overload signal. If this output rises above approximately 10 V, the multiplier gate pulse cuts out, thus limiting the current to an "in-range" value. The gate ring on the oscilloscope will cut in and out when this occurs, thus indicating the overload condition.

14.6   TYPICAL PULSE DELAY CIRCUIT

Figure 14.6-1 shows the pulse delay circuit of the scanner MA-001. The trigger pulse input to the pulse delay circuit is coincident with the leading edge of the ion drawout pulse (spectrum time = 0). A voltage ramp time delay circuit (R_{44} and C_{17}) determines the delay of the output. The ramp is started and retracted by a bistable multivibrator switch (Q_3, Q_4, and associated components).

Pulse Delay Circuit Operation

The pulse delay circuit generates pulses that act as the trigger for the gate pulse board shown in Figure 14.6-2. The gate pulse board furnishes the pulses that actually gate a portion of the spectrum into the electrometer, located in the scanner.

To assist in understanding the operation of this circuit, the sequence of events that constitute one cycle of operation will now be described. The quiescent state of the multivibrator (Q_3 and Q_4 both off) allows R_{43} to hold C_{17} at 0 V. The two matched temperature compensated diodes (DP4) determine the voltage

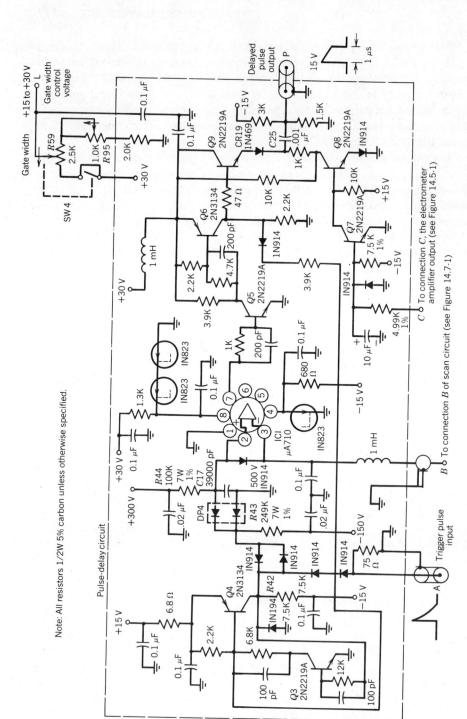

FIGURE 14.6-1 Pulse delay circuit of the output scanner Bendix MA-001. (Courtesy CVC Products, Inc.)

level on capacitor C_{17}. When the input trigger pulse occurs, the bistable multivibrator changes state and Q_4 back biases both diodes of DP4. The collector of Q_4 rises to approximately $+15$ V. The input trigger pulse also back biases DP4 directly. C_{17} will now charge through R_{44}, which is at a $+300$-V potential, to a maximum of 4 V at maximum delay. The ramp is approximately linear.

The voltage on C_{17} is applied to the noninverting input of level detector IC1. When this voltage reaches a level equal to the reference voltage on IC1, a positive pulse is generated in IC1. This pulse is amplified by Q_5, Q_6, and Q_9. The output pulse from Q_9 is fed through CR19 and C_{25} to the output line.

The output pulse from Q_6 is also fed back to the bistable multivibrator, resetting it to the quiescent state. C_{17} discharges through R_{42} and R_{43} to 0 V. After C_{17} is discharged equal current flows through each diode of DF4, resulting in a zero voltage on C_{17} and a termination of the output pulse. The circuit remains in this state until another trigger pulse is applied.

If the electrometer amplifier output exceeds approximately 10 V, the overload circuit prevents an output trigger pulse from occurring. Transistors Q_7 and Q_8 (and associated components) remove the emitter ground return on Q_9. Therefore, when the output trigger pulse is terminated, the gating pulse circuit will not produce a pulse, and the amount of anode current from the multiplier is effectively limited. All other stages of the pulse delay circuit will continue to operate as they did before the overload occurred.

Gate Width Control

This potentiometer controls the dc voltage fed to the gate pulse generator shown in Figure 14.6-2. The voltage level determines the time lag between the leading and trailing edges of the gating pulse and, therefore, the pulse width. In the minimum position, a switch (SW4) is shorted and acts as a detent, thus applying a minimum preset voltage to the gate width circuit. This minimum voltage can be adjusted by rotating potentiometer R_{95}.

Gate Pulse Circuit and Drawout Pulse Circuit

The gate pulse circuit (Figure 14.6-2) generates the actual pulse that is applied to one of the gates located in the multiplier portion of the mass spectrometer. The input trigger to the circuit is the output pulse of the pulse delay circuit (Figure 14.6-1). The gate pulse is used to divert a portion of the spectrum into the particular gate anode that is connected to the input of the electrometer.

The ion drawout pulse circuit is shown in Figure 14.6-3. Its input trigger pulse (about $+15$ to $+30$ V, peak) is furnished through a voltage divider shown in Figure 14.4-1. The output from the drawout pulse circuit is a negative pulse with a peak of about 100–150 V. This ion drawout pulse is applied to the first drawout grid of the ion source in the vacuum chamber. As illustrated in Figure 14.4-1, the leading edge of the drawout pulse is coincident with the trigger pulse input to the pulse delay circuit.

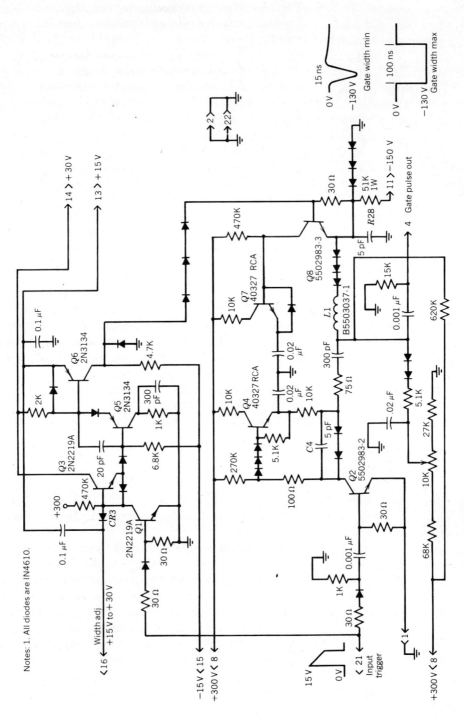

FIGURE 14.6-2 Typical gate-pulse circuit (C5502933, Bendix). (Courtesy CVC Products, Inc.)

480

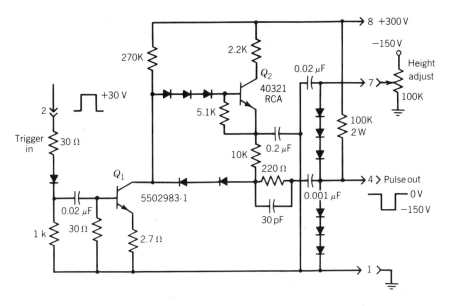

Notes:
1. All diodes are 1N4610.
2. All resistors are 1/2W, ±5%, unless otherwise specified.

FIGURE 14.6-3 Typical drawout pulse circuit Bendix C5502974). (Courtesy CVC Products, Inc.)

14.7 TYPICAL SCAN CIRCUIT

Three Scan Modes

The reference voltage used to control the delay time of the delay circuit (Figure 14.6-1) is provided by the scan generator circuit shown in Figure 14.7-1. The scan generator provides the manual, recurrent, and trigger modes. In the manual mode, a specific mass peak out of the spectrum can be selected for continuous monitoring. In the recurrent mode, the part of the spectrum selected is scanned repetitively. In the trigger mode only a single scan occurs when the circuit is triggered. In either the recurrent or trigger scan position, all or any part of the spectrum can be scanned at a variable rate.

Manual Scan

For manual scanning, the reference voltage is controlled only by the manual mass selector controls (R_{16} and R_{23}). The combined voltage from these controls is fed to the noninverting input of mixer-amplifier DP3 to bias the pulse delay circuit level detector. The amplifier has a very high input impedance and operates at a voltage gain of 1.4. The stability of the output voltage approximates

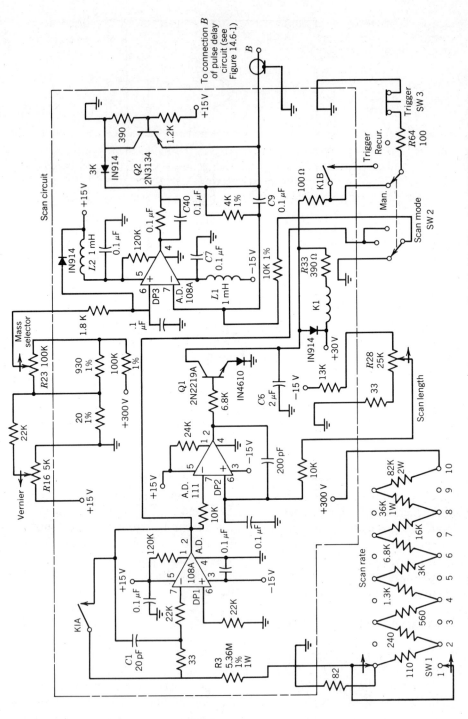

FIGURE 14.7-1 Scan generator circuit of the output scanner Bendix MA-001. (Courtesy CVC Products, Inc.)

the resistance stability of the high resolution, wire-wound mass selector. This output voltage is the inverting input to IC1 located in the pulse delay circuit. This voltage, therefore, acts as a reference voltage and determines at which portion of the spectrum the gate pulse fires.

Trigger Scan

In this mode, the reference voltage is determined by both the manual mass selector controls and by an automatic scan voltage fed to the inverting input of DP3. The automatic scan ramp is generated by amplifier DP1 and its associated components, which form a Miller integrator. Capacitor C_1, which is around amplifier DP1, is charged through resistor R_3 to a variable positive voltage determined by the scan rate control Sw_1.

The Miller integrator produces a voltage ramp with a negative slope starting at zero voltage when trigger switch Sw_3 is actuated. This opens switch K1A and allows C_1 to charge. This voltage ramp is applied to the inverting input of mixer-amplifier DP3 and also to the inverting input of level detector DP2. The voltage on the noninverting input of DP2 is controlled by the scan length control. When the ramp voltage reaches a level equal to the scan length voltage, DP2 will generate a positive pulse causing transistor Q_1 to conduct. This closes K1, resulting in the following actions:

1. Section K1A discharges C_1 back to 0 V, thus retracting the ramp to 0 V.
2. Section K1B acts as a holding contact. This section is in series with trigger push button Sw_3 and keeps K1 actuated until the trigger switch is again pushed. One scan is generated for each push of the trigger switch.

Recurrent Scan

This mode operates the same way as the trigger scan mode except that when the level detector (DP2) causes K1 to actuate and retrace the ramp to zero volts, the holding contact (K1B) is not connected. K1 will stay closed only long enough to retrace the ramp to 0 V and will then reopen. The RC network of C_6 and R_{33} (plus the K1 coil resistance) determines the length of time K1 is closed.

Procedure for Obtaining Output Scan

1. Select the desired portion of the spectrum at which the scan is to begin. This is achieved by adjusting the mass selector and observing the gate pulse ring on the oscilloscope.
2. Select any scan length.
3. Select any scan rate.
4. Select either the "trigger" or "recurrent" mode.
5. Depress the "trigger" switch.

It is advisable to make a few practice scans, not recorded, to observe the scan rate and scan length on the oscilloscope. The scan rate and length can now be adjusted to the desired positions, and a permanent record can be produced.

14.8   HIGH VACUUM PUMPS USED IN MASS SPECTROMETERS

Vacuum Requirements in Mass Spectrometers

The gas in a mass-spectrometer tube causes peak broadening, as the ions are diverted from their prescribed trajectories by the collisions. The vacuum should be only good enough to avoid intolerable peak broadening—an operating pressure in the tube of 1×10^{-6} mm Hg is usually sufficient. For high ion energies, as in mass spectrography (i.e., about 20 keV), a pressure of 1×10^{-5} mm Hg may be tolerated. If the instrument is used for accurate measurements of isotope ratio (as it is impossible to exclude coincidence of sample and background peaks), the vacuum must be as good as possible, unless one can use a type of ion source (a three-filament source) that will not ionize the residual gas. Under operating conditions a vacuum of 10^{-8} mm Hg may be considered normal.

A vacuum of the order of 10^{-8} to 10^{-7} mm Hg can be easily obtained from the ion pumps; such vacuums, Ultek[1] 5 to 25 L/s differential-ion (D-I) pumps, are described next.

Typical Ion Pumps

General.   The Ultek 5–25 L/s D-I pumps are multiple discharge, cold cathode ion pumps operating on the cold cathode discharge principle. These pumps utilize a discharge in a strong magnetic field to sputter reactive metal from a cathode plate. This reactive metal combines with gas particles in the system to form stable solid compounds, thus removing a gas from the system and reducing the pressure. The ion pump is not a pump in the usual sense of the word; it does not expel gas from the system but merely traps it within the system. The pump associated with its control unit, installed in a temperature-control room, operates entirely from electric power and does not require oil, water, or refrigerants of any kind. It also serves as its own vacuum gauge.

Ultek ion pumps have a conservatively rated life in excess of 32,000 hr at 10^{-6} torr (mm Hg). Pumping element life is inversely proportional to pressure during operation and will be 320,000 hr at 10^{-7} torr.

Hot Start Procedure.   The following is the hot start (isolated) procedure. It is used when there is an isolation valve between the ion pump and the

[1] Perkin-Elmer, Ultek Division, Box 10920, Palo Alto, California 94303.

system. Note that this valve is normally closed.

1. Check the ion pump control unit connection and be sure the system has been leak checked.
2. Turn on the roughing (rotary) pump and open the roughing valve.
3. Set the ion pump control unit RANGE SELECTOR switch at 200 mA and the mode switch to START.
4. When the system pressure has reached about 5 microns (5×10^{-3} torr), close the roughing valve.
5. Open the isolation valve slowly, keeping the ion pump at a pressure below 5×10^{-3} torr, with approximately 10 mA reading on the meter. This is maximum throughout range. Note that surface temperature of the pump must be less than 100°C.
6. As soon as the isolation valve is completely open, set the RANGE SELECTOR switch at the current range in which the pump is operating and the mode switch to RUN.
7. If the system is to be left unattended for some time, the ion pump control unit RANGE SELECTOR switch should be left on either the 200-mA or 4-kV position, to prevent the meter from reading off scale if a pressure rise should occur.

Low-Temperature Bakeout.   The low-temperature bakeout is a simple and very effective procedure for improving the pump-down speed and base pressure of a system that does not appear to be performing as well as it should. The principal reason for a decrease in performance is contamination of the system and/or pump with atmospheric water vapor. Water vapor enters the system and adheres to the walls when the system is open to atmosphere. It is not removed by subsequent roughing cycles and forms a significant additional gas source in the system, particularly in moist climates. The water vapor contamination can be removed by the following procedure:

1. Pump the system down and leave the ion pump in operation.
2. Arrange infrared heat lamps and/or heater tape around the vacuum chamber and ion pump and proceed to heat them. Make sure that Viton seals and the ion pump magnets are not heated above 250°C. This temperature is approximated by a 500-watt heat lamp at a 3-in distance.
3. Adjust the heat to prevent pressure in the system from rising above 5×10^{-5} torr (about 15 mA of current).
4. Continue heating the system for as long as is practicable with only the ion pump in operation. Generally an overnight bakeout will be sufficient, but longer bakeouts result in cleaner systems.

Typical Ion Pump Control Unit

The 150-mA ion pump control unit (Model 222-0400) is basically a high voltage current-limited operating power source for Ultrek ion pumps of 50 L/s or less. It also provides a means by which the operator can continuously monitor the operating conditions in the ion pump and thereby know, for example, the approximate pressure in the vacuum system. The control unit circuitry shown in Figure 14.8-1 consists of a transformer/rectifier circuit and three metering circuits for current, voltage, and pressure.

The high voltage transformer T_1 is of the high leakage inductance type. As the load decreases, the increase in primary current causes more magnetic flux to be shunted through the air gap of the transformer. Therefore less flux is linking the secondary winding and the output voltage drops. The ion pump control unit can operate indefinitely under any load condition.

The primary side of transformer T_1 is supplied with 115 Vac through circuit breaker CB2. Circuit breaker CB1 is on the return side of the transformer primary, but it functions as a result of certain current values in the secondary side of the transformer. Specifically CB1 will trip after the current on the output side of the transformer has sustained 50 mA for approximately 2 min for the run mode of operation, and thus protect the ion pump from excessive current. In the start mode of operation, however, CB1 is bypassed through switch S2 and only CB2 determines the maximum current in the primary side of T_1.

The secondary side of T_1 is rectified in the encapsulated diode bridge circuit B_1 and then is filtered through capacitor C_1 to give a nominal 4750 V dc output. Resistors R_1 and R_2 have low resistance and high voltage characteristics and serve to limit surge currents in the output of the ion pump control unit. The return side of the output circuit is completed through connector J_2 and RANGE SELECTOR switch S1A.

The three metering circuits all use the common 20-μA meter M_1, and the current and torr circuits use a common shunt diode D_2 to protect the meter against destructive overload currents. The diode will shunt less than 1 mA at full scale and 1 A at 1 V, limiting the meter current to five times the rated value. The 4.7-kΩ resistor in series with meter M_1 provides the impedance to bias the shunt diode D_2.

The current values indicated on the current scales of the meter are read through S1A and S1B in the return line of the power output circuit. The torr scale is also read on the return line, through diode D_3. The logarithmic characteristics of the semiconductor diode provide the logarithmic readout of the torr scale. Potentiometers R_3 and R_4 allow adjustment of this circuit for use with different pump sizes.

The voltage readings are obtained through the resistor bank RB1, the 1-kΩ potentiometer that acts as a voltage divider network and meter shunt. The voltage is connected to M_1 through S1B.

The RECORDER OUT connector J3A has potentiometer R_5 across it in order to shunt M_1 and to adjust the output of J3A. Diode D_1, in the connector

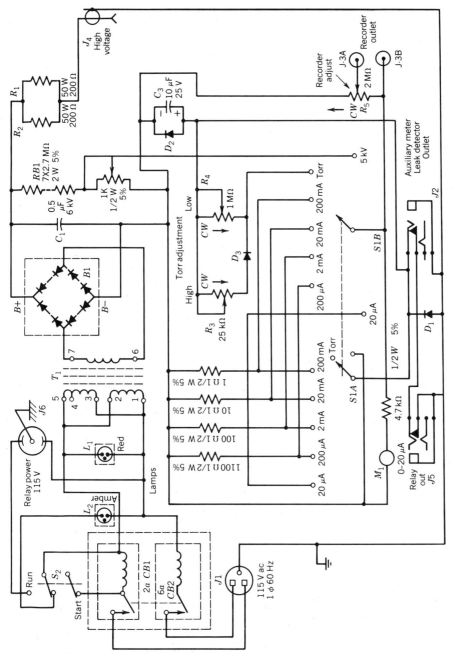

FIGURE 14.8-1 Circuitry of Ultek 150-mA ion pump control. (Courtesy Perkin-Elmer Vacuum Products.)

J_2 circuit, shunts the connector to provide protection to the leak detector and auxiliary meter output circuit.

REFERENCES

1. Kiser, R. W., *Introduction to Mass Spectrometry and Its Applications*, Prentice-Hall, Englewood Cliffs, New Jersey, 1965.

2. White, F. A., *Mass Spectrometry in Science and Technology*, John Wiley & Sons, New York, 1968.

3. McDowell, C. A., *Mass Spectrometry*, McGraw-Hill, New York, 1963.

4. Waldron, J. D., *Advances in Mass Spectrometry*, Pergamon Press, New York, 1959.

5. *Type MA-001 Analog Output Scanner*, Instruction Manual No. 20-34-A, Bendix Corporation, Scientific Instrument and Equipment Division, 1775 Mt. Read Blvd., Rochester, New York 14603, 1968.

6. *Operation and Maintenance Manual (C-1471) for High Vacuum Differential-Ion Pumps, 5, 11, 20, & 25 liters/second (Models 202-500, 202-800, 202-2000, 202-2500)*, Perkin-Elmer, Ultek Division, Box 10920, Palo Alto, California 94303, 1970.

7. *Operation and Maintenance Manual (C-1459A) for Ion Pump Control Unit (Models 222-0400, 222-0451, 222-0500, 222-0551, 222-0600, 222-0650)*, Perkin-Elmer, Ultek Division, Palo Alto, California, 1970.

8. Ducorps, A. M. and C. J. Yashinovitz, *CAMAC Controller for a Conventional and Pseudorandom Time-of-Flight System, Rev. Sci. Instrum.* Vol. 54, No. 4, April 1983, pp. 444–453.

QUESTIONS

14-1. Describe the uses of a mass spectrometer.

14-2. Name the different classes of mass spectrometers.

14-3. How does the time-of-flight (TOF) spectrometer work?

14-4. What is the relationship between time-of-flight and mass-to-charge ratio of the ions?

14-5. What sections are contained in the vacuum chamber of a TOF mass spectrometer?

14-6. Explain how the TOF mass spectrometer works.

14-7. Under which conditions can the electrostatic multiplier be successfully operated in a pulsed counting mode for a quantitative mass spectral work?

14-8. Describe the basic construction of dynodes in an electrostatic electron multiplier.

14-9. Describe the basic construction of the Wiley magnetic multiplier.

14-10. Briefly describe basic functions in TOF spectrometry.

14-11. What sections are contained in the anode current readout circuit of the output scanner Bendix MA-001?

14-12. What are the main features of the electrometer amplifier in the output scanner MA-001?

14-13. Describe the feedback circuit of the electrometer amplifier in the scanner MA-001. What will occur if the electrometer amplifier output rises above 10 V?

14-14. Explain the function of the main parts in the output amplifier of the scanner MA-001.

14-15. Discuss how the capacitor C_{17} charges and discharges in the circuit of Figure 14.6-1. What waveform will be produced from C_{17}.

14-16. How does the overload circuit (in Figure 14.6-1) work when the electrometer amplifier output exceeds 10 V? What will occur in the gating pulse circuit and the electron multiplier under the overload condition?

14-17. What is the purpose of the gate pulse used in the mass spectrometer?

14-18. What scan modes are provided by the scan generator of Figure 14.7-1?

14-19. Describe the circuit action for a push of the trigger switch Sw_3 in Figure 14.7-1?

14-20. What is the difference between the recurrent scan mode and the trigger scan mode?

14-21. Describe the procedure for obtaining output scan.

PROBLEMS

14-1. Derive an equation to express the relationship between time-of-flight and mass-to-charge ratio of the ions.

14-2. A TOF mass spectrometer has a flight length of 98 cm. The accelerating potential is 2700 V. How long will it take a PH_2^+ ion to traverse the spectrometer from ion source to detector?

TYPICAL STANDARD RESISTOR AND CAPACITOR VALUES

Standard Resistor Values

Ω	Ω	Ω	$k\Omega$	$k\Omega$	$k\Omega$	$M\Omega$	$M\Omega$
—	10	100	1	10	100	1	10
—	12	120	1.2	12	120	1.2	—
—	15	150	1.5	15	150	1.5	15
—	18	180	1.8	18	180	1.8	—
—	22	220	2.2	22	220	2.2	22
2.7	27	270	2.7	27	270	2.7	—
3.3	33	330	3.3	33	330	3.3	—
3.9	39	390	3.9	39	390	3.9	—
4.7	47	470	4.7	47	470	4.7	—
5.6	56	560	5.6	56	560	5.6	—
6.8	68	680	6.8	68	680	6.8	—
—	82	820	8.2	82	820	—	—

Standard Capacitor Values

pF	pF	pF	pF	μF	μF	μF	μF	μF	μF	μF
5	50	500	5000	0.05	0.5	5	50	500	5000	
—	51	510	5100	—	—	—	—	—	—	
—	56	560	5600	0.056	0.56	5.6	56	—	5600	
—	—	—	6000	0.06	–	6	—	—	6000	
—	62	620	6200	—	—	—	—	—	—	
—	68	680	6800	0.068	0.68	6.8	—	—	—	
—	75	750	7500	—	—	—	75	—	—	
—	—	—	8000	·—	—	8	80	—	—	
—	82	820	8200	0.082	0.82	8.2	82	—	—	
—	91	910	9100	—	—	—	—	—	—	
10	100	1000		0.01	0.1	1	10	100	1000	10000
—	110	1100		—	—	—	—	—	—	
12	120	1200		0.012	0.12	1.2	—	—	—	
—	130	1300		—	—	—	—	—	—	
15	150	1500		0.015	0.15	1.5	15	150	1500	
—	160	1600		—	—	—	—	—	—	
18	180	1800		0.018	0.18	1.8	18	180	—	
20	200	2000		0.02	0.2	2	20	200	2000	
22	220	2200		—	0.22	2.2	22	—	—	
24	240	2400		—	—	—	—	240	—	
—	250	2500		—	0.25	—	25	250	2500	
27	270	2700		0.027	0.27	2.7	27	270	—	
30	300	3000		0.03	0.3	3	30	300	3000	
33	330	3300		0.033	0.33	3.3	33	330	3300	
36	360	3600		—	—	—	—	—	—	
39	390	3900		0.039	0.39	3.9	39	—	—	
—	—	4000		0.04	—	4	—	400	—	
43	430	4300		—	—	—	—	—	—	
47	470	4700		0.047	0.47	4.7	47	—	—	

APPENDIX 2

ANALYSIS OF PERIODIC NONSINUSOIDAL WAVEFORMS

A2.1 FOURIER SERIES —THE GENERAL SOLUTION OF A PERIODIC WAVEFORM

Any periodic waveform is composed of a constant, a fundamental, and harmonics. The general solution of a periodic waveform is called a Fourier series and is expressed as

$$y = f(t) = a_0 + a_1 \cos \omega t + a_2 \cos 2\omega t + a_3 \cos 3\omega t + \cdots$$
$$+ b_1 \sin \omega t + b_2 \sin 2\omega t + b_3 \sin 3\omega t + \cdots$$
$$= a_0 + \sum_{n=1}^{\infty} (a_n \cos n\omega t + b_n \sin n\omega t), \qquad (A2.1\text{-}1)$$

where $f(t)$ may represent voltage, current, or power; a_0 is the constant or dc component; a_1, a_2, and a_3 represent the coefficients of the cosine terms of the fundamental, second, and third harmonic terms, respectively; b_1, b_2, and b_3 represent the coefficients of the sine terms of fundamental, second, and third harmonic terms, respectively; n is the integer sequence, 1, 2, 3, . . . ; and a_0, a_n, and b_n are called the Fourier coefficients and are calculated from $f(t)$. The term $\omega(= 2\pi/T)$ represents the fundamental frequency of the periodic function $f(t)$; 2ω, 3ω, . . . , and $n\omega$ are the second harmonic, the third harmonic, . . . , and the nth harmonic, respectively. The functions and integrals related to the Fourier coefficients are described as follows.

492

Even Function

A waveform that is symmetrical about the y axis is called an even function. Its amplitude at $+t$ (or $+\omega t$) equals that at $-t$ (or $-\omega t$). Thus $f(t)$ is an even function if

$$f(t) = f(-t). \qquad (A2.1\text{-}2)$$

The following is an example of even function:

$$f(t) = 2 + t^2 + t^4.$$

The cosine wave is also an even function since it can be expressed as the following series:

$$\cos x = 1 - \frac{x^2}{2!} + \frac{x^4}{4!} - \frac{x^6}{6!} + \cdots$$

Periodic waveforms shown in Figure A2.1-1 illustrate examples of even functions.

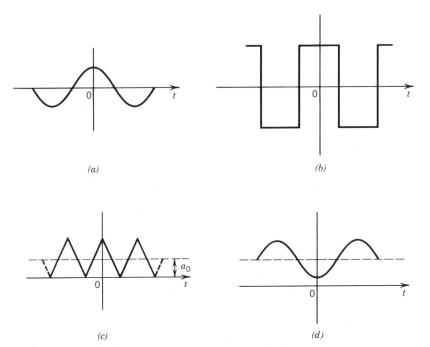

FIGURE A2.1-1 Periodic waveforms illustrating even functions: (a) cosine wave is an even function and its average value $a_0 = 0$; (b) even function and $a_0 = 0$; (c) even function and $a_0 \neq 0$; (d) even function and $a_0 \neq 0$.

Odd Function

A waveform that is symmetrical about an origin is called an odd function. Its amplitude at $+t$ (or $+\omega t$) is the negative of the amplitude at $-t$ (or $-\omega t$). Thus a function is odd if

$$f(t) = -f(-t). \qquad (A2.1-3)$$

The following is an example of odd function:

$$f(t) = t + t^3 + t^5.$$

The sine wave is also an odd function since it can be expressed as the following series:

$$\sin x = x - \frac{x^3}{3!} + \frac{x^5}{5!} - \frac{x^7}{7!} + \cdots$$

Periodic waveforms shown in Figure A2.1-2 illustrate examples of odd functions.

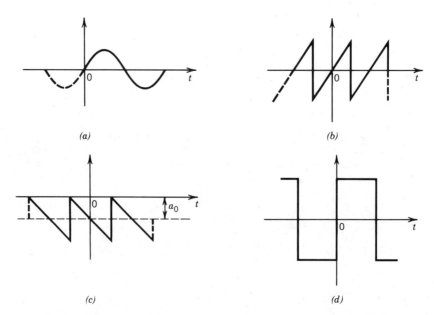

FIGURE A2.1-2 Periodic waveforms illustrating odd functions; (a) sine wave is an odd function and its average value $a_0 = 0$; (b) odd function and $a_0 = 0$; (c) odd function and $a_0 \neq 0$; (d) odd function and $a_0 = 0$.

Half-Wave Symmetry

A periodic function possesses half-wave symmetry if it satisfies the constraint

$$f(t) = -f\left(t + \frac{T}{2}\right). \qquad \text{(A2.1-4)}$$

Equation (A2.1-4) tells us that a periodic function has half-wave symmetry if after it is shifted one-half period ($T/2$) and inverted it is identical to the original function. The even harmonics of both the sine and cosine terms are zero. Thus waveforms having half-wave symmetry contain only odd harmonics. Two examples of such waveforms are shown in Figure A2.1-3.

Integral Relationships

The following integral relationships hold when m and n are integers:

$$\int_{-T/2}^{T/2} \cos m\omega t \, dt = 0 \quad \text{for all } m; \qquad \text{(A2.1-5)}$$

$$\int_{-T/2}^{T/2} \sin m\omega t \, dt = 0 \quad \text{for all } m; \qquad \text{(A2.1-6)}$$

$$\int_{-T/2}^{T/2} \cos m\omega t \cos n\omega t \, dt = \begin{cases} 0 & \text{for } m \neq n \\ T/2 & \text{for } m = n; \end{cases} \qquad \text{(A2.1-7)}$$

$$\int_{-T/2}^{T/2} \sin m\omega t \sin n\omega t \, dt = \begin{cases} 0 & \text{for } m \neq n \\ T/2 & \text{for } m = n; \end{cases} \qquad \text{(A2.1-8)}$$

$$\int_{-T/2}^{T/2} \sin m\omega t \cos n\omega t \, dt = 0 \quad \text{for all } m \text{ and } n. \qquad \text{(A2.1-9)}$$

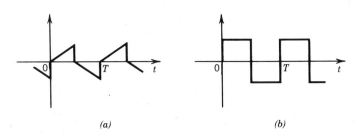

(a)   (b)

FIGURE A2.1-3 Periodic waveforms illustrating half-wave symmetry.

Derivation of Equation (A2.1-7). When $m \neq n$, use the formula

$$\cos A \cos B = \tfrac{1}{2}[\cos(A + B) + \cos(A - B)];$$

$$\omega t\big|_{t = \pm T/2} = \frac{2\pi}{T}\left(\pm \frac{T}{2}\right) = \pm\pi.$$

Then

$$\int_{-T/2}^{T/2} \cos m\omega t \cos n\omega t\, dt = \frac{1}{2}\int_{-T/2}^{T/2} \{\cos[(m + n)\omega t] + \cos[(m - n)\omega t]\}\, dt$$

$$= \frac{1}{2}\frac{1}{(m + n)\omega}\sin[(m + n)\omega t]\Big|_{-T/2}^{T/2} + \frac{1}{2}\frac{1}{(m - n)\omega}\sin[(m - n)\omega t]\Big|_{-T/2}^{T/2}$$

$$= \frac{1}{2}\frac{1}{(m + n)\omega}\{\sin[(m + n)\pi] - \sin[-(m + n)\pi]\}$$

$$+ \frac{1}{2}\frac{1}{(m - n)\omega}\{\sin[(m - n)\pi] - \sin[-(m - n)\pi]\}$$

$$= 0, \quad \text{if } m \neq n.$$

When $m = n$, use the formula

$$\cos^2 \theta = \tfrac{1}{2}(1 + \cos 2\theta).$$

Then

$$\int_{-T/2}^{T/2} \cos m\omega t \cos n\omega t\, dt = \int_{-T/2}^{T/2} \cos^2 m\omega t\, dt = \frac{1}{2}\int_{-T/2}^{T/2}(1 + \cos 2m\omega t)\, dt$$

$$= \frac{1}{2}t\Big|_{-T/2}^{T/2} + \frac{1}{4m\omega}\sin 2m\omega t\Big|_{-T/2}^{T/2} = \frac{T}{2}.$$

Derivation of Equation (A2.1-8). When $m \neq n$, use the formula

$$\sin A \sin B = \tfrac{1}{2}[\cos(A - B) - \cos(A + B)].$$

Then

$$\int_{-T/2}^{T/2} \sin m\omega t \sin n\omega t\, dt = \frac{1}{2}\int_{-T/2}^{T/2}[\cos(m - n)\omega t - \cos(m + n)\omega t]dt$$

$$= \frac{1}{2}\frac{1}{(m - n)\omega}\sin(m - n)\omega t\Big|_{-T/2}^{T/2} - \frac{1}{2}\frac{1}{(m + n)\omega}\sin(m + n)\omega t\Big|_{-T/2}^{T/2}$$

$$= \frac{1}{2} \frac{1}{(m-n)\omega} \{\sin(m-n)\pi - \sin[-(m-n)\pi]\}$$

$$- \frac{1}{2} \frac{1}{(m+n)\omega} \{\sin(m+n)\pi - \sin[-(m+n)\pi]\} = 0, \text{ if } m \neq n.$$

When $m = n$, use the formula

$$\sin^2 \theta = \tfrac{1}{2}(1 - \cos 2\theta).$$

Then

$$\int_{-T/2}^{T/2} \sin m\omega t \sin n\omega t \, dt = \int_{-T/2}^{T/2} \sin^2 m\omega t \, dt$$

$$= \frac{1}{2} \int_{-T/2}^{T/2} (1 - \cos 2m\omega t) \, dt = \frac{t}{2} \Big|_{-T/2}^{T/2} - \frac{1}{4m\omega} \sin 2m\omega t \Big|_{-T/2}^{T/2}$$

$$= \frac{T}{2}, \text{ if } m = n.$$

Derivation of Equation (A2.1-9).   When $m \neq n$, use the formula

$$\sin A \cos B = \tfrac{1}{2}[\sin(A+B) + \sin(A-B)].$$

Then

$$\int_{-T/2}^{T/2} \sin m\omega t \cos n\omega t \, dt = \frac{1}{2} \int_{-T/2}^{T/2} [\sin(m+n)\omega t + \sin(m-n)\omega t] \, dt$$

$$= \frac{1}{2} \left\{ \frac{-1}{(m+n)\omega} \cos(m+n)\omega t \Big|_{-T/2}^{T/2} + \frac{-1}{(m-n)\omega} \cos(m-n)\omega t \Big|_{-T/2}^{T/2} \right\}$$

$$= 0.$$

When $m = n$, use the formula

$$\sin \theta \cos \theta = \tfrac{1}{2}\sin 2\theta.$$

Then

$$\int_{-T/2}^{T/2} \sin m\omega t \cos n\omega t \, dt = \frac{1}{2} \int_{-T/2}^{T/2} \sin 2 \, m\omega t \, dt = - \frac{1}{4m\omega} \cos 2m\omega t \Big|_{-T/2}^{T/2}$$

$$= 0.$$

The coefficients a_n and b_n in Equation (A2.1-1) can be found using the integrals of Equations (A2.1-5) to (A2.1-9).

A2.2 FOURIER COEFFICIENTS

In order to find coefficient a_n, multiply both sides of Equation (A2.1-1) by $\cos n\omega t$ and then integrate both sides over one period $(-T/2, T/2)$. Thus

$$\int_{-T/2}^{T/2} f(t) \cos n\omega t\, dt = \int_{-T/2}^{T/2} a_0 \cos n\omega t\, dt$$

$$+ \int_{-T/2}^{T/2} a_1 \cos \omega t \cos n\omega t\, dt + \cdots$$

$$+ \int_{-T/2}^{T/2} a_n \cos^2 n\omega t\, dt + \cdots$$

$$+ \int_{-T/2}^{T/2} b_1 \sin \omega t \cos n\omega t\, dt + \cdots$$

$$+ \int_{-T/2}^{T/2} b_2 \sin 2\omega t \cos n\omega t\, dt + \cdots. \qquad \text{(A2.2-1)}$$

The definite integral terms on the right-hand side of Equation (A2.2-1) are zero except the term

$$\int_{-T/2}^{T/2} a_n \cos^2 n\omega t\, dt = \frac{a_n T}{2}. \qquad \text{(A2.1-7a)}$$

Hence

$$a_n = \frac{2}{T} \int_{-T/2}^{T/2} f(t) \cos n\omega t\, dt. \qquad \text{(A2.2-2)}$$

If both sides of Equation (A2.1-1) are multiplied by $\sin n\omega t$ and then integrated over one period $(-T/2, T/2)$, then the coefficient b_n can be expressed as

$$b_n = \frac{2}{T} \int_{-T/2}^{T/2} f(t) \sin n\omega t\, dt. \qquad \text{(A2.2-3)}$$

To find the first term (a_0) of Equation (A2.1-1), we simply integrate both sides of Equation (A2.1-1) over one period $(-T/2, T/2)$; thus

$$\int_{-T/2}^{T/2} f(t)\, dt = \int_{-T/2}^{T/2} \left[a_0 + \sum_{n=1}^{\infty} (a_n \cos n\omega t + b_n \sin n\omega t) \right] dt$$

$$= \int_{-T/2}^{T/2} a_0\, dt + \sum_{n=1}^{\infty} \int_{-T/2}^{T/2} (a_n \cos n\omega t + b_n \sin n\omega t)\, dt$$

$$= a_0 T + 0.$$

Hence

$$a_0 = \frac{1}{T} \int_{-T/2}^{T/2} f(t)\, dt. \qquad \text{(A2.2-4)}$$

Note that a_0 is the average value of $f(t)$, a_n is twice the average value of $f(t) \cos n\omega t$, and b_n is twice the average value of $f(t) \sin n\omega t$.

A2.3   FOURIER SERIES FOR SQUARE WAVEFORM

Even Quarter-Wave Symmetry

The square waveform shown in Figure A2.3-1 possesses even quarter-wave symmetry since the periodic function satisfies the conditions

$$f(t) = f(-t)$$

and

$$f(t) = -f(t + \tfrac{1}{2}T).$$

The Fourier series for an even quarter-wave symmetrical function contains only odd harmonics in cosine terms. We can see how the fundamental and odd

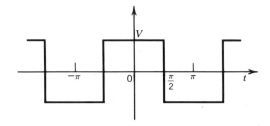

FIGURE A2.3-1   Square waveform illustrating even quarter-wave symmetrical function.

harmonics add to form a square wave from Figure 1.9-3c. Thus the Fourier coefficients and $f(t)$ are given by the following equations:

$$a_0 = 0;$$

$$a_{2n} = 0, \text{ for } 2n \text{ even};$$

$$b_n = 0 \text{ for all } n, \text{ since the function is even};$$

$$f(t) = \sum_{n=1}^{\infty} a_{2n-1} \cos(2n - 1)\omega t; \tag{A2.3-1}$$

$$a_{2n-1} = \frac{8}{T} \int_0^{T/4} f(t) \cos(2n - 1)\omega t \, dt \quad \text{for } 2n - 1 \text{ odd}. \tag{A2.3-2}$$

Derivation of Equation (A2.3-2). (a) Since the even function $f(t)$ is a symmetrical wave,

$$\int_{-a}^0 f(t)dt = \int_0^a f(-t)dt = \int_0^a f(t)dt. \tag{A2.3-3}$$

Hence

$$\int_{-a}^a f(t)dt = \int_{-a}^0 f(t)dt + \int_0^a f(t)dt = 2\int_0^a f(t)dt, \tag{A2.3-4}$$

and from Equation (A2.3-4),

$$a_{2n-1} = \frac{2}{T} \int_{-T/2}^{T/2} f(t) \cos(2n - 1)\omega t \, dt$$

$$= \frac{4}{T} \int_0^{T/2} f(t) \cos(2n - 1)\omega t \, dt$$

$$= \frac{4}{T} \left[\int_0^{T/4} f(t) \cos(2n - 1)\omega t + \int_{T/4}^{T/2} f(t) \cos(2n - 1)\omega t \right] dt. \tag{A2.3-5}$$

(b) If $t + T/2$ is used instead of the variable t in the second term on the right-hand side of Equation (A2.3-5), then

$$a_{2n-1} = \frac{4}{T} \left[\int_0^{T/4} f(t) \cos(2n - 1)\omega t \, dt \right.$$

$$\left. + \int_{-T/4}^0 f\left(t + \frac{T}{2}\right) \cos(2n - 1)\omega\left(t + \frac{T}{2}\right) dt \right]. \tag{A2.3-6}$$

Note that $f(t) = -f(t + T/2)$; if this relation is applied to Equation (A2.3-6) again, then

$$a_{2n-1} = \frac{4}{T}\left[\int_0^{T/4} f(t)\cos(2n-1)\omega t\, dt + \int_{-T/4}^0 f(t)\cos(2n-1)\omega t\, dt\right].$$

(A2.3-7)

Note also that $f(-t) = f(t)$ and $f(t)\cos(2n-1)\omega t$ is an even function. Thus Equation (A2.3-2) will be obtained from Equation (A2.3-7), referring to Equation (A2.3-4).

Series for Square Waves

From Equation (2.3-2),

$$a_{2n-1} = \frac{8}{T}\int_0^{T/2} f(t)\cos(2n-1)\omega t\, dt$$

$$= \frac{4}{\pi}\int_0^{\pi/2} V\cos(2n-1)\omega t\, d\omega t$$

$$= \frac{4V}{\pi}\frac{1}{2n-1}\sin(2n-1)\omega t\Big|_0^{\pi/2}$$

$$= \frac{4V}{\pi}\frac{1}{2n-1}\sin(2n-1)\frac{\pi}{2}$$

$$= \begin{cases} \dfrac{4V}{(2n-1)\pi}, & 2n-1 = 1, 5, \dots \\[2mm] -\dfrac{4V}{(2n-1)\pi}, & 2n-1 = 3, 7, \dots. \end{cases}$$

Hence

$$v = f(t) = \sum_{n=1}^{\infty} a_{2n-1}\cos(2n-1)\omega t$$

$$= \frac{4V}{\pi}\left(\cos\omega t - \tfrac{1}{3}\cos 3\omega t + \tfrac{1}{5}\cos 5\omega t - \tfrac{1}{7}\cos 7\omega t + \cdots\right). \quad (1.7\text{-}2)$$

The frequency spectrum of the square waveform is shown in Figure 1.7-3b.

A2.4   FOURIER SERIES FOR SAWTOOTH WAVEFORMS

Positive Sawtooth Wave

The Fourier series for the positive sawtooth wave of Figure 1.7-2 is given by

$$v = \frac{V}{2} - \frac{V}{\pi} \sum_{n=1}^{\infty} \frac{1}{n} \sin n\omega t,$$

or

$$v = \frac{V}{2} - \frac{V}{\pi} (\sin \omega t + \tfrac{1}{2} \sin 2\omega t + \tfrac{1}{3} \sin 3\omega t + \cdots), \qquad (1.7\text{-}1)$$

Where the first term $V/2$ is the average value. This equation is derived as follows. The expression for v between 0 and T is

$$v = \left(\frac{V}{T}\right) t = f(t).$$

The equation for a_0 is

$$a_0 = \frac{1}{T} \int_0^T \left(\frac{V}{T}\right) t \, dt = \tfrac{1}{2} V.$$

The equation for the value of a_n is

$$a_n = \frac{2}{T} \int_0^T \left(\frac{V}{T}\right) t \cos n\omega t \, dt$$

$$\left(\text{Use} \int u \, dv = uv - \int v \, du; \text{take } u = t \text{ and } dv = \frac{1}{n\omega} \cos n\omega t \, d n\omega t \right)$$

$$= \frac{2V}{T^2} \left\{ \frac{1}{n^2 \omega^2} \cos n\omega t \Big|_0^T + \frac{t}{n\omega} \sin n\omega t \Big|_0^T \right\}$$

$$= \frac{2V}{T^2} \left\{ \frac{1}{n^2 \omega^2} (\cos 2\pi n - 1) \right\} = 0 \quad \text{for all } n.$$

The equation for the nth value of b_n is

$$b_n = \frac{2}{T} \int_0^T \left(\frac{V}{T}\right) t \sin n\omega t \, dt$$

$$\left(\text{Use } u \, dv = uv - \int v \, du; \text{ take } u = t \text{ and } dv = \frac{1}{n\omega} \sin n\omega t \, dn\omega t \right)$$

$$= \frac{2V}{T^2} \left\{ \frac{1}{n^2\omega^2} \sin n\omega t \bigg|_0^T - \frac{t}{n\omega} \cos n\omega t \bigg|_0^T \right\}$$

$$= \frac{2V}{T^2} \left\{ 0 - \frac{T}{n\omega} \cos 2\pi n \right\} = \frac{-V}{n\pi}.$$

Thus Equation (1.7-1) is obtained.

Sawtooth Wave as Example of Odd Function

The sawtooth waveform of Figure A2.4-1 shows that

$$v = f(t) = -f(-t).$$

This waveform has zero average value and is an example of odd function. Since the coefficients of the cosine terms equal zero for odd functions, the Fourier series for this waveform only contains sine terms. Thus

$$f(t) = \sum_{n=1}^{\infty} b_n \sin n\omega t.$$

The product of two odd functions $[f(t) \sin \omega t]$ is an even function. According to Equation (A2.3-4), b_n given by Equation (A2.2-3) can also be written as follows:

$$b_n = \frac{2}{T} \int_{-T/2}^{T/2} f(t) \sin n\omega t \, dt$$

$$= \frac{4}{T} \int_0^{T/2} f(t) \sin n\omega t \, dt. \tag{A.2.4-1}$$

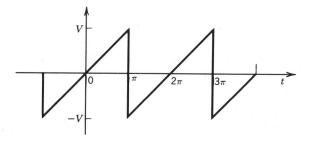

FIGURE A2.4-1 Sawtooth waveform with zero average value.

In the half-cycle from 0 to π,

$$f(t) = (V/\pi)\omega t. \qquad (A2.4-2)$$

Substituting Equation (A2.4-2) into Equation (A2.4-1), we obtain

$$b_n = \frac{4}{T} \int_0^{T/2} \frac{V}{\pi} \omega t \sin n\omega t \, dt$$

$$= \frac{4}{T} \int_0^{T/2} \frac{V}{\pi n} t \sin n\omega t \, dn\omega t$$

$$= \frac{2}{\pi} \int_0^{\pi} \frac{V}{\pi n} \omega t \sin n\omega t \, dn\omega t$$

$$= \frac{2V}{\pi^2} \int_0^{\pi} \frac{\omega t}{n} \sin n\omega t \, dn\omega t$$

$$\left(\text{Use formula} \int u \, dv = uv - \int v \, du; \text{ take } u = \omega t/n \text{ and} \right.$$

$$\left. dv = -d \cos n\omega t = \sin n\omega t \, dn\omega t \right)$$

$$= \frac{2V}{\pi^2} \left(-\frac{\omega t}{n} \cos n\omega t \Big|_0^{\pi} + \int_0^{\pi} \frac{\omega}{n} \cos n\omega t \, dt \right)$$

$$= \frac{2V}{\pi^2} \left(-\frac{\omega t}{n} \cos n\omega t \Big|_0^{\pi} + \int_0^{\pi} \frac{1}{n^2} d \sin n\omega t \right)$$

$$= \frac{2V}{\pi^2} \left[-\frac{\omega t}{n} \cos n\omega t + \frac{1}{n^2} \sin n\omega t \right]_0^{\pi} = \frac{2V}{\pi^2} \left[-\frac{\pi}{n} \cos n\pi \right]$$

or

$$b_n = -\frac{2V}{n\pi} \cos n\pi. \qquad (A2.4-3)$$

when n is an odd number, $\cos n\pi$ is negative; when n is an even number, $\cos n\pi$ is positive. Thus the Fourier series for the sawtooth wave of Figure A2.4-1 is given by

$$f(t) = \sum_{n=1}^{\infty} b_n \sin n\omega t$$

$$= \frac{2V}{\pi} (\sin \omega t - \tfrac{1}{2} \sin 2\omega t + \tfrac{1}{3} \sin 3\omega t - \cdots). \qquad (A2.4-4)$$

A2.5 FOURIER SERIES FOR TRIANGULAR WAVEFORMS

Triangular Waveform with Even Function Symmetry

The triangular waveform of Figure A2.5-1a shows that

$$f(t) = f(-t).$$

This waveform is an even function whose average value is $V/2$. The Fourier series for the even function contains only cosine terms. The product of $f(t) \cos n\omega t$ is an even function. From Equations (A2.2-2) and (A2.3-4),

$$a_n = \frac{2}{T} \int_{-T/2}^{T/2} f(t) \cos n\omega t \, dt = \frac{4}{T} \int_0^{T/2} f(t) \cos n\omega t \, dt. \qquad (A2.5\text{-}1)$$

In the interval $0 < \omega t < \pi$, the expression for $f(t)$ is

$$f(t) = V - (V/\pi)\omega t. \qquad (A2.5\text{-}2)$$

Substitution of Equation (A2.5-3) into Equation (A2.5-4) yields

$$a_n = \frac{4}{T} \int_0^{T/2} [V - (V/\pi)\omega t] \cos n\omega t \, dt$$

$$= \frac{2V}{n\pi} \int_0^{\pi} \cos n\omega t \, dn\omega t - \frac{2V}{n\pi^2} \int_0^{\pi} \omega t \cos n\omega t \, dn\omega t$$

$$= \frac{2V}{n\pi} \int_0^{\pi} d \sin n\omega t - \frac{2V}{n\pi^2} \int_0^{\pi} \omega t \, d \sin n\omega t$$

$$= \frac{2V}{n\pi} [\sin n\omega t]_0^{\pi} - \frac{2V}{n\pi^2} [\omega t \sin n\omega t]_0^{\pi} + \frac{2V}{(n\pi)^2} \int_0^{\pi} \sin n\omega t \, dn\omega t$$

$$= 0 - 0 - \frac{2V}{(n\pi)^2} [\cos n\omega t]_0^{\pi} = -\frac{2V}{(n\pi)^2} [\cos n\pi - 1];$$

or

$$a_n = \begin{cases} \dfrac{4V}{(n\pi)^2}, & n = \text{odd number } 1, 3, 5, \ldots, \\ 0, & n = \text{even number } 2, 4, 6, \ldots. \end{cases} \qquad (A2.5\text{-}3)$$

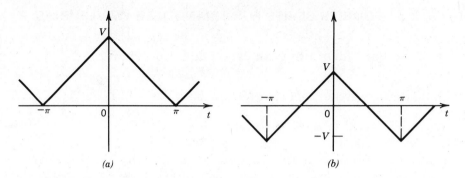

FIGURE A2.5-1   Triangular waveforms with (a) even-function symmetry and (b) even quarter-wave symmetry.

The Fourier series for the triangular waveform of Figure A2.5-1a is

$$v = a_0 + \sum_{n=1}^{\infty} a_n \cos n\omega t. \qquad (A2.5\text{-}4)$$

Substitution of Equation (A2.5-3) into Equation (A2.5-4 yields

$$v = \frac{V}{2} + \frac{4V}{\pi^2} \cos \omega t + \frac{4V}{(3\pi)^2} \cos 3\omega t + \frac{4V}{(5\pi)^2} \cos 5\omega t + \cdots. \qquad (A2.5\text{-}5)$$

Triangular Waveform with Even Quarter-Wave Symmetry

The triangular waveform of Figure A2.5-1b shows that

$$f(t) = f(-t)$$

and

$$f(t) = -f(t + T/2).$$

This waveform with zero average value has even quarter-wave symmetry. In the interval $0 < \omega t < \pi/2$,

$$f(t) = V - \frac{2V}{\pi}\omega t. \qquad (A2.5\text{-}6)$$

Substitution of Equation (A2.5-6) into Equation (A2.3-2) yields

$$a_{2n-1} = \frac{8}{T} \int_0^{T/4} f(t) \cos(2n-1)\omega t\, dt$$

$$= \frac{8}{T} \int_0^{T/4} \left(V - 2V \frac{\omega t}{\pi} \right) \cos(2n - 1)\omega t \, dt$$

$$= \frac{4V}{\pi(2n - 1)} \int_0^{\pi/2} \left(1 - \frac{2\omega t}{\pi} \right) \cos(2n - 1)\omega t \, d(2n - 1)\omega t$$

$$= \frac{4V}{\pi(2n - 1)} \left[\int_0^{\pi/2} d \sin(2n - 1)\omega t - \int_0^{\pi/2} \frac{2\omega t}{\pi} d \sin(2n - 1)\omega t \right]$$

$$= \frac{4V}{\pi(2n - 1)} \left\{ [\sin(2n - 1)\omega t]_0^{\pi/2} - \left[\frac{2\omega t}{\pi} \sin(2n - 1)\omega t \right]_0^{\pi/2} \right.$$

$$\left. + \int_0^{\pi/2} \sin(2n - 1)\, \omega t \frac{2}{\pi(2n - 1)} d(2n - 1)\omega t \right\}$$

$$= \frac{4V}{\pi(2n - 1)} \left\{ \sin(2n - 1)\pi/2 - \sin(2n - 1)\pi/2 \right.$$

$$\left. - \frac{2}{\pi(2n - 1)} \cos(2n - 1)\omega t \bigg|_0^{\pi/2} \right\}$$

$$= \frac{4V}{\pi(2n - 1)} \left\{ \frac{2}{\pi(2n - 1)} \right\} = \frac{8V}{\pi^2(2n - 1)^2}$$

or

$$a_{2n-1} = \frac{8V}{\pi^2(2n - 1)^2}. \qquad \text{(A2.5-7)}$$

Substituting Equation (A2.5-7) into Equation (A2.3-1), we obtain

$$v = f(t) = \sum_{n=1}^{\infty} a_{2n-1} \cos(2n - 1)\omega t$$

$$= \frac{8V}{\pi^2} \left[\cos \omega t + \frac{1}{9} \cos 3\omega t + \frac{1}{25} \cos 5\omega t + \frac{1}{49} \cos 7\omega t + \cdots \right]. \quad \text{(A2.5-8)}$$

This is the Fourier series for the triangular waveform of Figure A2.5-1b.

A2.6 FOURIER SERIES FOR HALF-WAVE RECTIFIED SINE WAVE

The ac sine wave after half-wave rectification becomes a half-wave rectified waveform, as shown in Figure A2.6-1. Since this waveform does not possess

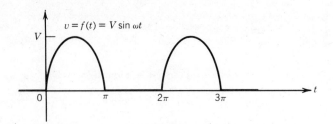

FIGURE A2.6-1 Half-wave rectified sine wave. Its average value is V/π.

any symmetry, its Fourier series representation must contain cosine and sine terms in addition to a dc component. From Equation (A2.2-2), the expression for a_n is

$$a_n = \frac{2}{T} \int_{-T/2}^{T/2} f(t) \cos n\omega t \, dt = \frac{1}{\pi} \int_0^\pi V \sin \omega t \cos n\omega t \, d\omega t.$$

Using $\sin A \cos B = \frac{1}{2}[\sin(A + B) + \sin(A - B)]$,

$$a_n = \frac{V}{2\pi} \int_0^\pi [\sin(1 + n)\omega t + \sin(1 - n)\omega t] d\omega t$$

$$= \frac{V}{2\pi} \left[\int_0^\pi -\frac{1}{1 + n} d \cos(1 + n)\omega t + \int_0^\pi -\frac{1}{1 - n} d \cos(1 - n)\omega t \right]$$

$$= -\frac{V}{2\pi} \left[\frac{\cos(1 + n)\omega t}{1 + n} + \frac{\cos(1 - n)\omega t}{1 - n} \right]_0^\pi$$

$$= -\frac{V}{2\pi} \left[\frac{\cos(1 + n)\pi}{1 + n} + \frac{\cos(1 - n)\pi}{1 - n} - \frac{1}{1 + n} - \frac{1}{1 - n} \right]$$

$$= -\frac{V}{2\pi} \left[\frac{(1 - n) \cos(1 + n)\pi + (1 + n) \cos(1 - n)\pi - (1 - n) - (1 + n)}{(1 + n)(1 - n)} \right]$$

$$= -\frac{V}{2\pi} \left[\frac{2 \cos \pi \cos n\pi + 2n \sin \pi \sin n\pi - 2}{1 - n^2} \right]$$

$$= \frac{V}{\pi} \left[\frac{-\cos \pi \cos n\pi - n \sin \pi \sin 2\pi + 1}{1 - n^2} \right]$$

or

$$a_n = \frac{V}{\pi(1 - n^2)} [\cos n\pi + 1]. \qquad (A2.6\text{-}1)$$

a_n is zero when n is an odd number but indefinite when $n = 1$. Thus a_1 must be determined from the following:

$$a_1 = \frac{1}{\pi} \int_0^\pi V \sin \omega t \cos \omega t \, d\omega t$$

$$= \frac{V}{\pi} \int_0^\pi \sin \omega t \, d \sin \omega t = 0. \qquad (A2.6\text{-}2)$$

From Equation (A2.2-3), the expression for b_n is

$$b_n = \frac{2}{T} \int_{-T/2}^{T/2} f(t) \sin n\omega t \, dt$$

$$= \frac{1}{\pi} \int_0^\pi V \sin \omega t \sin n\omega t \, d\omega t.$$

From Equation (A2.1-8), we know

$$b_n = 0 \quad \text{when } n \neq 1.$$

But when $n = 1$,

$$b_1 = \frac{1}{\pi} \int_0^\pi V \sin^2 \omega t \, d\omega t$$

$$= \frac{V}{\pi} \int_0^\pi \frac{1}{4} (1 - \cos 2\omega t) d2\omega t$$

$$= \frac{V}{\pi} \left[\frac{\omega t}{2} - \frac{\sin 2\omega t}{4} \right]_0^\pi = \frac{V}{2}.$$

Thus the Fourier series for the half-wave rectified sine wave of Figure A2.6-1 is given by

$$f(t) = a_0 + \sum_{n=1}^{\infty} (a_n \cos n\omega t + b_n \sin n\omega t)$$

$$= \frac{V}{\pi} \left(1 + \frac{\pi}{2} \sin \omega t - \frac{2}{3} \cos 2\omega t - \frac{2}{15} \cos 4\omega t - \frac{2}{35} \cos 6\omega t - \cdots \right).$$

$$(A2.6\text{-}3)$$

A2.7 EXPONENTIAL FORM OF FOURIER SERIES

Derivation of Exponential Form

The trigonometric Fourier series is

$$f(t) = a_0 + \sum_{n=1}^{\infty} (a_n \cos n\omega t + b_n \sin n\omega t), \tag{A2.1-1}$$

where the cosine and sine functions can be replaced with their exponential equivalents

$$\cos n\omega t = \frac{1}{2}(e^{jn\omega t} + e^{-jn\omega t}) \tag{A2.7-1}$$

and

$$\sin n\omega t = \frac{1}{2j}(e^{jn\omega t} - e^{-jn\omega t}). \tag{A2.7-2}$$

Substitution of Equations (A2.7-1) and (A2.7-2) into Equation (A2.1-1) yields

$$f(t) = a_0 + \sum_{n=1}^{\infty} \left[a_n \frac{1}{2}(e^{jn\omega t} + e^{-jn\omega t}) + b_n \frac{1}{2j}(e^{jn\omega t} - e^{-jn\omega t}) \right]$$

$$= a_0 + \sum_{n=1}^{\infty} \left[\frac{1}{2}(a_n - jb_n)e^{jn\omega t} + \frac{1}{2}(a_n + jb_n)e^{-jn\omega t} \right]. \tag{A2.7-3}$$

Let

$$C_0 = a_0; \tag{A2.7-4}$$

$$C_n = \tfrac{1}{2}(a_n - jb_n); \tag{A2.7-5}$$

$$C_{-n} = \tfrac{1}{2}(a_n + jb_n). \tag{A2.7-6}$$

The integral relationships for C_0, C_n, and C_{-n} can be obtained from Equations (A2.2-4), (A2.2-2), and (A2.2-3). Substituting Equations (A2.7-4), (A2.7-5), and (A2.7-6) into Equation (A2.7-3), we get

$$f(t) = C_0 + \sum_{n=1}^{\infty} (C_n e^{jn\omega t} + C_{-n} e^{-jn\omega t})$$

$$= \sum_{n=0}^{\infty} C_n e^{jn\omega t} + \sum_{n=1}^{\infty} C_{-n} e^{-jn\omega t}. \tag{A2.7-7}$$

Observe that the second summation on the right side of Equation (A2.7-7) is equivalent to summing $C_n e^{jn\omega t}$ from -1 to $-\infty$; that is,

$$\sum_{n=1}^{\infty} C_{-n} e^{-jn\omega t} = \sum_{n=-1}^{-\infty} C_n e^{jn\omega t}. \tag{A2.7-8}$$

Since the summation from -1 to $-\infty$ is the same as the summation from $-\infty$ to -1, we can use Equation (A2.7-8) to rewrite Equation (A2.7-7) as

$$f(t) = C_0 + \sum_{n=1}^{\infty} C_n e^{jn\omega t} + \sum_{n=-1}^{-\infty} C_n e^{jn\omega t}$$

$$= \sum_{n=0}^{\infty} C_n e^{jn\omega t} + \sum_{-\infty}^{-1} C_n e^{jn\omega t}$$

$$= \sum_{n=-\infty}^{\infty} C_n e^{jn\omega t}. \tag{A2.7-9}$$

Equation (A2.7-9) is the exponential form of the Fourier series. The coefficients C_n and C_{-n} can be obtained from the integral expressions of a_n and b_n.

$$C_0 = a_0 = \frac{1}{T} \int_{-T/2}^{T/2} f(t)dt; \tag{A2.7-10}$$

$$C_n = \tfrac{1}{2}(a_n - jb_n)$$

$$= \frac{1}{T} \left[\int_{-T/2}^{T/2} f(t) \cos n\omega t\, dt - j \int_{-T/2}^{T/2} f(t) \sin n\omega t\, dt \right]$$

$$= \frac{1}{T} \left[\int_{-T/2}^{T/2} f(t)(\cos n\omega t - j \sin n\omega t)dt \right]$$

$$= \frac{1}{T} \int_{-T/2}^{T/2} f(t)e^{-jn\omega t}\, dt. \tag{A2.7-11}$$

$$C_{-n} = \frac{1}{2}(a_n + jb_n) = \frac{1}{T} \int_{-T/2}^{T/2} f(t)e^{jn\omega t}\, dt. \tag{A2.7-12}$$

Combining Equations (A2.7-10), (A2.7-11), and (A2.7-12), we can obtain the following expression:

$$C_n = \frac{1}{T} \int_{-T/2}^{T/2} f(t)e^{-jn\omega t}\, dt, \qquad n = 0, \pm 1, \pm 2, \ldots. \tag{A2.7-13}$$

Since $f(t)e^{-jn\omega t}$ is a function with period T, Equation (A2.7-13) can be expressed as

$$C_n = \frac{1}{T} \int_0^T f(t)e^{-jn\omega t} \, dt \qquad \text{(A2.7-14)}$$

or

$$C_n = \frac{1}{2\pi} \int_0^{2\pi} f(t)e^{-jn\omega t} \, d\omega t. \qquad \text{(A2.7-15)}$$

From Equations (A2.7-4), (A2.7-5), and (A2.7-6), we see the following conclusion: If the series contains only sine terms, then $a_n = 0$ and the values of C_n are imaginary for all n. If the coefficients of the sine terms are zero or $b_n = 0$, then the values of C_n are real for all n. If the trigonometric series does not contain even harmonics, or if the function has half-wave symmetry, then $C_n = 0$ for n even.

Complex-Number Transition from Rectangular to Polar Form

In the rectangular form, a complex number (n) is written in terms of its real component (a) and imaginary component (b); thus

$$n = a + jb; \qquad \text{(A2.7-16)}$$

$$j = \sqrt{-1}.$$

In polar form, a complex number is written in terms of its magnitude (c), or modulus, and angle (θ), or argument; thus

$$n = ce^{j\theta}, \qquad \text{(A2.7-17)}$$

where e is the base of the natural logarithm. In the literature, the symbol $\underline{/\theta}$ is frequently used in place of $e^{j\theta}$; that is, the polar form is written

$$n = c\underline{/\theta}. \qquad \text{(A2.7-18)}$$

The transition from the rectangular to the polar form makes use of Euler's identity:

$$\cos\theta \pm j\sin\theta = e^{\pm j\theta}. \qquad \text{(A2.7-19)}$$

A complex number in polar form can be put in rectangular form by writing

$$ce^{j\theta} = c[\cos\theta + j\sin\theta]$$

$$= c\cos\theta + jc\sin\theta$$

$$= a + jb. \tag{A2.7-20}$$

The transition from rectangular to polar form makes use of the geometry of the right triangle; that is,

$$a + jb = (\sqrt{a^2 + b^2})e^{j\theta} = ce^{j\theta}, \tag{A2.7-21}$$

where

$$\tan\theta = \frac{b}{a} \quad \text{and} \quad \sqrt{a^2 + b^2} = c. \tag{A2.7-22}$$

Example A2.7-1. Determine the exponential Fourier series for the sawtooth waveform shown in Figure A2.7-1.

Solution. In the interval $0 < \omega t < 2\pi$,

$$f(t) = \frac{10}{2\pi}\omega t.$$

From Equation (A2.7-15),

$$C_n = \frac{1}{2\pi}\int_0^{2\pi}\frac{10}{2\pi}\omega t e^{-jn\omega t}\, d\omega t = \frac{10}{(2\pi)^2}\int_0^{2\pi}\frac{\omega t}{-jn}de^{-jn\omega t}$$

$$= \frac{10}{(2\pi)^2}\left[\frac{e^{-jn\omega t}}{(-jn)^2}(-jn\omega t - 1)\right]_0^{2\pi}$$

$$= \frac{10}{(2\pi)^2}\left[\frac{\cos n\omega t - j\sin n\omega t}{(-jn)^2}(-jn\omega t - 1)\right]_0^{2\pi} = j\frac{10}{2\pi n}.$$

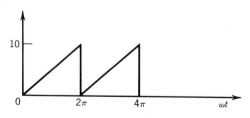

10

0   2π 4π ωt

FIGURE A2.7-1

$$C_0 = a_0 = \frac{1}{T} \int_0^T f(t)dt$$

$$= \frac{1}{2\pi} \int_0^{2\pi} \frac{10}{2\pi} \omega t \, d\omega t$$

$$= \frac{10}{8\pi^2} (\omega t)^2 \Big|_0^{2\pi} = 5.$$

Substituting the values of C_0 and C_n into Equation (A2.7-9), we obtain

$$f(t) = 5 + \sum_{n=1}^{\infty} j \frac{10}{2\pi n} e^{jn\omega t} + \sum_{n=-1}^{-\infty} j \frac{10}{2\pi n} e^{jn\omega t}.$$

This is the exponential form of the Fourier series for the sawtooth wave of Figure A2.7-1.

A2.8   RMS VALUE OF PERIODIC NONSINUSOIDAL WAVEFORM

From Equations (A2.1-1) and (A2.7-20),

$$f(t) = a_0 + a_1 \cos \omega t + a_2 \cos 2\omega t + a_3 \cos 3\omega t + \cdots$$

$$+ b_1 \sin \omega t + b_2 \sin 2\omega t + b_3 \sin 3\omega t + \cdots$$

$$= a_0 + \sum_{n=1}^{\infty} (a_n \cos n\omega t + b_n \sin n\omega t)$$

$$= a_0 + \sum_{n=1}^{\infty} (C_n^* \cos \theta_n \cos n\omega t + C_n^* \sin \theta_n \sin n\omega t)$$

or

$$f(t) = a_0 + \sum_{n=1}^{\infty} C_n^* \cos(n\omega t - \theta_n). \qquad \text{(A2.8-1)}$$

By definition, the rms (root-mean-square) value of a periodic function can be expressed as

$$F_{\text{rms}} = \sqrt{\frac{1}{T} \int_0^T f(t)^2 \, dt} \qquad \text{(A2.8-2)}$$

or

$$F_{rms} = \sqrt{\frac{1}{T} \int_0^T \left[a_0 + \sum_{n=1}^{\infty} C_n^* \cos(n\omega t - \theta_n) \right]^2 dt}. \qquad \text{(A2.8-3)}$$

The integral of the squared time function is simplified because the only terms to survive the integration over a period will be the product of the dc term and the harmonic products of the same frequency. All other products will integrate to zero. Hence Equation (A2.8-3) reduces to

$$F_{rms} = \sqrt{\frac{1}{T} \left[a_0^2 T + \sum_{n=1}^{\infty} \frac{T}{2} C_n^{*2} \right]} = \sqrt{a_0^2 + \sum_{n=1}^{\infty} \frac{C_n^{*2}}{2}}$$

$$= \sqrt{a_0^2 + \sum_{n=1}^{\infty} \left[\frac{C_n^*}{\sqrt{2}} \right]^2}. \qquad \text{(A2.8-4)}$$

Equation (A2.8-4) tells us that the rms value of a periodic function is the square root of the sum obtained by adding the square of the rms value of each harmonic to the square of the dc value. For example, assume that a periodic voltage is represented by the finite series

$$v = 15 + 27.01 \cos(400\pi t - 45°) + 19.10 \cos(800\pi t - 90°)$$

$$+ 9.00 \cos(1200\pi t - 135°) + 5.40 \cos(2000\pi t - 45°).$$

The rms value of the voltage is

$$V_{rms} = \sqrt{15^2 + \left(\frac{27.01}{\sqrt{2}}\right)^2 + \left(\frac{19.10}{\sqrt{2}}\right)^2 + \left(\frac{9.00}{\sqrt{2}}\right)^2 + \left(\frac{5.40}{\sqrt{2}}\right)^2}$$

$$= 28.76 \text{ V}.$$

GLOSSARY

Absolute address. A memory address that is given exactly, not computed or inferred from information.

Access time. The time needed to gain access to stored information in an electronic memory.

Accumulator. (a) A parallel array of flip-flops in the CPU of a computer that serves as the source and destination of most arithmetic operations. (b) A register in a microprocessor in which the result of a given operation is stored temporarily.

Accuracy. The degree to which a device is calibrated against a known standard.

Active edge. The triggering edge that causes an edge-sensitive device to sample its input(s).

Active filter. An electronic filter that uses passive circuit elements with active devices, such as operational amplifiers. In general, resistors and capacitors, but not inductors, are used.

Active level. The logic state in which a device is enabled.

Actuator. A device that converts a voltage or current input into a mechanical output.

ADC. Abbreviation for analog-to-digital converter.

A/D converter. Analog-to-digital converter. A circuit that converts an analog (continuous) voltage or current into an output digital code.

Addend. In arithmetic addition, the quantity added to the augend.

Address. The location of a given storage cell in a memory.

Address bus. A parallel array of conductors capable of transmitting address information from the CPU (central processing unit) of a computer to external elements such as memory and I/O circuitry. It is a unidirectional bus.

AFT. Automatic fine tuning.

AGC. Automatic gain control.

Algorithm. A step-by-step procedure that outlines the sequence of actions necessary to solve a problem.

Analog multiplexer. An array of switches with a common output connection for selecting one of a number of analog inputs. The output signal follows the selected input within a small error.

AND gate. Has a 1 output if and only if all inputs are at the 1 level.

Argument. The values upon which a function operates.

Arithmetic and logic unit (ALU). A complex combinational logic block in the CPU of a computer responsible for all arithmetic and logic operations.

Assembler. A computer program that translates binary-coded alphanumeric symbols that represent instruction mnemonics into executable instruction codes.

Astable blocking oscillator. An oscillator that conducts for a short period of time and is blocked out (cut off) for a much longer period.

Astable multivibrator. An oscillator circuit whose output is a binary signal. Typically an astable produces a pulse train or a square wave.

Asynchronous. Having no fixed time relationship.

Audio frequencies. Frequencies that can be heard by the human ear (approximately between 15 and 20,000 Hz).

Augend. In arithmetic addition, the quantity to which the addend is added.

Bandwidth. The range of frequencies over which a given device is designed to operate within specified limits.

Base. Same as *Radix*.

Baud. In general, a unit of signaling speed. In data processing, it is a group of tracks on a magnetic drum or on the side of a magnetic disk. In data communication the band is the frequency spectrum between two defined limits.

BCD code. *Binary-coded-decimal* code is a modification of the decimal number system different only in that the decimal digits are independently coded as four-bit binary numbers.

BCD counter. A ten-state natural binary counter module whose sequence begins at 0000 and ends at 1001, cascadable to any length to produce a multidigit binary-coded-decimal count sequence.

Bidirectional bus. A parallel array of conductors capable of transmitting information in two directions, requiring three-state driver gates at each end.

Binary counter. An interconnection of flip-flops whose outputs progress through a natural binary sequence when a periodic signal is applied to its clock input.

Bipolar junction transistor (BJT).   A three-layer silicon (or germanium) device consisting of either two *p*- and one *n*-type layers of material (*pnp*) or two *n*- and one *p*-type layers of material (*npn*). The three portion of a transistor are called the emitter (*E*), base (*B*), and collector (*C*).

Bistable multivibrator.   An alternative name for a flip-flop.

Bit.   A binary digit (either 0 or 1).

Boolean algebra.   A mathematics of logic.

Borrow digit.   A digit produced from the MSB stage in subtraction when the subtrahend is larger than the minuend.

Bubble memory.   A memory that uses tiny magnetic bubbles to store 1s and 0s.

Bus.   One or more conductors used for transmitting signals or power. These buses include a microprocessor interface bus, address bus, data bus, and control bus.

Byte.   An eight-bit binary number.

Carry digit.   A digit produced from the MSB stage in arithmetic addition when the sum cannot be expressed with the alloted number of digits.

Cascade.   A configuration in which one device drives another.

Cascode pair.   A pair of the same type of transistors with one emitter connected to the other device's collector. (A similar meaning applies to FETs or vacuum tubes.)

Channel.   A channel is part of a communication system with a frequency bandwidth sufficient for a one-way system. A channel in computers is a circuit, link, or path for the flow of information.

Character.   A member of a set of elements for which agreement on meaning has been reached and that is used for the organization, control, or representation of data. Characters may be letters, digits, punctuation marks, other symbols.

Chopper.   An electromechanical or electronic device used to interrupt a dc or low-frequency ac signal at regular intervals to permit amplification of the signal by an ac amplifier.

Chrominance (C) signal.   A modulated subcarrier of 3.58 MHz.

Class A operation.   A transistor operated in the class A mode draws current for the full ac cycle (360°).

Class AB operation.   A transistor operated in the class AB mode draws current for more than half the ac cycle but less than a full ac cycle.

Class B operation.   A transistor operated in the class B mode draws current for exactly 180° of the full ac cycle.

Class C operation.   A transistor operated in the class C mode draws current for less than 180° of the full ac cycle.

Clear direct.   An asynchronous input found on flip-flops, registers, and counters that causes the internal state of such devices to become 0 (or all 0's).

Clock. A digital circuit that provides a continuous pulse train.

Clock rate. The frequency of the timing pulses of the clock circuit.

Clocked D latch. An *S-R* latch with $S = D$, and $R = \bar{S}$. When enabled, Q follows D. When disabled, Q is held steady at its last correct value.

Clocked S-R latch. A type of latch that contains set, reset, and enable (or clock) inputs and requires an enable signal to load data.

CMOS. Complementary symmetry metal oxide semiconductor.

Combinational logic. Memoryless logic circuitry that can be unambiguously specified by a truth table, and in which no feedback exists.

Common-mode rejection ratio (CMRR). For an amplifier, the ratio of differential voltage gain to common-mode voltage gain, generally expressed in dB.

$$\text{CMRR} = 20 \log_{10} \frac{A_d}{A_c},$$

where A_d is the differential voltage gain and A_c is the common-mode voltage gain.

Common mode signal. A signal that appears simultaneously at both amplifier input terminals with respect to a common point.

Comparator. Compares the magnitudes of two digital quantities and produces an output indicating the relationship of the quantities.

Complement. The value obtained by reversing the state of a binary digit.

Computer peripherals. Auxiliary machines under the control of the computer.

CPS. Counts per second.

Crosstalk. A dynamic fault condition that exists when two parallel conductors are physically too close and become electromagnetically coupled.

Crystal oscillator. An astable circuit whose frequency-determining element is a quartz crystal; it is characterized by frequency stability over time and over variations in temperature.

DAC. Abbreviation for digital-to-analog converter. A circuit that converts a digital code word into an output analog (continuous) voltage or current.

Damping. An electrical, mechanical, or magnetic force that opposes oscillation of a body or system capable of free oscillation.

Data acquisition system. A system consisting of analog multiplexers, sample holds, A/D converters, and other circuits that process one or more analog signals and convert them into digital form for use by a computer.

Data bus. A parallel array of conductors capable of transmitting and receiving data between the CPU of a computer and external elements, such as memory and I/O circuitry. It is often a bidirectional bus.

Data converter. An A/D or D/A converter.

Data word. A digital code word that represents data to be processed.

DC offset. An unchanging voltage value added to a signal, causing the O state of a digital to be offset from zero volts.

Decibel (abbreviated dB). Ten times the logarithm of the ratio between two amounts of power P_1 and P_2 existing at two points or at two instants in time. By definition,

$$\text{number of dB} = 10 \log_{10} \frac{P_2}{P_1}$$

$$= 20 \log_{10} \frac{V_2}{V_1} + 10 \log_{10} \frac{Z_1}{Z_2},$$

assuming the power factors of the two impedances are equal. If the impedances themselves are equal, the right term becomes zero and

$$\text{number of dB} = 20 \log_{10} \frac{V_2}{V_1}.$$

Decode. To determine the meaning of coded information.

Decoder. A communications term for digital-to-analog converter.

Decoupling capacitor. A capacitor connected locally across the power supply pins of an IC to prevent high-frequency variations in load from affecting nearby ICs. Usually a decoupling capacitor is of the disc variety and between 0.01 and 0.1 μF.

Demodulator. A device that extracts the modulation information from a modulated carrier. In most cases it rectifies the incoming modulated carrier frequency and separates the desired modulation signal from the carrier.

Demultiplexer. A logic circuit in which a single input is gated onto one of a multitude of output lines.

Differential amplifier. A device that amplifies the difference between two signals.

Differential comparator. An analog integrated circuit to compare the magnitudes of two input voltages, producing a logic-compatible output whose value depends upon which input is greater.

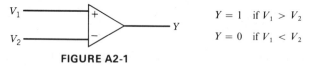

$$Y = 1 \quad \text{if } V_1 > V_2$$
$$Y = 0 \quad \text{if } V_1 < V_2$$

FIGURE A2-1

Differential transducer. A device that is capable of following simultaneously the voltages across or from two separate sources and providing a final output proportional to the difference between the two signals.

Digital computer. A digital device whose function and operation depend upon a stored sequence of binary numbers.

Digital-to-analog (D/A) converter. See DAC.

DIP. Dual in-line package. A type of integrated circuit package.

Disable. To inhibit the passage of a signal from *A* to *B* by applying the appropriate signal to the inhibit line.

Discriminator. A device in which the properties of a signal such as frequency or phase are converted into amplitude variations.

Distortion. An unwanted change in waveform. Principal forms of distortion are inherent nonlinearity of the device, nonuniform response at different frequencies, and lack of constant proportionality between phase shift and frequency.

Distributed capacitance. Capacitance evenly distributed over the entire length of a signal cable. Includes capacitance between signal conductors and from each conductor to ground.

DMA. Direct memory access.

Dynamic memory. A type of memory design utilizing a semiconductor capacitor matrix and requiring periodic refreshing to maintain data.

Electrocardiogram (ECG). Essentially an electromyogram of the heart muscle. All muscular activity in the body is characterized by the discharge of polarized cells; the aggregate current flow from it causes a voltage drop that can be measured on the skin. A changing emf will appear between electrodes connected to the arms, legs, and chest that rises and falls with heart action such that the period of the resulting waveform is the time between heartbeats. Various positive and negative peaks within one cycle of this waveform have been lettered *P, Q, R, S,* and *T,* a notation that aids in subsequent analysis and diagnosis.

Electroencephalogram (EEG). A waveform obtained by plotting brain voltages (available between two points on the scalp) against time. An EEG is not necessarily a periodic function, although it can be—particularly if the patient is unconscious. These voltages are of extremely low level and require recording apparatus that displays excellent noise rejection.

Electromyogram (EMG). Classically, a waveform of the contraction of a muscle as a result of electrical stimulation. Usually this stimulation comes from the nervous system (normal muscular activity). The record of potential difference between two points on the surface of the skin resulting from the activity or action potential of a muscle.

Electrostatic coupling. Coupling by means of capacitance so that charges on one circuit influence the other circuit owing to the capacitance between the two.

Enable. To activate or put into an operational mode.

Enable input. An input usually found on counter and register modules that, when asserted, permits normal operations to occur. When deasserted, normal operation is disabled.

Encode. To convert information into coded form.

Encoder. A communications term for an A/D converter.

Erasable programmable read only memory (EPROM). Programmed by a programmable device called the PROM burner. The burning or programming process takes a few minutes to store a charge in an essentially perfect insulator. The EPROM retains the stored data indefinitely. If the user decides to reprogram the EPROM chip, the previously stored data can be erased by shining a very strong ultraviolet light through a tiny window on top of the chip for about 20 minutes.

Execute process. The second phase of an instruction cycle during which the instruction code (or, alternatively, op code) is interpreted.

Fan in. The number of unit loads the input of a gate will present to the output of another gate.

Fan out. The number of unit loads the output of a gate can safely drive.

FCC. Federal Communication Commission.

Feedback path. A connection between an output and input of the same circuit.

Fetch process. The first phase of an instruction cycle during which an instruction code (or, alternatively, op code) is fetched from memory.

Field-effect transistor (FET). A unipolar device. It operates as a voltage-controlled device with either electron current in an n-channel FET or hole current in a p-channel FET.

Flag. (a) A one-bit register used as an indicator of status (of a device or program). (b) The carry and the overflow bits are called flags.

Flip-flop. A bistable multivibrator. A logical device capable of assuming one of two stable states, often equipped with two complementary outputs.

Flow chart. A graphical means of expressing an algorithm. Circles are used to denote *start* or *stop*, rectangles are used for *processing*, and diamonds are used for *decisions*.

Frequency response. The portion of the frequency spectrum that can be covered by a device within specified limits of amplitude error.

Full adder. A digital circuit that adds two binary digits and an input carry to produce a sum and an output carry.

Gain-bandwidth product. The product of gain and small signal bandwidth for an operational amplifier or other circuit. This product is constant for a single-pole response.

Glitch. A voltage or current spike of short duration and usually unwanted.

Half adder. A digital circuit that adds two bits and produces a sum and an output carry. It cannot handle input carries.

Hazard. A condition that exists in combinational logic when two or more inputs change simultaneously, but because of unequal path propagation delays in the logic, cause improper excitations and momentary false outputs to exist.

Hexadecimal code. A simplified means of representing binary numbers in a radix-sixteen number system.

High-pass filter. A filter that transmits alternating current above a given cutoff frequency and substantially attenuates all other currents.

Hold capacitor. A high quality capacitor used in a sample-hold circuit to store the analog voltage. The capacitor must have low leakage and low dielectric absorption. Types commonly used include polystyrene, teflon, polycarbonate, polypropylene, and MOS.

Hysteresis. A condition in which the threshold between two discrete states varies and is determined by the state that is currently occupied. Hysteresis operates so as to allow the widest possible analog variation within a discrete state prior to switching. Having switched to the next state, the threshold of that new state widens to include points that were previously stable in the last state.

IC chip. A single integrated circuit package.

Increment. To increase the value of a counter by a known amount.

Inhibit. To prevent an action from taking place by applying an appropriate signal to the proper input.

Instruction code. The binary-encoded form of a computer instruction that can be interpreted by the instruction decoder and sequencer in the CPU.

Instruction cycle. The detailed sequence of steps required to carry out one instruction in a computer.

Instruction mnemonic. A three- or four-letter abbreviation that represents a computer instruction. All mnemonics should be assembled into binary code before execution on a computer.

Instruction register (IR). A parallel array of flip-flops in the CPU of a computer used to hold the instruction code after the fetch process is complete.

Instruction set. The set of basic operations that a computer can execute, specified by assembly-language mnemonics and coded as binary numbers.

Intercarrier sound signal. The 4.5-MHz signal.

Interface. A shared boundary. For example, the interface can be the boundary that links two devices.

Interrupt. A signal sent to the microprocessor that may request service at any time and is asynchronous to the program.

Inverter. NOT gates are often referred to as inverters.

Laser. An acronym for *light amplification by stimulated emission of radiation*.

LCD. Abbreviation for *liquid-crystal display*. The LCD has a lower power requirement than the LED.

LED (light-emitting diode). A diode that will give off visible light when it is energized.

Low-pass filter. A filter that transmits alternating current below a given cutoff frequency and substantially attenuates all other currents.

LSB. An acronym for *least significant bit*, which is the digit of a binary number weighted least heavily.

LSI. Large-scale integration.

Lumped delay. The delay of a combinational network consolidated into one block and shown in the feedback path of an asynchronous sequential circuit for the purpose of analysis.

Master-slave flip-flop. A clocked flip-flop consisting of two serial S-R latches. Input data is transferred to the master when the clock is asserted, and from the master to slave when the clock is deasserted.

Memory address register (MAR). A parallel array of flip-flops in the CPU of a computer that is used to hold the address of the RAM cell currently being accessed. The MAR outputs drive the address bus.

Memory data register (MDR). A parallel array of flip-flops in the CPU of a computer used to hold data that is being received from or transmitted to RAM.

Microprocessor. The central processing unit (CPU) of a computer, implemented as a single integrated circuit containing 5000 or more transistors.

Modem. A device that modulates and then demodulates signals transmitted over communication facilities.

Modulation. The process of placing information onto an information carrier.

Modulus. Abbreviated mod. The maximum number of representable states in a device that has a specific number of digits. A four-digit binary counter is mod-16.

Monostables. Multivibrators with only one stable state.

MSB. An acronym for *most significant bit*, which is the digit of a binary number weighted most heavily.

MSI (medium-scale digital integrated circuits). Chips usually containing between 10 and 100 gates.

Multiplexer. A logic circuit in which one of a multitude of input lines is gated to a single output line.

Multivibrators. Sequential circuits that provide both an output (Q) an its complement ($\bar{Q}$).

NAND gate. A gate that has a 1 output if an only if one or more of the inputs are at the 0 level.

Natural frequency. The frequency at which a system with a single degree of freedom will oscillate upon momentary displacement from rest position by a transient force in the absence of damping.

Negative feedback. A circuit connection in which a proportion of the output signal is returned to and subtracted from the input to bring about a reduction in output.

Negative logic. The convention in electronic logic by which logic 0 values are represented by a more positive voltage than logic 1 values. Thus the assertion level for logic 0 is high.

Nibble. A group of four bits.

NMOS. *n*-channel metal oxide semiconductor.

Noise immunity. The maximum value of noise that can be tolerated without causing false outputs.

NOR gate. A gate that has a 1 output if and only if all the inputs are at the 0 level.

NOT gate. A gate that has a 1 output if and only if the single input is at the 0 level.

Notch filter. An electronic filter that attenuates or rejects a specific frequency or narrow band of frequencies with a sharp cutoff on either side of the band.

npn. Referring to the junction structure of a bipolar transistor.

Octal code. A simplified means of representing binary numbers in a radix-eight number system.

One's complement. The value obtained by reversing the logic state of each bit in a given binary number. The one's complement of 0110111 is 1001000, for example.

One shot. Same as monostable multivibrator.

Operand. Data to be operated on by a computer instruction. Most arithmetic operations require two operands.

Operation code. That part of a computer instruction that specifies the operation to be performed on the operands.

Operational amplifier. A high-gain differential amplifier used in feedback circuits for linear amplification, current summation, and a variety of nonlinear applications.

OR gate. A gate that has a 1 output if and only if one or more inputs are at the 1 level.

Overflow. A condition that occurs in an arithmetic operation when the result is greater than that which can be expressed with the permissible number of bits.

Parity. The oddness or evenness of the number of 1's in a specified group of bits.

Passive filter. A filter circuit using only resistors, capacitors, and inductors.

Period. The time required for a periodic waveform to repeat itself.

Periodic waveform. A sequence of changing signal levels that occurs during a fixed interval of time and repeats itself over and over.

Phase. In a periodic function or wave, the fraction of the period that has elapsed measured from some fixed origin. If the time for one period is represented as $360°$ along a time axis, the phase position is called the phase angle.

Phase shift. A change in the phase relationship between two periodic functions.

Phonocardiogram.   A graphic recording of the sounds produced by the heart and its associated parts—for example, of its mitral and aortic valves.

PMOS.   *p*-channel metal oxide semiconductor.

pnp.   A junction structure of a bipolar transistor.

Positive logic.   The system of logic where a high represents a 1 and a low represents a 0.

Precision.   The degree of repeatability or reproducibility of a series of successive measurements. Precision is affected by the noise, hysteresis, time, and temperature stability of a data converter or other device.

Preset.   To initialize a digital circuit to a predetermined state.

Program.   A list of instructions that are arranged in a specified order to control the operation of a microprocessor system or computer. The program tells the machine what to do on a step-by-step basis.

Program counter.   A counter in a microprocessor that keeps up with the place in the program. It acts as a bookmark that tells the computer the next instruction to be executed.

Programmer.   The person who understands the task and can "talk" to the computer.

Programming.   The arranging of a sequence of steps that will solve a given problem upon execution.

Propagation delay.   The time interval between the occurrence of an input transition and the corresponding output transition.

Pulse.   A sudden change from one level to another followed by a sudden change back to the original level.

Pulse duration.   The time interval that a pulse remains at its high level (positive-going pulse) or at its low level (negative-going pulse). Typically measured between the 50 percentage points on the leading and trailing edges of the pulse.

Pulse width.   Pulse duration.

PUT.   Programmable unijunction transistor.

QRS complex.   That portion of the waveform in an electrocardiogram extending from point Q to point S; it includes the maximum amplitude shown in an ECG trace.

Quotient.   The result of a division.

Race.   A condition in a logic network where the differences in propagation times through two or more signal paths in the network can produce an erroneous output.

Radix.   The base of a number system. The number of digits in a given number system.

RAM.   Random access memory.

Raster.   The rectangular area scanned by the electron beam as it is deflected horizontally and vertically.

Read. The process of retrieving information from a memory.

Read cycle. The sequence of signal changes needed to constitute a read operation in a memory.

Refresh. The process of renewing the contents of a dynamic memory.

Regenerative. Having feedback so that an initiated change is automatically continued, such as when a multivibrator switches from one state to the other.

Register. A set of flip-flops used for the temporary storage of a digital word.

Relay. An electromechanical switch in which the contact positions can be changed by applying an electrical signal.

Reset. The state of a flip-flop, register, or counter when 0's are stored. Equivalnet to the clear function.

Resolution. The capacity to distinguish difference between closely spaced values.

RFC. Radio frequency choke.

Ring counter. A digital circuit made up of a series of flip-flops in which the contents are continuously recirculated.

Ringing. A damped sinusoidal oscillation.

Ripple counter. A digital counter in which each flip-flop is clocked with the output of the previous stage.

Rise time. The time required for the positive-going edge of a pulse to go from 10% of its full value to 90% of its full value.

Roll over. For a collection of multivibrators this is the transition from a maximum value to a minimum value or vice versa.

ROM. Read only memory.

Sample and hold. A device used in data converter circuits to sample the analog input and then store and hold the sampled value while conversion takes place.

SCR. Silicon controlled rectifier.

Sequential circuits. Circuits that have outputs that depend not only on the present inputs but also on some memory of past inputs.

Serial. An in-line arrangement where one element follows another, such as in a serial shift register.

Set. The state of a flip-flop when it is in the binary 1 state.

Shift. To move binary data within a shift register or other storage device.

Shift register. A digital circuit capable of storing and shifting binary data.

SSI. Small scale integration.

Stack. A LIFO (last-in-first-out) memory consisting of registers or memory locations.

Static memory. A memory composed of flip-flops or magnetic cores that are capable of retaining information indefinitely.

Strobe. A pulse used to sample the occurrence of an event at a specified point in time in relation to the event.

Subroutine. A program that is normally used to perform specialized or repetitive operations during the course of a main program. A subprogram.

Subtractor. One of the operands in a subtraction.

Subtrahend. The other operand in a subtraction.

Synchronous counter. A counter circuit in which all flip-flops are triggered simultaneously, permitting all output bits to change at the same instant.

Telemetry. The complete measuring, transmitting, and receiving apparatus for indicating, recording, or integrating at a distance, by electrical translating means, of a quantity.

T flip-flop. A type of flip-flop that toggles or changes state on each clock pulse.

Three-state logic. A type of logic circuit having the normal two-state (high, low) output and, in addition, an open state in which it is disconnected from its load.

Threshold voltage. A value where the circuit output changes from one level to the other level.

Time-division multiplexing. The sharing of a single transmission channel (wire) by two or more signals at specific and periodic times. No two signals are ever on line at the same time.

Transducer. A device that converts one form of energy to antoher, usually electrical.

Transition. A change from one level to another.

Transmission line. A cable or other physical medium over which data is sent from one point to antoher.

Trigger. A pulse used to initiate a change in the state of a logic circuit.

Two's complement. The value obtained by adding one to the one's complement of a given binary number. The two's complement of 0101100 is 1010100.

UJT. Unijunction transistor.

Up-count. A counter sequence in which each binary state has a successively higher value.

UV EPROM. Ultraviolet erasable programmable read only memory.

Video frequencies. From 30 Hz to 4 MHz.

Video terminal. An interface device between the computer and a human operator consisting of a typewriter-like keyboard and a television display capable of displaying alphanumeric characters.

Volatile. The characteristic of a memory whereby it loses stored information if power is removed.

Wave trap. A resonant circuit tuned to attenuate a specific frequency.

Weight. The value of a digit in a number based on its position in the number.

Word. A multibit binary quantity.

Write. The process of storing information in a memory.

Write cycle. The sequence of signal changes needed to constitute a write operation in a memory.

XOR. Exclusive OR, a logic function that is true if one but not both of the variables is true.

Zener diode. A semiconductor diode designed to operate in the reverse breakdown region of its characteristics.

INDEX